Sustainable Development Goals Series

The **Sustainable Development Goals Series** is Springer Nature's inaugural cross-imprint book series that addresses and supports the United Nations' seventeen Sustainable Development Goals. The series fosters comprehensive research focused on these global targets and endeavours to address some of society's greatest grand challenges. The SDGs are inherently multidisciplinary, and they bring people working across different fields together and working towards a common goal. In this spirit, the Sustainable Development Goals series is the first at Springer Nature to publish books under both the Springer and Palgrave Macmillan imprints, bringing the strengths of our imprints together.

The Sustainable Development Goals Series is organized into eighteen subseries: one subseries based around each of the seventeen respective Sustainable Development Goals, and an eighteenth subseries, "Connecting the Goals", which serves as a home for volumes addressing multiple goals or studying the SDGs as a whole. Each subseries is guided by an expert Subseries Advisor with years or decades of experience studying and addressing core components of their respective Goal.

The SDG Series has a remit as broad as the SDGs themselves, and contributions are welcome from scientists, academics, policymakers, and researchers working in fields related to any of the seventeen goals. If you are interested in contributing a monograph or curated volume to the series, please contact the Publishers: Zachary Romano [Springer; zachary.romano@springer.com] and Rachael Ballard [Palgrave Macmillan; rachael.ballard@palgrave.com].

Abdo Hassoun
Editors

War on Gaza

Consequences on Sustainability and Global Security

Springer

Editor
Abdo Hassoun
Sustainable AgriFoodtech Innovation & Research (SAFIR)
Arras, France

Faculty of Agricultural Engineering
University of Aleppo
Aleppo, Syria

ISSN 2523-3084 ISSN 2523-3092 (electronic)
Sustainable Development Goals Series
ISBN 978-3-031-88499-3 ISBN 978-3-031-88500-6 (eBook)
https://doi.org/10.1007/978-3-031-88500-6

This Springer imprint is published by the registered company Springer Nature Switzerland AG.
The registered company address is: Gewerbestrasse 11, 6330 Cham, Switzerland

To the indomitable people of Gaza

This is for you! The unyielding hearts, strong, and true. The sacred bond with the soil beneath your feet. Echoes of sacrifice no force can defeat. Rooted deep in the ancient olive tree. A symbol of strength, of hope, to be free.

You have stood defiant, through night and through day. Against the storms that tear and betray. Refusing to yield, refusing to flee. Claiming your right to live and to be. To breathe and to rise, through shadow and light. On the land that holds your history tight.

Through loss and devastation, you rise. Unbowed, unbroken, under the skies. A testament to roots that never die. To courage that no force can deny. To the fire in your soul that burns bright. Guiding you through the darkest night.

May your sacrifice light the way. For generations to rise and stay. To stand with strength, to claim their place. To live with freedom, with pride, with grace.

And may that peace finally arrive. So Palestine may bloom and thrive. Free and sovereign, proud and strong. A homeland restored, where you belong— This book is dedicated to the great people of Gaza—its children, women, elderly, and all the Palestinian people who continue to resist and persevere with unwavering courage—as well as to the nations who have stood with the Palestinian cause and rights, especially Spain and Norway; and to the free people of Syria, who deserve a future beyond the grip of the Assad regime—one of the deadliest dictatorships of our time.

Editor: Abdo Hassoun

Contents

List of Figures

List of Tables

Introduction to Gaza's War: Unprecedented Toll on Life and Nature

1

Abdo Hassoun

Abstract

"*War on Gaza: Consequences on Sustainability and Global Security*" explores the impacts of the recent war (October 2023–January 2025) on Gaza, highlighting its disastrous effects on human life, the environment, and economic stability. The war has severely threatened sustainability and undermined progress toward achieving the Sustainable Development Goals. Poverty has deepened under the tight blockade, forcing people into destitution and dependency, while hunger has escalated due to famine and malnutrition. The collapse of healthcare has led to widespread disease and trauma, increasing casualties by overwhelming facilities and denying critical care. The destruction of schools and universities has resulted in educide, while gender inequality has heightened vulnerability and exploitation. Water shortages and contamination threaten sanitation, and energy blackouts make life unbearable, exacerbated by the weaponization of necessities, such as food, water, and medical supplies. Economic collapse and soaring unemployment have worsened conditions, while displacement and exclusion have increased suffering, particularly for women and children. Environmental destruction, including ecocide and pollution, intensifies the crisis as marine ecosystems suffer from overfishing and contamination, while land degradation accelerates due to habitat and biodiversity loss from bombardment and toxic munitions. The war, rooted in Israeli apartheid policies and genocidal practices, has fueled violence and lawlessness, further weakening governance and institutions. International cooperation remains hindered, leaving Gaza isolated and fragmented, making sustainable recovery increasingly challenging. Despite these immense challenges, the resilience of Gazans stands as a testament to their determination to resist and rebuild. The book underscores the urgent need for global intervention to address the humanitarian catastrophe, hold those responsible accountable for genocide and crimes against humanity, and pave the way for a sustainable and just future for Gaza and beyond.

Keywords

Blockade · Conflict · Ecocide · Educide · Gaza war · Genocide · Humanitarian crisis · October 2023 · Palestine · Peace · Resilience · Sustainability · Sustainable development goal

A. Hassoun (✉)
Sustainable AgriFoodtech Innovation & Research (SAFIR), Arras, France

Faculty of Agricultural Engineering, University of Aleppo, Aleppo, Syria
e-mail: a.hassoun@saf-ir.com

A. Hassoun (ed.), *War on Gaza*, Sustainable Development Goals Series,
https://doi.org/10.1007/978-3-031-88500-6_1

1.1 Background of the Recent War (October 2023–January 2025) on Gaza

On October 7, 2023, Hamas and other Palestinian resistance groups launched a large-scale attack on Israeli military camps and settlements bordering the Gaza Strip. The attack reportedly resulted in 1200 fatalities and the capture of more than 240 Israelis. In response, Israel launched a punitive war on Palestinians in Gaza on the same day, cutting off essential supplies of water, food, and fuel and carrying out relentless bombardments targeting not only suspected Hamas military positions but also key civilian infrastructure, including hospitals, schools, mosques, and churches. This military response caused unprecedented catastrophe and a disastrous human toll, with Gaza's civilian population suffering immensely under the relentless assault (Boisen 2024; Fraihat and Hedaya 2024; Ramadan and Abouzid 2024).

Over the following 15 months of aerial, maritime, and ground-based military campaign, this brutal war led to the widespread destruction of Gaza, displacing millions and devastating the region's infrastructure and economy. The scale and intensity of the war prompted international concern, drawing widespread condemnation and calls for a ceasefire. Although the reasons behind the conflict are complex and multifaceted, the UN Secretary-General António Guterres stated that the attack by Hamas on October 7 "did not happen in a vacuum." He emphasized that "the Palestinian people have been subjected to 56 years of suffocating occupation", highlighting the deep-rooted historical context and longstanding grievances that fueled the eruption of violence (Agha et al. 2024; Fraihat and Hedaya 2024; Gwaya 2024).

The program "*The Hidden is More Immense*" by Aljazeera provided an in-depth analysis of the Al-Aqsa Flood operation carried out by Palestinian groups, outlining that the events of October 7, 2023, were driven by five key reasons (AlJazeera 2025).

First, the increasingly extreme Israeli government has openly opposed the two-state solution, denying Palestinians their fundamental right to self-determination and sovereignty. This persistent refusal has fueled frustration and despair, leaving Palestinians with little hope for a peaceful resolution. Second, systematic Israeli military operations and settler violence in the occupied West Bank and Jerusalem, including frequent raids on Al-Aqsa Mosque, have further intensified tensions. The repeated violations of Palestinian religious and cultural sites have provoked widespread outrage and resistance. Third, the longstanding blockade imposed on Gaza has created unbearable living conditions by severely restricting access to basic necessities such as food, water, medical supplies, and economic opportunities. This suffocating siege has fueled resentment and desperation among Gazans. Fourth, the continued detention of thousands of Palestinians in Israeli prisons, many held without trial under administrative detention, along with reports of deaths under torture and mistreatment, has further exacerbated tensions and fueled calls for resistance. Fifth, Israel had reportedly planned to launch a surprise air attack targeting all resistance factions, followed by a large and destructive ground invasion. This looming threat contributed to the sense of urgency and the timing of the operation.

These factors, combined with years of systematic oppression and the absence of meaningful diplomatic progress, culminated in the events of October 7, 2023, highlighting the urgent need for a just and lasting solution to this conflict.

1.2 Unprecedented Crisis

The recent war on Gaza (October 2023–January 2025) has resulted in an unprecedented humanitarian crisis, with levels of destruction surpassing any previous conflicts in scale and intensity. The sheer scale of bombardment on Gaza has surpassed that of major wars, with over 25,000 tons of explosives dropped on Gaza within just the first weeks of the war. This relentless assault has resulted in unprecedented levels of destruction and mass casualties, devastating entire neighborhoods and critical infrastructure. With unceasing attacks across the entire territory, nowhere in Gaza remains safe, leaving its population trapped in an unrelenting cycle of fear, displacement, and suffering (Asi et al. 2024; Wells et al. 2024).

The war on Gaza has resulted in an unprecedented human toll, with approximately 47,000 people killed as ceasefire begins in January 2025 (ReliefWeb 2025), while other studies suggest that the actual number of casualties could be significantly higher than official estimates (Khatib et al. 2024; Jamaluddine et al. 2025). Entire families have been wiped out, and mass graves continue to be uncovered, raising serious concerns about the true extent of the devastation. The scale of civilian deaths, coupled with the deliberate targeting of densely populated areas, has led to growing accusations that Israel's military campaign constitutes genocide (Nijim 2023, 2024; Boisen 2024; Repo 2024).

In response to the escalating crisis, South Africa filed a case against Israel at the International Court of Justice (ICJ), accusing it of violating the 1948 Genocide Convention. The case argues that Israel's military actions—including mass killings, the destruction of civilian infrastructure, and the blockade-induced deprivation of essential services—amount to genocide against the Palestinian people. Since January 2024, the ICJ has issued three orders of provisional measures, demanding Israel take immediate action to provide urgently needed humanitarian aid and basic services in Gaza, including food, water, fuel, and medical supplies. Despite these legally binding orders, Israel has continued to impose severe restrictions on aid delivery. The ICJ's May 2024 ruling specifically required Israel to keep the Rafah crossing open to allow unhindered humanitarian assistance, yet it has remained closed since Israeli forces took control of it (Gwaya 2024; Human Rights Watch 2024).

Food insecurity in Gaza has reached catastrophic levels due to the blockade and relentless attacks, leaving hundreds of thousands without access to basic necessities. The destruction of agricultural infrastructure, supply chain disruptions, and restrictions on aid have exacerbated the crisis, pushing the population to the brink of famine. Malnutrition rates are soaring, with children and vulnerable groups facing severe health risks (de Waal 2024; Hassoun et al. 2024). Urgent international intervention is needed to prevent a full-scale humanitarian disaster.

The healthcare system in Gaza has been pushed beyond its limits, with hospitals and clinics repeatedly targeted and destroyed, rendering them nonfunctional and leaving the wounded with little to no access to life-saving treatment. Thousands of medical personnel have been killed or injured, while the remaining facilities are overwhelmed and critically undersupplied (Zughbur 2023; Kunichoff et al. 2024; Abed Alah 2024). Despite intensive efforts by humanitarian organizations, the blockade and continuous airstrikes have obstructed aid deliveries, exacerbating the humanitarian catastrophe and further increasing the death toll.

The war also led to widespread educide—the systematic destruction of Gaza's educational institutions. Thousands of schools and universities were bombed or rendered inoperable, denying an entire generation their right to education. Students and teachers have faced significant psychological and social challenges, with many forced to abandon their education amid the chaos and displacement (Dader et al. 2024; Iriqat et al. 2025).

The unprecedented scale of destruction and human suffering inflicted upon Gaza underscores the dire need for international accountability and urgent efforts to mitigate the long-term consequences of the war. Addressing the environmental, social, and economic impacts will require coordinated global action, significant investment in reconstruction, and a commitment to justice and sustainable peace in the region.

1.3 Sustainability and Sustainable Development Goals

1.3.1 A Holistic Approach to Sustainable Development

The Sustainable Development Goals (SDGs), adopted by the United Nations in 2015, take a comprehensive and integrated approach to global development by addressing economic, social, and environmental challenges. The SDGs apply universally to all countries, emphasizing inclusivity, long-term sustainability, and accountability through measurable targets and indicators. They are structured around five core areas: people, planet, prosperity, peace, and partnership, highlighting the interconnectedness of global challenges and the need for collective action (Wong 2021; Yamaguchi et al. 2023).

A widely used framework to conceptualize the SDGs is the "Wedding Cake" model, developed by the Stockholm Resilience Centre, which presents development as a tiered structure. The model positions the environment as the foundation that supports social and economic goals, illustrating the critical dependency of human well-being and economic growth on a healthy biosphere. This holistic perspective underscores the necessity of integrated solutions for achieving sustainable development (Trappey et al. 2023; Schafer et al. 2025).

The SDGs are categorized into three interdependent pillars: social, environmental, and economic, each addressing specific global challenges that are particularly relevant in conflict-affected regions such as Gaza.

The *social pillar* encompasses goals related to poverty, hunger, health, education, gender equality, and access to clean energy. Chapters focusing on Gaza's humanitarian crisis highlight the devastating impact of war on SDG 1 (No Poverty), SDG 2 (Zero Hunger), SDG 3 (Good Health and Well-being), SDG 4 (Quality Education), SDG 5 (Gender Equality), and SDG 7 (Affordable and Clean Energy). The destruction of homes, healthcare systems, and educational institutions in Gaza demonstrates the urgent need for integrated social policies to address systemic inequalities and human suffering.

The *environmental pillar* focuses on preserving the planet's ecosystems and ensuring sustainability for future generations. Goals such as SDG 6 (Clean Water and Sanitation), SDG 13 (Climate Action), SDG 14 (Life Below Water), and SDG 15 (Life on Land) are critical in Gaza, where the war has led to severe environmental degradation. The weaponization of water resources, destruction of marine and terrestrial ecosystems, and pollution have exacerbated the region's ecological crisis, highlighting the need for environmental stewardship as an essential component of sustainable development.

The *economic pillar* addresses economic growth, infrastructure, industry, and reduced inequalities, aligning with SDG 8 (Decent Work and Economic Growth), SDG 9 (Industry, Innovation, and Infrastructure), SDG 10 (Reduced Inequalities), and SDG 12 (Responsible Consumption and Production). The war has caused severe economic devastation in Gaza, undermining local industries, disrupting supply chains, and exacerbating socioeconomic disparities. The challenges faced by Gaza emphasize the importance of resilient economic systems that can support livelihoods and sustainable development even in times of crisis. Achieving these goals will require strengthened global partnerships and multi-stakeholder collaboration, as outlined in SDG 17 (Partnerships for the Goals), to mobilize resources, share knowledge, and build capacity for sustainable recovery and long-term development.

These three pillars of sustainability are deeply interconnected, with progress in one area reinforcing advancements in the others. The war in Gaza demonstrates how disruptions across social, environmental, and economic domains can create long-term challenges that hinder development efforts, making a holistic and integrated approach essential to achieving the SDGs in conflict-affected regions.

1.3.2 Impact of War on the Three Pillars of Sustainability

The war has severely impacted the social pillar of sustainability, exacerbating poverty, hunger, health crises, gender inequality, and access to education and essential services (Fig. 1.1).

Economic disruption and widespread infrastructure damage have led to increased poverty levels and unemployment, leaving many households without income or access to basic services

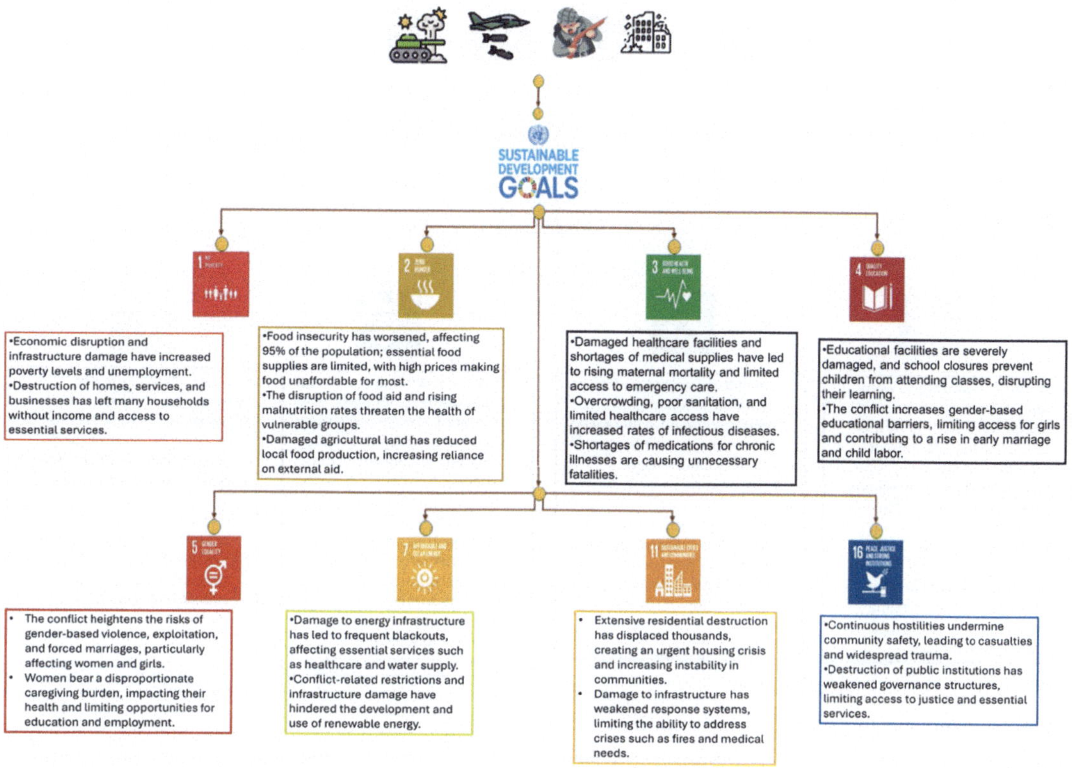

Fig. 1.1 Social impact of the war on Gaza

(ESCWA/UNDP 2024). Food insecurity has drastically worsened, with soaring prices and extensive damage to agricultural land, pushing the population to the brink of famine and increasing dependence on external aid for survival (Buheji and Hasan 2024; Hassoun et al. 2024). The collapse of healthcare systems, overcrowding, and shortages of medical supplies have resulted in rising mortality rates and preventable fatalities from chronic and infectious illnesses (Alnabih et al. 2024; Shorrab et al. 2024; Irfan et al. 2024). The war has also disrupted education, with damaged schools and closures preventing children from learning, increasing educational barriers, and leading to child labor. Women and girls face heightened risks of gender-based violence and exploitation, further limiting their opportunities for education and employment (Ramadan and Abouzid 2024; UN Women 2024; Jalambo et al. 2025).

Moreover, the destruction of energy infrastructure in Gaza has severely disrupted essential services such as healthcare and water supply, deepening the humanitarian crisis. Frequent power outages have severely disrupted hospital operations, leaving healthcare facilities unable to reliably power life-saving equipment, refrigerate critical medicines, or provide adequate patient care (Shellah 2024; Mohammed et al. 2024). Water treatment and distribution systems have also been heavily impacted, increasing the risk of waterborne diseases and sanitation crises. The war has left millions of people vulnerable, with service failures exacerbating health risks and placing immense strain on an already fragile healthcare system (Amnesty International 2024; Human Rights Watch 2024).

Having discussed the impact of the war on the social pillar, we now turn to its effects on the environmental pillar. This war has caused extensive environmental damage, severely impacting water resources, air quality, marine ecosystems, and terrestrial biodiversity (Fig. 1.2).

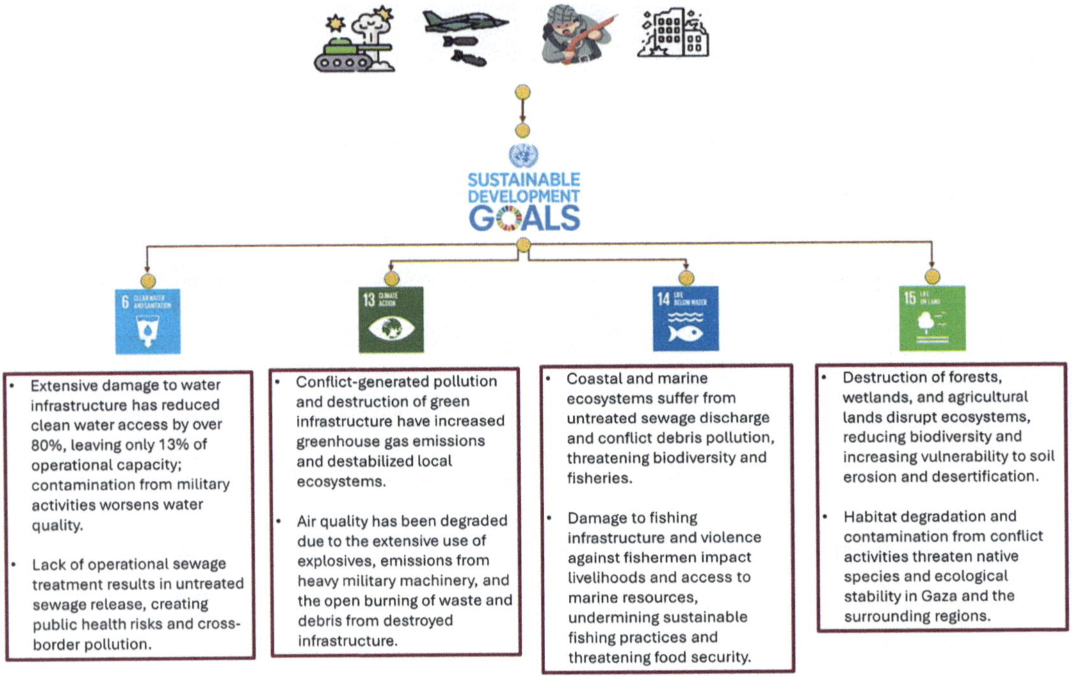

Fig. 1.2 Environmental impact of the war on Gaza

The extensive destruction of water infrastructure in Gaza has drastically reduced access to clean water. Contamination from military activities and untreated sewage discharge has worsened water quality, creating public health risks and cross-border pollution. Conflict-generated pollution and the destruction of infrastructure have led to increased greenhouse gas emissions and destabilized local ecosystems. Air quality has deteriorated due to the widespread use of explosives, emissions from heavy military machinery, and the burning of waste and debris. Marine ecosystems are under significant threat, suffering from untreated sewage discharge, debris pollution, and violence against fishermen, which undermines sustainable fishing practices and threatens food security. Damage to fishing infrastructure and restricted access to marine resources have further exacerbated the crisis. On land, the destruction of forests, wetlands, and agricultural lands has disrupted ecosystems, reduced biodiversity, and increased vulnerability to soil erosion and desertification. Habitat degradation and contamination from various military activities continue to threaten native species and the ecological stability of Gaza and its surrounding regions. The cumulative environmental degradation caused by the war underscores the urgent need for recovery efforts, as this ecocide accelerates ecosystem collapse, threatens biodiversity, and endangers long-term food and water security, further exacerbating the humanitarian crisis (Buheji and Al-Muhannadi 2023; Ahmed et al. 2024; Dardona et al. 2024; Kholoud 2024; Qumsiyeh 2024; UNEP 2024).

The war has also severely undermined Gaza's economic sustainability, leading to widespread unemployment, infrastructure devastation, and economic isolation (Fig. 1.3). The destruction of businesses and infrastructure has caused an economic collapse, disproportionately affecting youth and women, while constant instability has deterred investment and restricted the ability of small businesses to operate and grow. Severe damage to industrial areas and critical infrastructure has halted local production and stalled innovation, further limiting economic development.

Fig. 1.3 Economic impact of the war on Gaza

Telecommunications and digital infrastructure have also been heavily impacted, disrupting connectivity and hindering technological progress. Blockades and resource restrictions have deepened income disparities and poverty, exacerbating economic inequality by isolating Gaza's businesses and workforce from regional and global markets. Restrictions on movement and trade have significantly hindered economic resilience, making recovery efforts more challenging. Additionally, the war has caused severe scarcity and infrastructure damage, which has hindered sustainable resource use and effective waste management. The reliance on unsustainable materials and equipment due to war conditions further undermines efforts toward responsible consumption and production, making it difficult to achieve long-term economic sustainability (ESCWA/UNDP 2024; ILO 2024; The World Bank 2024). These economic challenges highlight the urgent need for rebuilding efforts that focus on restoring infrastructure, supporting small local businesses, and fostering sustainable economic growth to ensure long-term economic resilience.

1.4 Ceasefire in January 2025 and Israeli Resumption of Hostilities in March

On January 15, 2025, Hamas and Israel reached a ceasefire agreement, which came into effect on January 19, 2025, marking the first step toward ending the war in Gaza. The proposal follows a phased approach in three stages, starting with a six-week ceasefire, the release of all Israelis held hostage in Gaza in exchange for many of the Palestinians imprisoned by Israel, followed by a permanent ceasefire, Israel's withdrawal from Gaza, and a reconstruction process expected to last 3–5 years.

However, without addressing the root causes of the conflict, Israel may continue to undermine the agreement, justifying future military operations by citing Hamas's presence or even launching new assaults in other parts of Palestine, such as the West Bank. The sustainability of this agreement hinges on its comprehensiveness, the commitment of all parties involved, and the effective implementation of enforcement and monitoring mechanisms. A deal that fails to adopt a holistic approach is likely to be fragile and short-lived.

The long-term impact of this ceasefire will be shaped by the objectives outlined within it. If the agreement includes concrete steps to rebuild Gaza and improve living conditions, such as lifting the blockade and guaranteeing access to essential services like electricity, water, and healthcare, it could provide much-needed relief to the population. On the other hand, agreements limited to prisoner exchanges or temporary halts in fighting, without tackling core issues, are likely to result in an unstable truce, paving the way for renewed violence and further bloodshed. The recurring cycle of wars followed by fragile ceasefires demonstrates that resolving this conflict requires addressing critical issues, including the blockade, the occupation, and the recognition of Palestinian political rights. Without mechanisms to ensure compliance and accountability, any agreement remains at risk of collapse. Moreover, internal Palestinian unity and dialogue are vital, as divisions weaken stability and allow policies that undermine Palestinian rights and aspirations for self-determination.

A lasting and just resolution demands a political framework that promotes economic development, safeguards political rights, establishes effective governance in Gaza, and includes the withdrawal of the occupation. Without these fundamental elements, any agreement is unlikely to break the cycle of conflict and humanitarian suffering, instead perpetuating instability and violence. Additionally, a major shift in Israeli leadership is crucial. The far-right policies of Netanyahu's government have exacerbated tensions and obstructed peace efforts. A leadership transition toward moderate, peace-focused figures is essential to fostering an environment conducive to dialogue and reconciliation. A government that prioritizes coexistence and mutual respect can help build trust, encourage meaningful negotiations, and lay the foundation for a sustainable resolution that benefits both Palestinians and Israelis. On March 18, the Israeli military launched a series of coordinated strikes across the Gaza Strip, killing over 400 people whithin only a few hours and bringing an abrupt end to the temporary ceasefire. The scale and timing of the assault have drawn widespread condemnation, with many viewing it as a calculated move by Prime Minister Netanyahu to derail diplomatic efforts and escalate the conflict further. The renewed attacks have not only deepened the humanitarian crisis but have also exacerbated famine conditions, pushing an already devastated civilian population into even greater levels of suffering, deprivation, and despair.

1.5 The Structure and Chapters of the Book

This book is structured into three parts, each focusing on a critical pillar of sustainable development: social, environmental, and economic. Across its 18 chapters, the book provides an in-depth analysis of the profound impacts of the war on Gaza in relation to SDGs.

Part I, dedicated to the social pillar, encompasses Chaps. 2, 3, 4, 5, 6, 7, 8 and 9, addressing key issues such as poverty, food insecurity, healthcare, education, gender equality, energy access, housing, and governance challenges in war-torn Gaza.

After this introductory chapter, Chap. 2, led by Professor Mohamed Buheji, founder of the International Inspiration Economy, explores the evolving poverty dynamics in Gaza before and during the recent war. It provides a historical overview of socioeconomic challenges aggravated by the blockade and conflicts, framed within SDG 1. The chapter examines the impact of the war on food insecurity, unemployment, displacement, and access to essential services, highlighting the worsening multidimensional poverty. Drawing on authoritative reports and data, it evaluates the crisis with a long-term perspective and offers strategic recommendations for sustainable recovery and resilience in Gaza.

Chapter 3, led by Professor MoezAlIslam Ezzat Faris, explores the escalating food security crisis in Gaza, focusing on SDG 2. The chapter examines how the war has worsened acute food insecurity, pushing Gaza toward famine. It analyzes the destruction of agricultural infrastructure, the weaponization of food aid, and the blockade's impact on essential supplies, which have severely hindered self-sufficiency. The chapter also highlights rising malnutrition, especially among vulnerable groups such as children and women, and underscores the urgent need for humanitarian intervention and sustainable food system recovery in the region.

Chapter 4, led by Dr. Yara Ashour, a Palestinian medical doctor, examines the recent war's unprecedented humanitarian impact on healthcare in Gaza within the framework of SDG 3. It highlights the destruction of medical infrastructure, severe shortages of supplies, and the immense strain on healthcare workers. The chapter explores rising disease outbreaks, malnutrition, and mental health disorders, along with challenges in accessing care for displaced populations. Based on reports from local and international health organizations, it emphasizes the urgent need for humanitarian interventions and policy reforms to restore essential health services in Gaza.

Chapter 5, led by Dr. Tahani Aldahdouh, a Gazan Postdoctoral Research Fellow in Finland, examines the severe challenges the education in Gaza faces during the recent war, focusing on SDG 4. It explores the destruction of schools, displacement and killing of students and teachers, and the broader concept of "scholasticide." Despite these hardships, the chapter highlights the resilience of students and educators, showcasing efforts to sustain learning through online platforms and community initiatives. Combining qualitative and quantitative research, it provides insights into psychological, social, and economic factors influencing educational persistence in Gaza and presents policy recommendations to support recovery in alignment with global sustainability goals.

Chapter 6, led by Dr. Dunya Ahmed, Assistant Professor at the University of Bahrain and Strategic Planning & Development Adviser at the Supreme Council for Women, explores the profound gendered impacts of the recent war on Gaza, focusing on SDG 5. The chapter examines how the war has deepened existing gender inequalities—worsened by long-standing occupation and blockade—leading to increased gender-based violence, limited access to essential services such as healthcare and education, and major obstacles to economic empowerment. The widespread destruction of homes and infrastructure has further heightened the vulnerability of women and girls. Despite these challenges, the chapter highlights the resilience of Gazan women and their critical role in supporting families during crises.

Chapter 7, led by Nadia Mahdi, a Gazan Senior Electrical Engineer and leader of the Power Systems Studies Department at the Gaza Electricity Distribution Company, examines the war's devastating impact on energy infrastructure within the framework of SDG 7. The chapter explores the destruction of power plants, grids, and fuel supplies, leading to chronic electricity shortages and reliance on unsafe, unsustainable alternatives. Disruptions to essential services such as healthcare, water supply, and education have worsened energy poverty. The chapter emphasizes the urgent need for investment in renewable energy and international cooperation to rebuild infrastructure and ensure sustainable, reliable, and affordable access for all.

Chapter 8, led by Dr. Abdo Hassoun and a team of multidisciplinary experts from Gaza and beyond, examines the catastrophic impact of domicide—the deliberate destruction of homes—within the framework of SDG 11. It explores how systematic urban destruction has left vast areas uninhabitable, displaced nearly the entire population, and devastated critical infrastructure, including housing, healthcare, and education. The chapter analyzes the physical, social, and environmental consequences, highlighting psychological trauma, economic collapse, and weakened social cohesion. It underscores the urgent need for sustainable urban development and international intervention to rebuild resilient communities in Gaza while addressing long-term humanitarian and justice-related challenges.

Chapter 9, led by Bayan Arouri, a Gazan Doctoral Researcher at Tampere Peace Institute, examines critical challenges to achieving peace, justice, and effective governance in Gaza within the framework of SDG 16. The chapter explores the war's devastating impact on governance, the erosion of institutional trust, and growing instability driven by systemic human rights violations, forced displacement, and violence. It highlights persistent breaches of international law, apartheid policies, and the disproportionate harm affecting vulnerable populations, especially children. Advocacy efforts focus on enforcing international legal frameworks, protecting human rights, and developing inclusive governance mechanisms to restore justice and rebuild resilient institutions.

Part II, focusing on the environmental pillar, spans Chaps. 10, 11, 12 and 13, shedding light on water and sanitation challenges, environmental degradation, and the destruction of terrestrial and marine ecosystems.

Chapter 10, led by Dr. Rami Zurayk, Professor at the Faculty of Agricultural and Food Sciences at the American University of Beirut, examines the systematic weaponization of water and sanitation in Gaza within the framework of SDG 6. The chapter explores how repeated conflicts and prolonged blockades have deprived Gazans of clean water and adequate sanitation, severely impacting public health, environmental stability, and economic resilience. It discusses the destruction of critical infrastructure, including water treatment plants and desalination facilities, leading to widespread contamination and severe health crises such as cholera and dysentery. The analysis addresses the broader effects of water scarcity on food security and livelihoods, stressing the urgent need for international intervention to restore infrastructure, ensure accountability, and establish sustainable water management practices in Gaza.

Chapter 11, led by Dr. Khawla Al-Muhannadi, the President of the Environment Friends Society, examines the severe environmental impact of the recent war on Gaza within the framework of SDG 13. The chapter highlights how widespread destruction from military operations has caused extensive air, soil, and water pollution, worsening climate vulnerabilities. The use of chemical weapons like white phosphorus and the destruction of infrastructure have led to unprecedented environmental degradation, with long-term effects on public health and ecosystem stability in the region. It also addresses the role of the war in escalating greenhouse gas emissions and weakening climate resilience in Gaza, underscoring the urgent need for global intervention to support recovery and sustainability.

Chapter 12, led by Dr. Abdo Hassoun and a team of experts, examines the devastating impact of the recent war on coastal and marine ecosystems in Gaza within the framework of SDG 14. It explores the destruction of marine habitats, biodiversity loss, and severe pollution from debris, oil spills, and chemical contaminants released during military operations. The chapter highlights socioeconomic consequences for coastal populations, such as the collapse of fisheries, rising poverty, and food insecurity due to restricted fishing access and damaged infrastructure. It stresses the urgent need for environmental monitoring, targeted restoration efforts, and stronger international support to rehabilitate marine ecosystems in the region. The importance of sustainable management and global cooperation is also emphasized to ensure long-term resilience.

Chapter 13, led by Professor Alaa El-Din Ahmed Bekhit from the University of Otago's Department of Food Science, examines the devastating impact of the recent war on terrestrial ecosystems in Gaza within the framework of SDG 15. The chapter explores the destruction of agricultural land, soil degradation, and biodiversity loss caused by the war, which has severely affected food security and ecological balance. It highlights deforestation, habitat destruction, and contamination of soil and water due to military activities, creating long-term environmental challenges. The urgent need for ecological restoration, sustainable land management, and international collaboration is emphasized to address the crisis and support ecosystem recovery in Gaza.

Part III, which examines the economic pillar, includes Chaps. 14, 15, 16, 17 and 18, analyzing the war's impact on work and economic growth, industry, innovation, infrastructure, and sustainable consumption and production.

Chapter 14, led by Professor Mohammed Migdad, from the Faculty of economics and managerial sciences at the Islamic University of Gaza, explores the catastrophic impact of the recent war on the economy within the framework of SDG 8. It examines extensive job losses, widespread poverty, and the collapse of industries caused by the destruction of critical infrastructure and prolonged blockade. The chapter highlights the severe consequences of movement restrictions and military operations, which have disrupted trade and humanitarian aid, deepening economic instability. Unemployment has soared, especially among youth and women, while unsafe working conditions and rising child labor reflect the dire socioeconomic situation. Despite these challenges, the chapter also showcases resilience through entrepreneurial initiatives and community-driven efforts, offering hope for economic recovery.

Chapter 15, led by Dr. Samer Abuzerr, Assistant Professor at the University College of Science and Technology in Khan Younis in Gaza, examines the war's devastating impact on industry, innovation, and infrastructure in Gaza within the framework of SDG 9. The chapter explores the widespread destruction of critical infrastructure, including power plants, health facilities, and communication networks, which has crippled industrial activities and halted technological progress. Local production capabilities have been severely affected, resulting in economic decline and massive job losses. The destruction of research centers and educational institutions has further hindered innovation and technological growth, deepening isolation from global advancements. The chapter underscores the urgent need for international support to rebuild infrastructure, restore industrial capacity, and promote innovation for economic recovery and long-term sustainability in Gaza.

Chapter 16, led by Dr. Dalal Iriqat, Associate Professor at the Arab American University in Palestine, explores the widening social and economic inequalities in Gaza within the framework of SDG 10. The recent war has worsened existing disparities, hitting vulnerable groups such as women, children, and persons with disabilities the hardest. Destruction of infrastructure, businesses, and essential services, combined with the blockade, has driven unemployment and poverty to critical levels, drastically lowering living standards. The chapter discusses rising social exclusion, restricted access to education and healthcare, and long-term economic marginalization. It also reviews the international response to the crisis and offers policy recommendations to reduce inequalities and support sustainable recovery efforts that enhance social inclusion and resilience in Gaza.

Chapter 17, led by Dr. Hussein F. Hassan, Professor of Food Science and Technology, and Dr. Imane Dankar, Instructor at the Lebanese American University, examines the severe disruption of sustainable consumption and production systems in Gaza within the framework of SDG 12. The chapter explores the destruction of essential infrastructure, including manufacturing facilities, supply chains, and waste management systems, which has increased reliance on unsustainable practices, worsening resource scarcity and environmental degradation. It also highlights the war's social and economic impacts, such as rising unemployment and poverty, which fuel unsustainable consumption and deepen reliance on humanitarian aid. Policy recommendations are provided to rebuild production and consumption systems sustainably, emphasizing international collaboration and long-term recovery strategies for economic and environmental stability in Gaza.

Finally, Chap. 18, authored by Rasha Abousalem, a humanitarian aid worker and adjunct professor at the University of Missouri, USA, explores the weakening of global partnerships due to the recent war on Gaza within the framework of SDG 17. The chapter highlights how the war has undermined international cooperation, diplomacy, and economic collaboration, creating major obstacles to achieving global goals. It examines the growing crisis of credibility and trust among international actors, complicating peace efforts and humanitarian interventions. The destruction of infrastructure and economic instability in Gaza have further strained partnerships, limiting resource mobilization and sustainable development efforts. The chapter emphasizes the need for renewed international engagement and policy-driven strategies to rebuild trust, promote peace, and revitalize commitments to the SDGs.

1.6 Conclusion

The recent war on Gaza has resulted in unprecedented human suffering, environmental destruction, and economic collapse. The relentless bombardment and blockade have devastated critical infrastructure, leaving Gazans without access to basic necessities such as food, water, healthcare, and shelter. The destruction of energy and water infrastructure has deepened the humanitarian crisis, with frequent blackouts and severe water shortages exacerbating public health risks and increasing dependency on scarce humanitarian aid. The healthcare system has been pushed beyond its limits, overwhelmed by the sheer number of casualties and the widespread spread of preventable diseases. In addition to the immediate loss of life, the long-term consequences of the war continue to erode Gaza's social fabric, pushing people into deeper poverty and despair.

Beyond the humanitarian toll, the environmental impact of the war has been catastrophic, with widespread destruction of marine and terrestrial ecosystems, pollution of water resources, and significant land degradation. The extensive damage to agricultural land and infrastructure has worsened food insecurity and further hindered economic recovery. The economic collapse has resulted in soaring unemployment, leaving Gaza isolated from regional and global markets. Governance structures have been severely weakened, with institutional capacity diminished and access to justice and public services increasingly limited. The war has not only intensified existing vulnerabilities but also created new challenges that threaten Gaza's long-term sustainability and resilience. Addressing these challenges requires urgent international intervention, accountability for crimes committed, and a commitment to rebuilding Gaza with a focus on sustainable development, resilience, and justice for its people.

References

Abed Alah M (2024) Echoes of conflict: the enduring mental health struggle of Gaza's healthcare workers. Confl Heal 18:1–6. https://doi.org/10.1186/S13031-024-00577-6/METRICS

Agha Z, Esson J, Griffiths M, Joronen M (2024) Gaza: A decolonial geography. Trans Inst Br Geogr 49:e12675. https://doi.org/10.1111/tran.12675

Ahmed K, Gayle D, Mousa A (2024) 'Ecocide in Gaza': does scale of environmental destruction amount to a war crime? Guard

AlJazeera (2025) The hidden is more immense. https://www.aljazeera.net/program/the-hidden-is-more-immense/. Accessed 25 Jan 2025

Alnabih A, Alnabeh N, Aljeesh Y, Aldabbour B (2024) Food insecurity and weight loss during wartime: a mixed-design study from the Gaza strip. J Health Popul Nutr 43:222

Amnesty International (2024) 'You feel like you are subhuman'—Israel's genocide against Palestinians in Gaza

Asi Y, Mills D, Greenough PG et al (2024) 'Nowhere and no one is safe': spatial analysis of damage to critical civilian infrastructure in the Gaza Strip during the first phase of the Israeli military campaign, 7 October to 22 November 2023. Confl Heal 18:1–13. https://doi.org/10.1186/s13031-024-00580-x

Boisen C (2024) Israel's punitive war on Palestinians in Gaza. J Genocide Res 1–22. https://doi.org/10.1080/14623528.2024.2406098

Buheji M, Al-Muhannadi K (2023) Mitigating risks of environmental impacts on Gaza - review of precautions & solutions post (2023 war). Int J Adv Res Eng Technol 14:15–47

Buheji M, Hasan A (2024) Beyond famine and chaos—case of Gaza. Int J Manag. https://doi.org/10.17605/OSF.IO/B6SPQ

Dader K, Ghantous W, Masad D et al (2024) Topologies of scholasticide in Gaza: education in spaces of elimination. Fenn—Int J Geogr 202:1–12. https://doi.org/10.11143/FENNIA.147002

Dardona Z, Amane M, Dardona A, Boussaa S (2024) Health and environmental impacts of Gaza conflict (2023-2024): a review. One Heal Bull. https://doi.org/10.4103/ohbl.ohbl

de Waal A (2024) Famine in Gaza: an example of the global humanitarian crisis. Am J Clin Nutr 119:1383–1385. https://doi.org/10.1016/j.ajcnut.2024.04.015

ESCWA/UNDP (2024) Gaza war: expected socioeconomic impacts on the State of Palestine—October 2024

Fraihat I, Hedaya M (2024) Understanding the October 7th Hamas attack on Israel and the war on Gaza. In: Elkahlout G (ed) Gaza's cycle of destruction and rebuilding. Palgrave Macmillan, Cham, pp 77–103

Gwaya KR (2024) The authority of the ICC to institute criminal proceedings against Israeli and Hamas leaders: the Gaza conflict. Beijing Law Rev 15:2202–2217. https://doi.org/10.4236/blr.2024.154123

Hassoun A, Al-Muhannadi K, Hassan HF et al (2024) From acute food insecurity to famine: how the 2023/2024 war on Gaza has dramatically set back sustainable development goal 2 to end hunger. Front Sustain Food Syst 8:1402150. https://doi.org/10.3389/fsufs.2024.1402150

Human Rights Watch (2024) "Hopeless, starving, and besieged" Israel's forced displacement of Palestinians in Gaza

ILO (2024) A year of war in Gaza: impacts on employment and livelihoods in the West Bank and Gaza strip. Bulletin No 5

Irfan B, Shammala AA, Saleh K (2024) Will there be a future for newborns in Gaza? Lancet 404:1725–1726. https://doi.org/10.1016/S0140-6736(24)02249-9

Iriqat D, Alousi R, Aldahdouh TZ et al (2025) Educide amid conflict: the struggle of the Palestinian education system. Qual Educ All 2:81–99. https://doi.org/10.1108/QEA-10-2024-0120

Jalambo MO, Al-masri N, Somaya M (2025) Resilient Palestinian Women facing gender-based conflict in the Gaza strip. In: Yohannes HT, Phipps A, Sitholé T (eds) Cultures of sustainable peace: conflict transformation, gender-based violence, decolonial praxes. Bristol & Blue Ridge Summit: Multilingual Matters, pp 138–157

Jamaluddine Z, Abukmail H, Aly S et al (2025) Traumatic injury mortality in the Gaza strip from Oct 7, 2023, to June 30, 2024: a capture—recapture analysis. Lancet 2024:1–9. https://doi.org/10.1016/S0140-6736(24)02678-3

Khatib R, McKee M, Yusuf S (2024) Counting the dead in Gaza: difficult but essential. Lancet 404:237–238. https://doi.org/10.1016/S0140-6736(24)01169-3

Kholoud M (2024) The impact of Gaza war on environmental security. Serbian J Eng Manag 9:1–5. https://doi.org/10.5937/sjem2401001m

Kunichoff D, Mills D, Asi Y et al (2024) Are hospitals collateral damage? Assessing geospatial proximity of 2000 lb bomb detonations to hospital facilities in the Gaza strip from October 7 to November 17, 2023. PLOS Glob Public Heal 4:1–14. https://doi.org/10.1371/journal.pgph.0003178

Mohammed F, Elgailani USA, Ibrahim Ali SY et al (2024) Defending the right to health in Gaza: a call to action by health workers. Confl Heal 18:1–4. https://doi.org/10.1186/S13031-024-00613-5/METRICS

Nijim M (2023) Genocide in Palestine: Gaza as a case study. Int J Hum Rights 27:165–200. https://doi.org/10.1080/13642987.2022.2065261

Nijim M (2024) Gazacide: Palestinians from Refugeehood to ontological obliteration. Crit Sociol 1944:1–23. https://doi.org/10.1177/08969205241286530

Qumsiyeh MB (2024) Impact of the Israeli military activities on the environment. Int J Environ Stud 81:977–992. https://doi.org/10.1080/00207233.2024.2323365

Ramadan A, Abouzid M (2024) Resilience beyond borders: Gaza's women and their silent battles. Women Birth 37:101623. https://doi.org/10.1016/j.wombi.2024.101623

ReliefWeb (2025) WeWorld flash update #25—War on Gaza: ceasefire after 465 days, January 21st, 2025. https://reliefweb.int/report/occupied-palestinian-territory/weworld-flash-update-25-war-gaza-ceasefire-after-465-days-january-21st-2025. Accessed 25 Jan 2025

Repo J (2024) Genocide and the destruction of the means of social reproduction in Gaza. Eur J Polit Gend:1–8. https://doi.org/10.1332/25151088Y2024D000000061

Schafer LJ, Nielsen AR, Inisan A et al (2025) Holistic advisory tool to evaluate and enhance sustainable food procurement practices in professional kitchens. Int J Gastron Food Sci 39:101087. https://doi.org/10.1016/j.ijgfs.2024.101087

Shellah D (2024) War on Gaza: the impossible duty to care for the critically ill. Intensive Care Med 50:311–313. https://doi.org/10.1007/S00134-023-07309-Z/METRICS

Shorrab AA, Nassef M, Subhi A et al (2024) Health in the crossfire-Analyzing and mitigating the multifaceted health risks of the 2023 war on Gaza. Public Heal Res 14:1–11. https://doi.org/10.5923/j.phr.20241401.01

The World Bank (2024) Gaza strip—interim damage assessment

Trappey CV, Trappey AJC, Lin HJ, Chang AC (2023) Comparative analysis of food related sustainable development goals in the North Asia Pacific region. Food Ethics 8:1–24. https://doi.org/10.1007/S41055-023-00132-8/TABLES/10

UN Women (2024) Gender alert: Gaza: a war on women's health

UNEP (2024) Environmental impact of the conflict in Gaza: preliminary assessment of environmental impacts. Nairobi

Wells K, Cortés-Morales S, Esson J et al (2024) Israel's war on Gaza and the violation of children's rights. Child Geogr 22:197. https://doi.org/10.1080/14733285.2024.2316752

Wong P-H (2021) Economic deprivation: approaches, causes, and consequences for violent conflicts. In: Leal Filho W, Marisa Azul A, Brandli L, Lange Salvia A, Özuyar PG, Wall T (eds) Peace, justice and strong institutions, encyclopedia of the UN sustainable development goals. Springer, Cham, pp 187–197

Yamaguchi NU, Bernardino EG, Ferreira MEC et al (2023) Sustainable development goals: a bibliometric analysis of literature reviews. Environ Sci Pollut Res 30:5502–5515. https://doi.org/10.1007/s11356-022-24379-6

Zughbur MR (2023) Protect civilians' lives and health care in Gaza. Lancet 402:1620–1621. https://doi.org/10.1016/S0140-6736(23)02402-9

From Crisis to Catastrophe: Poverty Dynamics in Gaza Before and During the Recent War

2

Mohamed Buheji, Mohammed Migdad, and Abdo Hassoun

Abstract

The humanitarian crisis in Gaza is unprecedented, marked by extensive destruction, significant loss of life, and the breakdown of critical infrastructure. Before the war (October 2023–January 2025), poverty in Gaza was already at alarming levels, with unemployment at 45.1% and food insecurity affecting 68.5% of the population. This war has drastically worsened these conditions, pushing multidimensional poverty to 97.9% and bringing the region to the brink of famine as agriculture and supply chains are devastated. Unemployment has surged to extreme levels, while healthcare, education, and housing systems have been severely damaged. Decades of development progress have been erased, setting Gaza back by 69 years and underscoring the urgent need for comprehensive recovery measures. This chapter analyzes the dynamics of poverty in Gaza before and during this recent war, framed within the context of Sustainable Development Goal 1 (ending poverty). It begins with a historical overview of poverty in Gaza, highlighting the multidimensional deprivation exacerbated by the blockade and recurring conflicts. The chapter then examines the devastating impacts of war on key poverty indicators, including food insecurity, unemployment, displacement, and access to essential services. Drawing on authoritative reports and peer-reviewed publications, it evaluates the escalation of poverty since October 2023 and its far-reaching implications for Gaza's future.

Keywords

Aid dependency · Blockade · Displacement · Humanitarian crisis · Multidimensional poverty · Resource scarcity · Resilience · SDG 1 · Sustainable development · Sustainable recovery · Unemployment

M. Buheji (✉)
Socioeconomic Institute for Advanced Studies (SIAS), Kigali, Rwanda

Inspiration Economy Institute, Manama, Bahrain

M. Migdad
Faculty of economics and managerial sciences, Islamic University of Gaza, Gaza, Palestine
e-mail: mmigdad@iugaza.edu.ps

A. Hassoun
Sustainable AgriFoodtech Innovation & Research (SAFIR), Arras, France

Faculty of Agricultural Engineering, University of Aleppo, Aleppo, Syria
e-mail: a.hassoun@saf-ir.com

A. Hassoun (ed.), *War on Gaza*, Sustainable Development Goals Series,
https://doi.org/10.1007/978-3-031-88500-6_2

2.1 Introduction

Poverty is a multidimensional phenomenon that extends beyond mere income deprivation, encompassing both material and non-material aspects of human well-being. According to the United Nations Development Programme (UNDP), poverty includes the deprivation of basic capabilities, such as access to education, healthcare, and essential services, which are necessary for living a life of dignity. The Multidimensional Poverty Index (MPI), developed by the Oxford Poverty and Human Development Initiative (OPHI) and UNDP, measures poverty across three dimensions, namely health, education, and living standards, using indicators such as nutrition, years of schooling, and access to clean water and sanitation (D'Attoma and Matteucci 2024; OPHI/UNDP 2024).

Globally, progress in poverty reduction has been uneven, with regions affected by wars and conflicts facing significant challenges. According to the World Bank and the United Nations Economic and Social Commission for Western Asia (ESCWA), extreme poverty, which is defined as living on less than $1.90 per day, remains a pressing issue in many developing countries, particularly those impacted by conflict, such as Palestine, Syria, Yemen, and Sudan. In 2021, 7.2% of the Arab world's population lived in extreme poverty. In 2024, an estimated 36% is projected to live below national poverty lines, underscoring the persistent impact of conflict and economic instability on the region (ESCWA 2022; World Bank Group 2024).

Gaza represents one of the most striking examples of multidimensional poverty in the Arab world. Before the most recent war, multidimensional poverty in Gaza affected 45.1% of the population (ESCWA 2023a). The United Nations Relief and Works Agency (UNRWA) reported that 80% of Gaza's population depends on humanitarian assistance (UNRWA 2023), while recent studies showed that 84% of households were receiving food assistance (Horino et al. 2024; Nasreddine and Jomaa 2024). These figures underscore the chronic socioeconomic challenges faced by Gazans due to prolonged conflicts, blockade, and systemic deprivation. The recent war has further compounded these challenges, causing unprecedented levels of destruction and displacement. A recent joint report by ESCWA and UNDP reveals that the war has set back Gaza's development progress by 69 years (ESCWA/UNDP 2024), further entrenching poverty and worsening living conditions.

This chapter aims to analyze the poverty dynamics in Gaza within the framework of Sustainable Development Goal 1 (SDG 1). It begins by examining the historical context of poverty in Gaza since the 2007 blockade, exploring how restrictions on trade, movement, and access to resources have shaped the socioeconomic landscape of the Strip. Drawing on observational data and authoritative reports from organizations such as UNDP, ESCWA, UNICEF, United Nations Office for the Coordination of Humanitarian Affairs—occupied Palestinian territory (OCHA), the World Bank, as well as academic literature, the chapter investigates the escalation of poverty during the recent war (October 2023–January 2025), its immediate impacts on infrastructure and livelihoods, and the long-term implications for this region. The chapter concludes by offering recommendations for recovery, focusing on early self-sufficiency programs, rebuilding infrastructure, and addressing the psychological and social consequences of the war. This comprehensive analysis highlights the need for coordinated humanitarian and development efforts to address the drivers of poverty in Gaza.

2.2 Sustainable Development Goal 1 (SDG 1): No Poverty

SDG 1 aims to "end poverty in all its forms everywhere" by 2030. This goal is integral to ensuring that every person, regardless of their socioeconomic background, has the opportunity to live a life of dignity and economic security. SDG 1 is a key priority for global development, addressing both the eradication of extreme poverty and the reduction of poverty rates in all its dimensions. The first target of the United Nations' SDG 1 focuses on eradicating extreme poverty for all people everywhere, currently defined as living on less than \$1.90 per day, based on the international poverty line. Additionally, it aims to reduce by at least half the proportion of people living in poverty in all its dimensions, as defined by national standards. The other main targets of SDG 1 include implementing appropriate social protection systems and measures to ensure substantial coverage of the poor and vulnerable, ensuring equal access to basic services, and building the resilience of low-income and marginalized populations to economic, social, and environmental shocks. These targets collectively aim to address the root causes of poverty and promote sustainable development (Nassar et al. 2023; Mhlanga and Ndhlovu 2024; Ajaj et al. 2024).

Achieving SDG 1 requires targeted poverty reduction policies, measured through key indicators such as income levels, access to basic services, and social protection systems. Effective strategies include robust economic policies, social safety nets, and international cooperation to promote inclusive growth. Addressing multidimensional poverty involves improving education, clean water access, and employment opportunities. Innovation and digital technologies can further enhance economic prospects, especially in low-income regions, while infrastructure development fosters a sustainable environment for poverty alleviation (Zavyalova and Krotova 2021).

The war on Gaza has severely set back progress on each target of SDG 1. Extreme poverty has escalated due to widespread destruction of homes, livelihoods, and infrastructure, leaving much of the population reliant on emergency aid. Access to basic services and economic opportunities has been further hindered by damaged infrastructure, near complete halt of economic activity, and prolonged socioeconomic stagnation (ESCWA 2023a; World Bank Group 2024). Achieving SDG 1 in Gaza requires urgent relief efforts, significant investments in rebuilding infrastructure, and programs to restore livelihoods and essential infrastructure, create jobs, and support local businesses. Expanding social safety nets and access to social protection services is also critical to building community resilience to economic and environmental shocks.

2.3 Poverty Before the Recent War on Gaza

2.3.1 Historical Overview of Poverty in Gaza

The Gaza Strip, a small and densely populated territory, has faced a persistent and worsening poverty crisis since 2007. This period marks a turning point in the socioeconomic history of Gaza, characterized by political division, international isolation, and repeated cycles of conflict. These factors have converged to create a multidimensional poverty crisis, severely impacting the livelihoods, living conditions, and overall well-being of the population in Gaza.

Israel has maintained its occupation of Gaza, the West Bank, and East Jerusalem since June 1967. Although Israel "withdrew" from Gaza in 2005, it continues to exercise control over the territory's airspace, as well as its land and sea borders. Additionally, it strictly regulates the movement of people and goods in and out of Gaza and oversees the supply of essential resources, including electricity, water, fuel, and gas. In 2007, following Hamas' takeover of Gaza, Israel significantly tightened its restrictions on the movement of people, goods, and resources in and out of the territory. These measures escalated the blockade into a comprehensive siege, severely limiting access to essential supplies and economic opportunities (Asi 2020; Amir 2021; Issa et al. 2021; UNCTAD 2024). The intensified restrictions have profoundly impacted economy, infrastructure, and humanitarian conditions in Gaza, effectively isolating the population and exacerbating poverty and deprivation on an unprecedented scale. This dire situation under blockade has been described as "de-development" reflecting the systematic destruction of Gaza's economy and infrastructure, and a "slow-motion genocide" highlighting the prolonged and deliberate infliction of widespread suffering and deprivation on its population (Roy 2016; Asi 2020; Nijim 2024).

The repeated wars on Gaza have inflicted devastating economic losses, significantly exacerbating poverty levels. Between 2008 and 2021, direct economic losses from these assaults have been estimated at billions of dollars, including USD 1.9 billion in 2008–2009, USD 370 million in 2012, USD 460 million in 2014, and USD 500 million in 2021 (Palestine Economic Policy Research Institute—MAS 2023). These losses, combined with prolonged conflict and the tight blockade, have crippled the economy of Gaza, destroying livelihoods, businesses, and critical infrastructure. The cumulative impact of these wars has made the population increasingly reliant on humanitarian aid, exacerbating poverty, obstructing any prospects for sustainable development, and effectively turning the Strip into a battleground of economic warfare (Alagha and Hussein 2024).

The blockade on Gaza has severely restricted the movement of goods and people, including essential items such as food, medicine, and construction materials, drastically limiting trade and access to international markets. These restrictions have contributed to a sharp rise in poverty levels, as the inability to import critical goods has disrupted local industries and inflated prices, reducing purchasing power for Gazans. The blockade has also stifled economic growth by restricting imports and exports, leaving economy in Gaza increasingly dependent on international aid to meet basic needs. Additionally, the inability to bring in construction materials has delayed or entirely prevented the repair of critical infrastructure, including roads, schools, and healthcare facilities, further compounding the humanitarian crisis in the Strip. Access to essential services such as electricity and clean water has been severely compromised, with frequent power shortages and limited water availability further degrading living standards. Restrictions on raw materials for agriculture and industry, as well as machinery needed for infrastructure rehabilitation, have crippled productive sectors in Gaza, deepening poverty and increasing reliance on external assistance (Ajluni 2014; ILO 2018; Romano et al. 2019; UNCTAD 2024).

Unemployment rates in Gaza have reached critical levels, reflecting the severe economic challenges facing the region. As of September 2023, the overall unemployment rate stood at a staggering 45.1%, highlighting the crippling impact of prolonged economic restrictions and stagnation on the job market. The situation is even more dire among Gaza's youth, with unemployment reaching 59.5%, signifying a lost generation deprived of economic opportunities and hope for the future (The World Bank 2024).

Food insecurity and poverty are closely intertwined, each reinforcing the other in a vicious cycle. Poverty limits access to adequate and nutritious food due to insufficient income and high food prices, while food insecurity exacerbates poor health, reduces productivity, and restricts educational opportunities, deepening poverty. In Gaza, decades of conflict and a prolonged blockade have devastated agricultural infrastructure and severely disrupted local food systems. By 2018, 68.5% of the population was food insecure, with the majority reliant on humanitarian assistance to meet their basic dietary needs (Romano et al. 2019; Asi 2020; El Bilbeisi et al. 2022; Buheji and Hasan 2024; Migdad and Buheji 2024a). This dependency underscores the chronic vulnerability and the inability of local systems in this region to sustain the population under such prolonged strain.

Between 2006 and 2022, Gaza's GDP per capita experienced a dramatic decline, shrinking by 27% from $1994 to $1257 (UNCTAD 2024). This sharp contraction reflects the devastating impact of prolonged conflict, the blockade, and economic stagnation, which have eroded livelihoods and restricted economic growth. The declining GDP per capita underscores the deteriorating living standards and growing challenges faced by the population in Gaza during this period.

Based on the preceding analysis, it is clear that even before the current war, poverty in Gaza had reached alarming levels, with approximately 65% of the population living below the poverty line (UCHA 2023; World Bank 2024). The blockade has systematically undermined livelihoods, disrupted economic activity, and limited access to essential resources. Combined with widespread unemployment and food insecurity, these challenges left much of the population struggling to meet basic needs, creating a reliance on humanitarian aid. This pre-existing crisis has only been exacerbated by subsequent escalations, deepening the humanitarian and economic challenges faced by Gazans.

2.3.2 Multidimensional Poverty

Poverty in Gaza extends far beyond income deprivation, encompassing severe deficits in health, education, and living standards, as highlighted by the MPI. The healthcare system in Gaza faced significant challenges, with hospitals frequently experiencing shortages of essential medical supplies and limited access to specialized care due to the blockade. Malnutrition was widespread, particularly among children, who suffer from high rates of stunting and wasting, reflecting the long-term effects of chronic food insecurity and inadequate healthcare services. The situation is exacerbated by a lack of clean water and proper sanitation, which increases the prevalence of preventable diseases and further strains an already overwhelmed health system (Waterston and Nasser 2017; Issa et al. 2021; Assaf et al. 2023; Horino et al. 2024).

Education in Gaza has been profoundly affected by poverty and conflict. Overcrowded classrooms, damaged schools, and limited resources undermine the quality of education. Many children are unable to attend school regularly, with displacement, economic hardships, and the psychological toll of conflict acting as significant barriers (Akesson 2015; Abusamra et al. 2023; Iriqat et al. 2025). This has long-term implications for the future of Gaza, as generations are deprived of the education necessary to break the cycle of poverty.

Living standards have also been devastated by the blockade, which restricts access to essential goods such as construction materials, clean water, and fuel. Families faced daily challenges with electricity available for only a few hours each day, leaving homes and schools in darkness and disrupting essential services. The lack of access to fundamental necessities created a constant struggle for survival, further entrenching the cycle of multidimensional poverty in Gaza (ESCWA 2023b; Buheji and Hasan 2024; Migdad and Buheji 2024b).

2.4 Poverty During the Recent War on Gaza

2.4.1 Escalation of Poverty and Its Interconnected Challenges

This war on Gaza has led to a catastrophic escalation of poverty and its associated indicators, pushing the besieged enclave into unprecedented humanitarian and economic turmoil. On October 9, 2023, Israel announced a "total" blockade, banning the entry of food, electricity, and fuel. This, combined with the destruction of infrastructure, homes, and businesses, has left the entire population in Gaza grappling with acute shortages of essential goods and services. The war has exacerbated an already dire situation, compounding pre-existing poverty and reliance on humanitarian aid, and leading to what was recently described as a second Nakba (Choonara 2024).

One of the most devastating effects of the conflict is the massive loss of lives and injuries. As of November 2024, nearly 43,000 people have been killed by Israeli forces during the first year of the war, with the majority of victims being women and children. Tens of thousands more have been injured or remain missing, underscoring the devastating human cost of the conflict (Amnesty International 2024; Hamad et al. 2024b; McMahan 2024; Samuel 2024). The loss of life has not only devastated families but has also had profound economic implications, as breadwinners and working-age individuals constitute a significant portion of those lost. The psychological and social impacts of these losses are immeasurable, with survivors facing immense trauma, grief, and a breakdown of community cohesion.

Entire families have been displaced, and countless children have been orphaned, further deepening the cycle of poverty.

The loss of women and children during the war deepens poverty in Gaza by exacerbating immediate and long-term socioeconomic challenges. Women often play vital roles in household stability and economic support, and their loss leaves families more vulnerable. The death or injury of children diminishes future human capital, undermines economic recovery, and places emotional and financial burdens on families, especially when long-term medical care is required for surviving children. Already in November 2023, the United Nations Secretary-General, António Guterres, has warned that Gaza was becoming 'a graveyard for children'. The disruption of education for children further perpetuates intergenerational poverty, limiting opportunities for growth and stability. These losses amplify dependence of Gazans on humanitarian aid and hinder prospects for recovery and resilience (Al Attar and Brik 2024; Ben Saad 2024; Hamad et al. 2024b; Klin and Clin 2024; UN Women 2024).

Displacement could be considered another critical indicator of poverty, especially in the context of the war on Gaza. Nearly 1.9 million Palestinians have been forcibly displaced, with many uprooted multiple times under dangerous and chaotic conditions (Huss and Altehe 2024; Kerbaj et al. 2024). The majority of those displaced now live in overcrowded tents or makeshift shelters, lacking access to basic necessities such as food, clean water, and sanitation. This forced displacement has stripped families of their homes, livelihoods, and access to essential services, further entrenching poverty across the region. Overcrowding in displacement areas has exacerbated resource scarcity, intensifying food insecurity and increasing dependency on humanitarian aid (Human Rights Watch 2024; Abed Alah 2024). This cycle of displacement not only perpetuates economic vulnerability but also deepens psychological and social hardships, making poverty an inescapable reality for much of the population in Gaza.

Food insecurity in Gaza has reached catastrophic levels during this war, becoming a defining indicator of the deepening poverty crisis. The Integrated Food Security Phase Classification (IPC) has classified the entire Gaza Strip in IPC Phase 4 (Emergency) acute food insecurity, with significant portions of the population nearing famine-like conditions (IPC Phase 5). The destruction of agricultural lands, livestock, and fisheries has decimated local food production, leaving households unable to access fresh produce or staples. Disrupted trade and supply chains have caused food prices to surge, making basic necessities unaffordable for most families. As a result, the majority of the population relies on emergency aid, while some resort to desperate measures, such as consuming non-edible resources like animal feed or foraging for weeds (Al Sharjabi et al. 2024; Hassoun et al. 2024; IPC 2024). These alarming IPC findings highlight the severe scarcity of food, which not only exacerbates immediate poverty but also undermines health, resilience, and economic productivity, further entrenching long-term socioeconomic challenges.

The widespread destruction of healthcare and medical infrastructure in Gaza during the war has significantly deepened poverty and exacerbated the humanitarian crisis. Hospitals and clinics, already underfunded and overburdened before the war, have been severely damaged or rendered non-functional due to repeated airstrikes and other military actions. Essential medical equipment, supplies, and pharmaceuticals are in critical shortage, while power outages and fuel scarcity have further hindered the ability of healthcare facilities to provide even basic services. Ambulances and emergency response systems are frequently unable to reach those in need, leaving countless injuries untreated and preventable deaths on the rise (Zughbur 2023; Salmiya 2023; Mohammed et al. 2024). This collapse of the healthcare system has profound implications for poverty in Gaza. Families with injured or chronically ill members face overwhelming financial burdens due to limited access to affordable medical care, pushing them further into destitution (Blanchet et al. 2024; Mohammed et al.

2024). The inability to address malnutrition and other preventable illnesses exacerbates the already dire food insecurity and reduces productivity, hindering any potential economic recovery. The destruction of medical infrastructure not only increases immediate suffering but also perpetuates long-term poverty by eroding public health, weakening resilience, and increasing reliance on humanitarian aid. Addressing these healthcare challenges is critical to alleviating the multifaceted poverty crisis in Gaza.

Since its establishment in 1949, UNRWA has played a vital role in supporting Palestinian refugees and alleviating poverty across Gaza, the West Bank, and other regions. The agency provides essential services such as education, healthcare, food aid, and job opportunities to millions of vulnerable individuals, many of whom have limited access to basic resources. In Gaza, where over 80% of the population relies on humanitarian aid, UNRWA's programs have been a lifeline, addressing food insecurity, improving public health, and fostering education to reduce intergenerational poverty. However, following the attacks of October 7, 2023, Israel accused some UNRWA members of being affiliated with Hamas, claiming their involvement in or support of the group (El-Shewy et al. 2024). These accusations have led to heightened scrutiny and calls to restrict or dismantle UNRWA, culminating in bans that severely disrupt its operations. The impact of these actions is profound, particularly in Gaza, where the absence of support of UNRWA exacerbates poverty, food insecurity, and unemployment, leaving millions without access to essential services and worsening an already dire humanitarian crisis (Hamad et al. 2024a).

The war has further compounded unemployment in Gaza, which was already among the highest in the world. The destruction of workplaces, factories, and businesses during the war has left countless families without any means of livelihood, driving unemployment to unprecedented levels. Within just 3 months of the war, the unemployment rate soared to approximately 80%, and it has now approached nearly 100%, reflecting the near-total collapse of Gaza's economy (Palestine Economic Policy Research Institute—MAS 2023; UNCTAD 2024). This alarming rise underscores the devastating impact of the war on livelihoods and the inability of families to meet even their most basic needs. The absence of economic opportunities has created a cycle of dependency and despair, leaving little room for recovery.

The war and its far-reaching effects on livelihoods, displacement, and essential services have caused the poverty rate in Gaza to soar. By the 16th day of the war, over 96% of the population was classified as multidimensionally poor, according to ESCWA estimates (ESCWA 2023a; ESCWA/UNDP 2023). Widespread unemployment, severe food insecurity, forced displacement, and the destruction of healthcare and infrastructure have left nearly the entire population living in poverty (UNCTAD 2024; World Bank 2024). This alarming level of deprivation highlights not only immediate economic hardships but also profound long-term socioeconomic challenges, requiring substantial international intervention for recovery.

2.4.2 Long-Term Consequences of Poverty Created by the War

The war on Gaza has created far-reaching long-term consequences for the population, particularly in the areas of education, child food poverty, public health, and female labor force participation. These interconnected challenges not only reflect the depth of the humanitarian and socioeconomic crisis in Gaza but also exacerbate poverty in the long term. Disrupted education limits future opportunities for economic mobility, while worsening public health crises reduce workforce productivity and increase household financial burdens. Similarly, child food poverty undermines the physical and cognitive development of the next generation, perpetuating cycles of deprivation, and restricting household income and economic resilience.

In more details, the destruction of schools and the displacement of families have severely disrupted the education of children in the Gaza Strip. Many students are unable to continue their studies, resulting in significant learning gaps and delayed cognitive development for those who survived. This interruption in education not only hampers knowledge acquisition but also impacts the ability of children to process information, develop reasoning skills, and make sound decisions. Furthermore, displaced students often exhibit poor cognitive functions, negative emotions, and disruptive behaviors, creating additional barriers to effective learning (Buheji and Buheji 2024; Phusavat and Buheji 2024; Iriqat et al. 2025).

The war has exacerbated the prevalence of chronic and infectious diseases in Gaza, placing an overwhelming burden on the already devastated healthcare system. The destruction of healthcare facilities and limited access to clean water, sanitation, and essential medicines have worsened the spread of diseases like diabetes, hypertension, and respiratory infections, alongside outbreaks of preventable infectious diseases such as cholera and measles. Poor living conditions in overcrowded shelters further heighten the risks of disease transmission, while malnutrition weakens immune systems, particularly among children and the elderly (Marou et al. 2024; Taha et al. 2024).

UNICEF has introduced the concept of "child food poverty" to highlight the issues of dietary deprivation and poor-quality nutrition in early childhood, aiming to prioritize these challenges in global efforts to meet the nutrition targets of SDGs for children under the age of five (UNICEF 2024a). After only 6 months from the recent war, UNICEF (2024b) mentioned that nine out of ten children in the Gaza Strip are experiencing severe food poverty, surviving on two or fewer food groups per day, based on five rounds of data collected between December 2023 and April 2024. In a recent study, it was revealed that the war on Gaza has severely impacted the nutritional status of children under five, with 16.6% experiencing acute malnutrition. Among them, 6.7% suffer from severe acute malnutrition and 9.7% from moderate acute malnutrition (Albelbeisi et al. 2024). Displacement, overcrowded shelters, and disrupted infrastructure were identified as key contributors to this crisis. Children in North Gaza were found to be at the highest risk, reflecting the devastating impact of intensified blockades and military operations.

The war has also had profound implications for female labor force participation in Gaza. The destruction of businesses and workplaces has further marginalized women, many of whom now serve as the primary caregivers in displaced households. This increased caregiving burden, combined with the lack of employment opportunities, has severely restricted the ability of women to contribute to the labor market, exacerbating household poverty (ILO 2024; UN Women 2024).

The interconnected effects of disrupted education, a collapsing healthcare system, and declining female labor force participation underscore the need for comprehensive and sustained interventions to address the long-term consequences of poverty in Gaza.

2.4.3 Implications of the War on Gaza for SDG 1 Targets

The war on Gaza has had profound implications for achieving SDG 1. Each target under this SDG has been severely affected, further pushing the population of Gaza into extreme deprivation and exacerbating long-term challenges.

Target 1.1: Eradication of extreme poverty: The war has drastically increased extreme poverty in Gaza, with nearly the entire population now living below the poverty line. Widespread destruction of homes, livelihoods, and essential infrastructure has left families without stable sources of income or shelter, pushing them into deeper destitution (Human Rights Watch 2024; ILO 2024).

Target 1.2: Reduction of poverty in all its dimensions: The multidimensional nature of poverty in Gaza has worsened due to this war. Families face deprivations across multiple dimensions, including food security, health, education, and access to clean water and sanitation. From the early stages of the war, nearly 97.9% of Gazans were already categorized as living in multidimensional poverty (ESCWA 2023a; The

World Bank 2024). The destruction of schools and healthcare facilities, coupled with forced displacement, has left families unable to meet basic needs, further entrenching cycles of poverty.

Target 1.3: Implementation of social protection systems: Social protection systems in Gaza have been rendered nearly non-functional due to the war. The destruction of government facilities and service delivery systems has disrupted efforts to provide safety nets for vulnerable populations. Programs designed to assist displaced individuals, women, and children are unable to cope with the overwhelming demand (Al Attar and Brik 2024; The World Bank 2024; Alkhaldi and Alrubaie 2024). This collapse of social protection mechanisms leaves the most marginalized populations, especially women and children, without the support necessary to survive and recover.

Target 1.4: Equal access to economic resources and basic services: Access to economic resources and basic services has been severely disrupted by the war. The destruction of businesses, agricultural lands, and markets has eliminated economic opportunities for countless families. Furthermore, the blockade continues to limit access to essential goods, including food, fuel, and medical supplies, leaving Gazans reliant on external aid. Basic services such as electricity, water, and healthcare are either unavailable or extremely limited, further worsening the living conditions of the population in Gaza (Nassar et al. 2023; UNCTAD 2024).

Target 1.5: Building resilience to economic and environmental shocks: The war has significantly undermined the resilience of Gaza to economic and environmental shocks. The destruction of critical infrastructure and the displacement of almost all people have eroded the capacity of the Strip to recover from conflict-related damage. The war has also heightened vulnerabilities to environmental challenges, such as water scarcity and contamination, further compounding poverty (Buheji and Mushimiyimana 2023; Khalidi and Iwidat 2024). Without substantial international intervention, the ability of Gaza to rebuild and strengthen its resilience remains uncertain (Buheji and Al-Muhannadi 2023).

In conclusion, the war has set back progress on SDG 1 in Gaza by decades, creating a humanitarian and socioeconomic crisis that impacts every target of the goal. Urgent actions are required to address these challenges, including immediate humanitarian aid, the lifting of the blockade, and long-term strategies to rebuild infrastructure, restore livelihoods, and create sustainable development pathways.

2.5 Challenges and Recovery Efforts

Efforts to address the severe humanitarian situation require immediate and coordinated action. A significant increase in humanitarian needs is needed, with urgent requirements for food, medical care, and shelter, particularly for displaced individuals and those who have lost their homes and livelihoods.

The economy is likely to remain deeply disrupted, with extensive damage to businesses and significant job losses continuing to affect livelihoods (Khalidi and Iwidat 2024). Although international organizations and governments are anticipated to mobilize emergency relief efforts, such measures, including temporary shelters, food aid, and medical assistance, are typically insufficient to meet the full scope of needs. The destruction is so extensive that estimations suggest Gaza's development progress has been pushed back by 69 years, significantly exacerbating poverty and deteriorating living conditions (ESCWA/UNDP 2024). Addressing long-term economic recovery, which is pivotal for reducing poverty and rebuilding resilience, will require strategic and sustained interventions beyond immediate relief.

To rapidly establish early self-sufficiency networks and combat poverty in Gaza, a multifaceted approach focused on the prioritization and optimization of resource allocation is essential. Agricultural revitalization should be prioritized, with efforts to rehabilitate farmland and introduce urban farming initiatives to ensure food security. Supporting small enterprises and the private sector through microloans, grants, and technical training can help rebuild local businesses and create employment opportunities, particularly in critical sectors like construction and food production (ESCWA/UNDP 2024; PFI/FPCCIA 2024).

Reconstruction efforts should emphasize local engagement and capacity-building, with a focus on community-driven solutions. Emphasis should be placed on empowering local expertise, fostering equitable participation, and addressing systemic barriers like dependency on external actors. Sustainable development strategies should integrate local knowledge with practical solutions, promoting resilience and long-term self-sufficiency in Gaza (Barakat and Porteous 2024). In parallel, emergency social safety nets such as direct cash transfers, food vouchers, and subsidized goods should be established to ensure that vulnerable populations can meet their basic needs during the recovery period (Al Sharjabi et al. 2024). These strategies, rooted in local engagement and supported by international partnerships, can lay the groundwork for sustainable self-sufficiency in Gaza.

2.6 Proposed Framework for Mitigating the Risks of Poverty in Gaza

Addressing and mitigating the risks of poverty in Gaza after the October 2023–January 2025 war require a holistic approach that balances immediate humanitarian needs with long-term developmental strategies. The framework shown in Fig. 2.1 provides a comprehensive roadmap for tackling poverty and rebuilding resilience in the Gaza Strip.

The framework prioritizes immediate humanitarian assistance to alleviate the urgent needs of those affected by the war. This includes providing nutritious food to combat food insecurity and offering medical assistance to strengthen healthcare systems and address critical health needs. In addition, distributing emergency shelter and non-food essentials ensures that displaced and vulnerable populations have access to basic living necessities. Psychosocial support is also a key component, focusing on counseling and mental health services to help individuals cope with trauma. These efforts are complemented by programs designed to protect children from abuse, exploitation, and developmental delays, recognizing the heightened vulnerability of this group during times of crisis.

Beyond addressing immediate needs, the framework emphasizes long-term poverty mitigation strategies that focus on rebuilding and revitalization. Restoring critical infrastructure (such as electricity, clean water, sanitation systems, schools, and healthcare facilities) is identified as a cornerstone of recovery, enabling communities to regain stability and access essential services. Economic recovery is supported by initiatives to rebuild small local businesses, which play a vital role in generating income and employment opportunities. Improving transportation networks is also seen as essential for facilitating trade, movement, and access to resources.

Moreover, to enhance livelihoods and empower Gazans, the framework advocates for vocational and skills training to increase employability and help people integrate into the workforce. Strengthening social and child protection systems is essential to provide safety nets for the most vulnerable populations, ensuring that they are supported through this challenging period and beyond.

By addressing both immediate needs and systemic challenges, this proposed framework offers a path to not only alleviate poverty but also foster sustainable development and resilience in Gaza's recovery.

2.7 Discussion and Conclusion

The blockade of Gaza, imposed in 2007 and compounded by recurrent wars, has created a persistent humanitarian, economic crisis, and multidimensional poverty. The recent war has significantly worsened existing conditions, resulting in widespread destruction of infrastructure, mass displacement, and further economic decline. The impact of the war will continue to drive a sharp rise in multidimensional poverty,

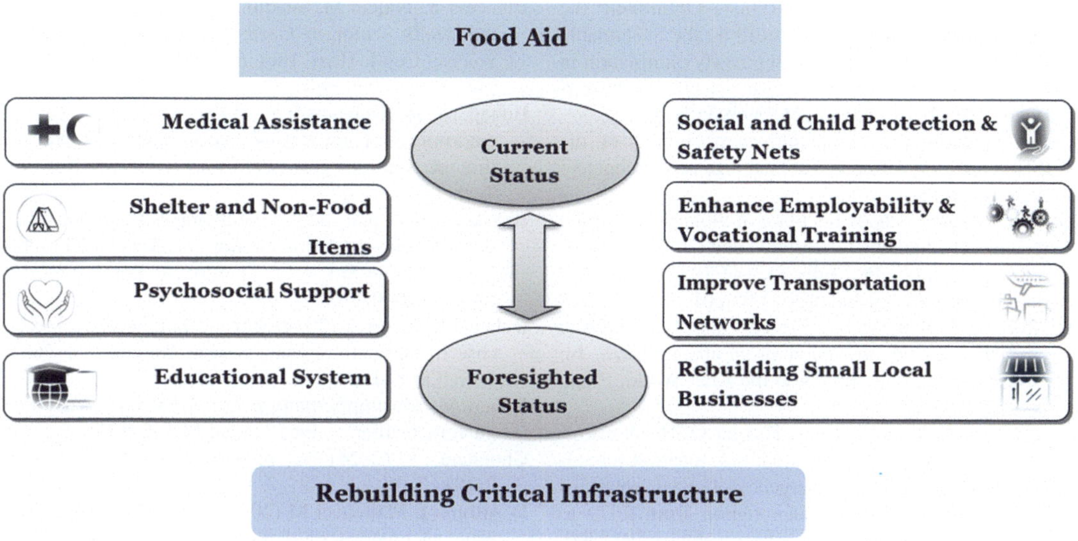

Fig. 2.1 Proposed framework for mitigating poverty risks and ensuring rapid recovery in Gaza

reflected in worsening indicators such as inadequate housing, loss of livelihoods, restricted access to healthcare, disruptions in education, and limited availability of clean water and sanitation. Displacement and the collapse of essential services have further intensified an already severe humanitarian crisis, affecting multiple dimensions of well-being.

The siege and restrictions on Gaza's trade, imports, exports, and infrastructure, such as the absence of an operational airport and seaport, have severely exacerbated poverty in this region. Ending these restrictions and enabling Gazans to access their undeniable economic and human rights, as emphasized by international resolutions, would significantly mitigate poverty. Gazans have demonstrated resilience and resourcefulness in rebuilding after previous wars, often reusing rubble as raw material for reconstruction and infrastructure development. With even minimal international support, rapid progress could be made to rebuild homes, businesses, and essential services and infrastructures, paving the way for economic revitalization and poverty reduction. This potential underscores the importance of lifting restrictions to allow Gaza's recovery and self-sufficiency.

Looking ahead, the poverty crisis in Gaza is expected to remain severe in the near future, with widespread humanitarian needs requiring urgent attention. The proposed framework underscores the critical importance of immediate relief efforts, including the provision of food, medical care, and emergency shelter, to address the most pressing needs. However, relying solely on short-term interventions will not be sufficient to tackle the root causes of poverty. A sustainable recovery demands a comprehensive approach that integrates immediate humanitarian assistance with long-term strategies for economic reconstruction, infrastructure rehabilitation, and the development of social safety nets. Crucially, achieving lasting progress and peace in the region requires recognizing the Palestinian people's right to establish their own sovereign state, which is fundamental to fostering resilience, restoring dignity, and paving the way for sustainable development in Gaza.

References

Abed Alah M (2024) Shattered hierarchy: how the Gaza conflict demolished Maslow's pyramid of needs. Curr Psychol 1–7. https://doi.org/10.1007/S12144-024-05784-1/FIGURES/1

Abusamra A, Suyanto, Wibawa S (2023) The principals' role in creating safe schools in war zones: Gaza strip as a case study. J Educ 203:367–380. https://doi.org/10.1177/00220574211031951

Ajaj R, Buheji M, Hassoun A (2024) Optimizing the readiness for industry 4.0 in fulfilling the Sustainable Development Goal 1: focus on poverty elimination in Africa. Front Sustain Food Syst 8:1393935. https://doi.org/10.3389/FSUFS.2024.1393935/BIBTEX

Ajluni S (2014) The 'Disemployment' impact of the 2014 conflict in Gaza: an ILO damage assessment and recovery strategy

Akesson B (2015) School as a place of violence and hope: tensions of education for children and families in post-intifada Palestine. Int J Educ Dev 41:192–199. https://doi.org/10.1016/j.ijedudev.2014.08.001

Al Attar HMJ, Brik AB (2024) The historical tragedy: evaluating social protection programs in Gaza. In: Brik AB (ed) Public policy in the Arab World. Elgar, pp 213–243

Al Sharjabi SJ, Al Jawaldeh A, Hassan OEH, Dureab F (2024) Understanding the food and nutrition insecurity drivers in some emergency-affected countries in the eastern Mediterranean region from 2020 to 2024. Nutrients 16:3853. https://doi.org/10.3390/nu16223853

Alagha M, Hussein N (2024) The political economy of Israeli economic warfare on the Gaza strip. In: Elkahlout G (ed) Gaza's cycle of destruction and rebuilding. Palgrave Macmillan, Cham, pp 125–145

Albelbeisi A, Zinszer K, Hamid A, Bilbeisi E (2024) The burden of acute malnutrition among children under five in conflict-afflicted Gaza strip: prevalence and associated factors. Front Nutr 1478485. https://doi.org/10.3389/fnut.2024.1478485

Alkhaldi M, Alrubaie M (2024) Roadmap for rebuilding the health system and scenarios of crisis path in Gaza. Int J Health Plann Manag 1–13. https://doi.org/10.1002/HPM.3861

Amir M (2021) Post-occupation Gaza: Israel's war on Palestinian futures. Geogr Ann Ser B Hum Geogr 103:283–300. https://doi.org/10.1080/04353684.2021.1958357

Amnesty International (2024) 'You feel like you are sub-human'—Israel's genocide against Palestinians in Gaza

Asi YM (2020) Achieving food security through localisation, not aid: "De-development" and food sovereignty in the Palestinian territories. J Peacebuild Dev 15:205–218. https://doi.org/10.1177/1542316620918555

Assaf EA, Al Sabbah H, Al-Jawadleh A (2023) Analysis of the nutritional status in the Palestinian territory: a review study. Front Nutr 10:1206090. https://doi.org/10.3389/FNUT.2023.1206090/BIBTEX

Barakat BS, Porteous P (2024) Gaza: leadership and reconstruction for the "day after"

Ben Saad H (2024) Urgent humanitarian appeal: protecting the lives of women and children in the Gaza strip (Palestine). J Public Health Res 13:1–2. https://doi.org/10.1177/22799036241229312

Blanchet K, Najem M, Shadid L et al (2024) Rebuilding the health sector in Gaza: alternative humanitarian voices. Confl Heal 18:4–6. https://doi.org/10.1186/s13031-024-00599-0

Buheji M, Al-Muhannadi K (2023) Mitigating risks of environmental impacts on Gaza—review of precautions & solutions post (2023 war). Int J Adv Res Eng Technol 14:15–47

Buheji M, Buheji B (2024) Mitigating risks of slow children development due to war on Gaza 2023. Int J Psychol Behav Sci 14:11–21. https://doi.org/10.5923/j.ijpbs.20241401.02

Buheji M, Hasan A (2024) Beyond famine and chaos—case of Gaza. Int J Manag. https://doi.org/10.17605/OSF.IO/B6SPQ

Buheji M, Mushimiyimana E (2023) "Gaza "—towards an agile resilience. Int J Manag 14:120–136

Choonara J (2024) A second Nakba? Int Social 2(181):3–26

D'Attoma I, Matteucci M (2024) Multidimensional poverty: an analysis of definitions, measurement tools, applications and their evolution over time through a systematic review of the literature up to 2019. Qual Quant 58:3171–3213

El Bilbeisi AH, Al-Jawaldeh A, Albelbeisi A et al (2022) Association of household food insecurity with dietary intakes and nutrition-related knowledge, attitudes, and practices among parents, aged $\geq$ 18 years in Gaza strip, Palestine: a descriptive study. Heliyon 8:e09582. https://doi.org/10.1016/j.heliyon.2022.e09582

El-Shewy M, Griffiths M, Jones C (2024) Israel's war on Gaza in a global frame. Antipode 1–21. https://doi.org/10.1111/ANTI.13094

ESCWA (2022) Survey of economic and social developments in the Arab region 2021–2022

ESCWA (2023a) War on Gaza 2023: an unprecedented and devastating impact

ESCWA (2023b) Social and economic situation of Palestinian women and girls: July 2020–June 2022

ESCWA/UNDP (2023) Gaza war: expected socioeconomic impacts on the State of Palestine: Preliminary estimations until 5 November 2023

ESCWA/UNDP (2024) Gaza war: Expected socioeconomic impacts on the State of Palestine—October 2024

Hamad BA, Jamaluddine Z, Aly S et al (2024a) UNRWA's work is at risk again. Lancet 404:2157. https://doi.org/10.1016/S0140-6736(24)02476-0

Hamad BA, Vintges J, Diab R, et al (2024b) Shattered lives and dreams: the toll of the war on Gaza on young people

Hassoun A, Al-Muhannadi K, Hassan HF et al (2024) From acute food insecurity to famine: how the 2023/2024 war on Gaza has dramatically set back sustainable development goal 2 to end hunger. Front Sustain Food Syst 8:1402150. https://doi.org/10.3389/fsufs.2024.1402150

Horino M, Zaqqout R, Habash R et al (2024) Food insecurity, dietary inadequacy, and malnutrition in the Gaza strip: a cross-sectional nutritional assessment of refugee children entering the first grade of UNRWA schools and their households before the conflict of 2023–24. Lancet Glob Heal 12:e1871–e1880. https://doi.org/10.1016/S2214-109X(24)00320-6

Human Rights Watch (2024) "Hopeless, starving, and besieged" Israel's forced displacement of Palestinians in Gaza

Huss M, Altehe S (2024) On spaciocide and resistance: between Bi'r as-Sab'a and Gaza. Polit Geogr 114:103186. https://doi.org/10.1016/j.polgeo.2024.103186

ILO (2018) The occupied Palestinian territory: an employment diagnostic study

ILO (2024) A year of war in Gaza: impacts on employment and livelihoods in the West Bank and Gaza strip. Bulletin No 5

IPC (2024) IPC famine review committee alert Gaza strip

Iriqat D, Alousi R, Aldahdouh TZ et al (2025) Educide amid conflict: the struggle of the Palestinian education system. Qual Educ All 2:81–99. https://doi.org/10.1108/QEA-10-2024-0120

Issa YY, Amro A, Rajabi RK (2021) Environmental and social determinants of health in Palestine. In: Handbook of healthcare in the Arab world. Springer, Cham, pp 607–649

Kerbaj M, Traboulsi J, Maasri Z, Toukan H (2024) Gaza in plain sight: witnessing in solidarity. J Vis Cult 22:1–19. https://doi.org/10.1177/14704129241227847

Khalidi R, Iwidat Q (2024) Assessing the economic and social impacts of Israel's war on Palestine. Arab Cent Res Policy Stud 7:80–90

Klin A, Clin A (2024) Behind the rubble: psychological trauma of wars and human rights abuses on women and children in Gaza. Anatol Clin J Med Sci 29:119–136. https://doi.org/10.21673/anadoluklin.1575372

Marou V, Vardavas CI, Aslanoglou K et al (2024) The impact of conflict on infectious disease: a systematic literature review. Confl Heal 18:1–32. https://doi.org/10.1186/s13031-023-00568-z

McMahan J (2024) Proportionality and necessity in Israel's invasion of Gaza, 2023-2024. Anal und Krit 46:387–407. https://doi.org/10.1515/auk-2024-2024

Mhlanga D, Ndhlovu E (2024) The Russia-Ukraine conflict and development in Africa. Springer

Migdad M, Buheji M (2024a) Resilience amidst devastation: qualitative insights from Gaza during the war. Int J Soc Sci Res Dev 6:13–27

Migdad M, Buheji M (2024b) Realizing the capacity of the internal front—case of Gaza 2024. Community Pract 21:544–565. https://doi.org/10.5281/zenodo.13622097

Mohammed F, Elgailani USA, Ibrahim Ali SY et al (2024) Defending the right to health in Gaza: a call to action by health workers. Confl Heal 18:1–4. https://doi.org/10.1186/S13031-024-00613-5/METRICS

Nasreddine L, Jomaa L (2024) Safeguarding the SDG promise to end hunger and leave no one behind: the plight of children in the Gaza strip. Lancet Glob Heal 12:e1748–e1749. https://doi.org/10.1016/S2214-109X(24)00367-X

Nassar S, Tóth ZN, Vasa L (2023) Economic empowerment as a result of achieving SDGs with resource access: a comparative research between Gaza strip and Hungary. J Int Stud 16:9–33. https://doi.org/10.14254/2071-8330.2023/16-2/1

Nijim M (2024) Gazacide: Palestinians from Refugeehood to ontological obliteration. Crit Sociol 1944:1–23. https://doi.org/10.1177/08969205241286530

OPHI/UNDP (2024) Global multidimensional poverty index

Palestine Economic Policy Research Institute—MAS (2023) Gaza war economy brief fundamental shocks pushed the Gazan economy to the brink of collapse on the eve of the war

PFI/FPCCIA (2024) Impact of the Gaza war on private sector and pathways for recovery

Phusavat K, Buheji M (2024) Mapping informal learning for displaced learners during the war on Gaza 2023-application of situated cognition. Int J Learn Dev 14:1–16. https://doi.org/10.5296/ijld.v14i1.21626

Romano D, Stefani G, Rocchi B, Fiorillo C (2019) The impact of assistance on poverty and food security in a fragile and protracted-crisis context: the case of west bank and Gaza strip. Bio Based Appl Econ 8:21–61. https://doi.org/10.13128/bae-8145

Roy S (2016) The Gaza strip: the political economy of de-development, 3rd edn. Institute for Palestine Studies, Washington, DC

Salmiya MA (2023) Gaza hospitals: military siege and bombings. Lancet 402:2191. https://doi.org/10.1016/S0140-6736(23)02566-7

Samuel MT (2024) The Israel-Hamas war one year later: mass violence and Palestinian dispossession. Middle East Policy 31:3–16. https://doi.org/10.1111/mepo.12792

Taha AM, Mahmoud H, Nada SA, Abuzerr S (2024) Controlling the alarming rise in infectious diseases among children younger than 5 years in Gaza during the war. Lancet Infect Dis 24:e211. https://doi.org/10.1016/S1473-3099(24)00067-7

The World Bank (2024) Gaza strip—interim damage assessment

UCHA (2023) Movement in and out of Gaza in 2022. https://www.ochaopt.org/content/movement-and-out-gaza-2022. Accessed 6 Dec 2024

UN Women (2024) Gender alert: the gendered impact of the crisis in Gaza

UNCTAD (2024) Preliminary assessment of the economic impact of the destruction in Gaza and prospects for economic recovery

UNICEF (2024a) Child food poverty. Nutrition deprivation in early childhood. Child nutrition report

UNICEF (2024b) 1 in 4 children globally live in severe child food poverty due to inequity, conflict, and climate crises. https://www.unicef.org/press-releases/1-4-children-globally-live-severe-child-food-poverty-due-inequity-conflict-and. Accessed 5 Dec 2024

UNRWA (2023) Gaza_15 years of blockade. https://www.unrwa.org/gaza15-years-blockade. Accessed 3 Dec 2024

Waterston T, Nasser D (2017) Access to healthcare for children in Palestine. BMJ Paediatr Open 1:1–6. https://doi.org/10.1136/bmjpo-2017-000115

World Bank (2024) Note on the impacts of the conflict in the Middle East on the Palestinian economy—September 2024. Washington, DC

World Bank Group (2024) Macro poverty outlook: country-by-country analysis and projections for the developing World

Zavyalova EB, Krotova TG (2021) Methods for achieving SDG 1, poverty eradication. In: Lecture notes in networks and systems. Springer, pp 1987–2001

Zughbur MR (2023) Protect civilians' lives and health care in Gaza. Lancet 402:1620–1621. https://doi.org/10.1016/S0140-6736(23)02402-9

From Acute Food Insecurity to Famine: The Struggle of Gazans for Food Security

3

MoezAlIslam Ezzat Faris, Tarek Ben Hassen, Carlos Parra-López, and Abdo Hassoun

Abstract

The recent war (October 2023–January 2025) on Gaza has triggered a profound humanitarian crisis, causing widespread destruction, deprivation, and suffering on a potentially unprecedented scale, with a grave risk of famine and severe food insecurity. This chapter examines the profound impact of this war in Gaza on Sustainable Development Goal 2 (Zero Hunger), emphasizing the acute food insecurity and escalating famine devastating the besieged region. The relentless bombardment and ground operations have devastated agricultural lands, livestock, the marine environment, and the fishing industry, destroying all critical food production infrastructure. The result has been severe shortages, exacerbating an already dire food crisis that some agencies deem unprecedented in this context. The blockade and continuous military actions have severely restricted the import of essential food supplies and fuel, hampering the ability to grow and process food locally. As a result, Gazans are left without sufficient access to basic nutrition, putting them at risk of famine. Moreover, there has been an alarming increase in the number of children succumbing to hunger and severe malnutrition in Gaza. Through detailed analysis of grey literature and scientific publications, this chapter sheds light on the devastating impact of this war on Gaza on food security, highlighting the multifaceted dimensions of this crisis. It highlights the urgent need for immediate international humanitarian intervention to address the crisis and implement sustainable solutions that will rebuild Gaza's food systems, ensuring all Gazans have access to a sufficient and nutritious diet.

M. E. Faris (✉)
Department of Clinical Nutrition and Dietetics, Faculty of Allied Medical Sciences, Applied Science Private University, Amman, Jordan
e-mail: m_faris@asu.edu.jo

T. B. Hassen
Department of International Affairs, College of Arts and Sciences, Qatar University, PO, Doha, Qatar
e-mail: thassen@qu.edu.qa

C. Parra-López
Department of Agrifood System Economics. Institute of Agricultural and Fisheries Research and Training (IFAPA), Centre 'Camino de Purchil', Granada, Spain
e-mail: carlos.parra@juntadeandalucia.es

A. Hassoun
Sustainable AgriFoodtech Innovation & Research (SAFIR), Arras, France

Faculty of Agricultural Engineering, University of Aleppo, Aleppo, Syria
e-mail: a.hassoun@saf-ir.com

A. Hassoun (ed.), *War on Gaza*, Sustainable Development Goals Series,
https://doi.org/10.1007/978-3-031-88500-6_3

Keywords

Blockade · Famine · Hunger · Humanitarian crisis · Malnutrition · War on Gaza · Sustainable Development Goal 2

3.1 Introduction

Hunger, food insecurity, and famine remain some of the most pressing challenges in the modern world. Despite significant advancements in agricultural technology and global efforts to combat hunger, millions of people across the globe continue to suffer from insufficient access to food. According to the Food and Agriculture Organization (FAO), International Fund for Agricultural Development (IFAD), the United Nations Children's Fund (UNICEF), the World Food Programme (WFP), and the World Health Organization (WHO), in 2023, approximately 733 million people globally (i.e., 1 in 11) faced hunger, with Africa experiencing the highest burden at 1 in 5. Food insecurity remains widespread, affecting 2.33 billion individuals, including 864 million in severe food insecurity, largely due to conflicts, economic challenges, and climate change. The lack of access to healthy diets persists, with over 2.8 billion people unable to afford nutritious food, particularly in low-income countries where affordability challenges are most severe. Despite some progress in child stunting reduction, malnutrition, including wasting, anemia, and obesity, continues to rise, underscoring the complex and multifaceted nature of the global food crisis (FAO et al. 2024).

Food insecurity, as defined by the FAO, is a condition where individuals lack consistent access to enough nutritious food for an active and healthy life. This issue often precedes famine, the most extreme manifestation of food insecurity, characterized by widespread hunger, acute malnutrition, and high mortality rates. Famine typically arises in the context of systemic crises, including war, natural disasters, and economic collapse. Projections in 2023 indicate that global progress toward Sustainable Development Goal 2 (SDG 2) has reversed, leaving millions vulnerable to the devastating impacts of food insecurity (Béné and Devereux 2023; Doocy et al. 2023; FAO, IFAD, WHO, UNICEF 2024).

The situation in conflict zones is particularly alarming. War and political instability disrupt food production, supply chains, and access to resources, creating environments where food becomes a weapon of war. Across regions like Yemen, Somalia, South Sudan, and Syria, the convergence of conflict and hunger has led to humanitarian crises of unprecedented scale. These regions underscore the complex relationship between conflict and food insecurity, where instability not only disrupts livelihoods but also entrenches long-term vulnerabilities in local populations (Martin-Shields and Stojetz 2019; Brück and d'Errico 2019; de Waal 2024a).

Against this global backdrop, the Gaza Strip represents one of the most acute and tragic cases of food insecurity escalating into famine. While hunger in Gaza is not new, the scale and severity of the current crisis are unparalleled. According to reports by various international agencies, decades of geopolitical conflict, exacerbated by a 17-year blockade, have drastically reduced Gaza's ability to sustain itself. Agricultural production has been stifled, water resources weaponized, and humanitarian aid obstructed in this region, leaving its population of over 2.3 million people in an unrelenting state of dependency and deprivation (Alnabih et al. 2024; de Waal 2024b; Hassoun et al. 2024; Insight Insecurity 2024; Whyte 2024). Hunger in Gaza reflects a multifaceted crisis. Unlike natural disasters or economic downturns, which can sometimes be mitigated through swift international aid, the crisis in Gaza is deeply rooted in political and military actions. Systematic policies restricting movement, limiting access to critical resources, and targeting essential infrastructure have compounded the vulnerability of the Gaza Strip. These factors have transformed what was once a condition of chronic food insecurity into a full-blown famine, with devastating consequences for the population

(B'TSELEM 2024a; Buheji and Hasan 2024; Devereux 2024; Iriqat et al. 2025).

The commitment of the international community to ending hunger is embodied in SDG 2, which aims to eliminate hunger, achieve food security, and promote sustainable agriculture by 2030. Yet, the situation in Gaza reveals how conflict and political instability can undermine these goals. The blockade and recurrent military offensives have destroyed most of the cropland, decimated its fishing industry, and rendered much of its food production infrastructure unusable. Consequently, Gazans are left heavily reliant on international aid, which is often delayed or insufficient due to logistical and political barriers (Hassoun et al. 2024; Nasreddine and Jomaa 2024). The implications of this crisis are dire. Recent reports indicate alarming rates of malnutrition among children, with rising instances of wasting, stunting, and severe hunger-related illnesses. These conditions not only threaten immediate survival but also compromise the long-term health and development of youth in Gaza. Women, children, and other vulnerable groups bear the brunt of these conditions, with far-reaching implications for social stability and resilience (Global Nutrition Cluster 2024; Samuel 2024; Nasreddine and Jomaa 2024).

This chapter examines the food security crisis in Gaza, beginning with an analysis of SDG 2 and a comprehensive overview of food security. It then provides an overview of food security in Gaza before the recent war, highlighting preexisting challenges and widespread food insecurity. The chapter delves into the devastating impact of the October 2023–January 2025 Gaza war on its food systems, detailing the escalation from acute food insecurity to famine. The implications of this war on SDG 2 targets are also briefly explained. It concludes with actionable recommendations for mitigating the crisis and rebuilding food systems, as well as a summary that underscores the urgent need for international intervention and sustainable solutions.

3.2 Sustainable Development Goal 2 (SDG 2): Zero Hunger and Food Security

SDG 2 of the United Nations' 2030 Agenda for sustainable development represents a global commitment to ending hunger, achieving food security, improving nutrition, and promoting sustainable agriculture. This goal reflects the understanding that hunger and malnutrition are not simply outcomes of food scarcity but are deeply tied to systemic issues, including poverty, inequality, and environmental degradation. SDG 2 comprises several ambitious targets that guide international efforts. These include ending hunger and ensuring access to sufficient, safe, and nutritious food for all, particularly for vulnerable groups such as women and children (Target 2.1); eliminating all forms of malnutrition by 2030, with a specific focus on addressing stunting and wasting in children under 5 years (Target 2.2); and doubling the agricultural productivity and incomes of small-scale food producers (Target 2.3). Additionally, it emphasizes sustainable food production systems (Target 2.4) and maintaining the genetic diversity of seeds, plants, and animals (Target 2.5) (United Nations Sustainable Development 2015; Arora and Mishra 2022; Guo 2024; Sporchia et al. 2024).

Achieving SDG 2 is pivotal for ensuring global well-being and resilience, as hunger affects not only individual health but also societal stability and economic productivity. However, global progress toward this goal has been uneven, with conflict-affected regions, such as Gaza, facing significant setbacks. In these contexts, understanding and addressing food security require a comprehensive framework that extends beyond the mere availability of food to encompass its multifaceted dimensions (Clapp et al. 2022; Manikas et al. 2023; Mockshell and Nielsen Ritter 2024).

Food security is a complex concept that integrates six key pillars: availability, access, utilization, stability, agency, and sustainability. Availability refers to the physical presence of sufficient quantities of food, either through local production, imports, or aid. Access involves the ability of individuals to obtain food, influenced by factors such as income, market structures, and social safety nets. Utilization pertains to the way food is prepared and consumed, as well as the ability of the body to absorb nutrients, which is heavily reliant on clean water, sanitation, and health services. Stability addresses the consistency of food availability and access over time, accounting for factors such as economic shocks, climate events, and conflict. The pillars of agency and sustainability further expand the traditional food security framework. Agency emphasizes the empowerment of individuals and communities to make their own decisions regarding food production and consumption, while sustainability focuses on ensuring that food systems do not compromise the needs of future generations (HLPE 2020; Clapp et al. 2022; Hassoun et al. 2024).

The interdependence of these pillars underscores the importance of a holistic approach to achieving food security, particularly in regions like Gaza, where systemic disruptions undermine all aspects of the framework. By exploring SDG 2 alongside the six pillars of food security, this chapter highlights the multidimensional nature of the crisis in Gaza and the necessity of integrated strategies to address it effectively.

3.3 Overview of Food Security in Gaza Before the Recent War: Acute Food Insecurity

The food security crisis in Gaza was deeply rooted in a history of geopolitical conflicts, blockades, and repeated military actions. Since the 1948 Arab-Israeli War, the 1967 Six-Day War, and subsequent occupation, Gaza endured systemic challenges that eroded its economic and agricultural capacities. The blockade imposed in 2007, combined with recurrent conflicts, exacerbated these vulnerabilities by restricting access to essential resources and tightly controlling the flow of goods, often calibrated to meet only a "humanitarian minimum." This policy, aimed at limiting resources to avoid malnutrition barely, left a significant portion of the population in Gaza dependent on humanitarian aid and rendered its food security system fragile and lacking resilience (UNCT 2012; Brück et al. 2019; ESCWA 2023a; Alshwaikh 2024; Iriqat 2024).

Throughout the years, multiple military operations, including the Israeli attacks in 2008–2009, 2012, 2014, and 2021, resulted in devastating damage to infrastructure, homes, and agricultural lands. The destruction of essential facilities has dramatically impaired the capacity of Gaza to produce food locally, exacerbating its fragile food security situation. Moreover, the blockade severely restricted the availability of food in Gaza according to international monitoring bodies. Imports of agricultural inputs, such as fertilizers, seeds, and essential farming materials, were heavily limited, while strict restrictions on fishing zones curtailed fishing activities. Approximately 30% of farmland in Gaza was removed from circulation due to Israeli security buffer zones (Smith and Isleem 2017), further reducing local food production in one of the most densely populated regions in the world. Repeated military offensives destroyed critical agricultural infrastructure, greenhouses, and irrigation systems. Fishing restrictions forced Gazan fishermen into polluted waters close to shore, leading to overfishing and diminished yields (Roy 2016; Hussein et al. 2022). These systemic constraints on food availability left Gaza heavily reliant on external imports and aid.

Access to food was constrained by widespread poverty and one of the highest unemployment rates globally, reaching about 80% among youth and women (Ridao-Cano et al. 2017). Soaring food prices, driven by disrupted supply chains and limited imports, left many households unable to afford sufficient food. By 2022, an estimated 64.3% of the population in Gaza experienced moderate or severe food insecurity, and 77% reported reducing the number of daily meals to cope with shortages (Abu-Ras and Mohamed 2018; Think Global Health 2024). Utilization was further undermined by poor water and sanitation infrastructure. Over 97% of the freshwater supply in Gaza was unfit for human consumption (Shomar and Rovira 2023) due to contamination from seawater and untreated sewage. Chronic energy deficits (Juaidi et al. 2016) caused by restrictions on dual-use materials, such as fuel and concrete, made refrigeration and food storage unreliable. These conditions contributed to widespread malnutrition, anemia, and other nutrient deficiencies, particularly among children (Radi et al. 2013; Horino et al. 2020; United Nations in Palestine 2022).

The severe food insecurity in Gaza reflected systemic challenges in achieving SDG 2. Progress toward Target 2.1, ensuring access to safe and nutritious food for all, was hindered by high poverty levels and restricted food availability. The inability to provide sufficient nutrition for vulnerable groups, particularly children, also highlighted the failure to meet Target 2.2. Targets 2.3 and 2.4, focusing on improving agricultural productivity and ensuring sustainable food systems, were undermined by the destruction of farmland, the unavailability of agricultural inputs, and environmental degradation caused by overexploitation and repeated military offensives. Therefore, even before October 2023, about 80% of the population in Gaza depended on food assistance to meet their basic needs (Butt and Butt 2016; Brück et al. 2019). These pre-existing vulnerabilities, rooted in the blockade, systemic poverty, and recurring conflicts, created a uniquely fragile food security environment that worsened with each crisis. The reliance on aid and the collapse of local food systems underscored the challenges Gaza faced in achieving SDG 2 targets and addressing its chronic food insecurity.

Decades of cumulative challenges left Gaza in a state of acute food insecurity, with rising malnutrition and hunger levels threatening the well-being of its population. These conditions underscored the urgent need for systemic interventions to address both immediate needs and the structural factors perpetuating the crisis. The following section examines how the recent escalation of hostilities has further intensified these vulnerabilities.

3.4 Impact of the Recent War on Gaza: Famine

Numerous recent reports have highlighted the deliberate destruction of food production facilities in Gaza and severe restrictions on the entry of essential food supplies. These actions have been described by some observers as part of a broader strategy that effectively uses food as a weapon, further exacerbating the humanitarian crisis in the region. The systematic targeting of agricultural infrastructure, including farms, fisheries, and food storage facilities, combined with the blockade on critical imports, has severely affected the lives of people in Gaza, plunging the region into an unprecedented humanitarian catastrophe marked by widespread hunger, forcible displacement of 90% of Gazans, and a collapsing ability to sustain even the most basic survival needs (ESCWA 2023b; Human Rights Watch 2023; B'TSELEM 2024b; Médecins Sans Frontières 2024).

3.4.1 Restrictions on Food Imports and Weaponization of Food Aid

The recent war on Gaza has brought unprecedented restrictions on food imports and intensified the weaponization of food aid, exacerbating an already dire humanitarian crisis. Israel has imposed a near-total blockade, with a systematic denial of humanitarian aid, preventing the entry of essential goods, including food, as part of its military strategy. This blockade has effectively cut off the residents of Gaza from vital resources, leading to a considerable decline in food availability (Checchi et al. 2024; Médecins Sans Frontières 2024), plunging the region into an unprecedented food crisis. A significant drop in the average daily truck entries into Gaza to just 58 in October 2024 marks the lowest level recorded since November 2023, highlighting the worsening severity of the humanitarian crisis (WFP Palestine Food Security Analysis 2024).

As early as mid-November 2023, the WFP raised urgent alarms, warning that food and water supplies in Gaza were nearly exhausted and that civilians faced an imminent risk of starvation (WFP 2023). On October 17, 2024, the Integrated Food Security Phase Classification (IPC) Famine Review Committee (FRC) published an updated report on the situation in Gaza. This report indicated that the humanitarian situation in the Gaza Strip is highly grave and rapidly deteriorating, classifying the entire population in the Strip in IPC Phase 4 (Emergency) acute food insecurity, including nearly 133,000 people facing catastrophic food insecurity (IPC Phase 5). According to this report, famine is highly likely to occur imminently in parts of the northern Gaza Strip (IPC 2024a, b; Samuel 2024).

Even when limited food aid manages to enter Gaza, its distribution has become a dangerous endeavor. Several nongovernmental organizations (NGOs) and human rights organizations report that civilians attempting to collect aid have been attacked and individuals seeking food have been targeted, resulting in injuries and deaths. Aid organizations have struggled to navigate this hostile environment, with numerous food warehouses destroyed and aid workers killed during their efforts to provide relief. For example, on April 1, 2024, Israeli drone strikes killed seven World Central Kitchen aid workers distributing food in Gaza (Yang 2024). On November 30, 2024, an airstrike on a World Central Kitchen vehicle in Khan Yunis killed five people, including three aid workers (Al Jazeera 2024). Incidents like the "flour massacre" (Gordon and Haddad 2024; OHCHR 2024; Whyte 2024), where civilians attempting to collect food aid were attacked, further highlight the deliberate targeting of humanitarian efforts. These actions disrupt the delivery of critical resources and create an atmosphere of fear, discouraging both aid distribution and access. Such attacks exacerbate the humanitarian crisis, leaving already vulnerable populations at greater risk of starvation and malnutrition. Some experts and international organizations have expressed concern that Israel's alleged use of food as a weapon has led to avoidable hunger, widespread starvation, and loss of life. They argue that these actions may constitute war crimes and crimes against humanity, considering the devastating impact on civilians and the potential violations of international law (Devereux 2024; Mahomed 2024; Sultany 2024).

The deliberate restriction of food imports and the targeting of aid distribution systems during this war have severely undermined the pillars of food security in Gaza. The blockade has decimated availability; access has been obstructed through attacks on civilians and aid convoys, while the unpredictable and violent environment has disrupted stability. This weaponization of food aid deepens the suffering of the population in Gaza. It emphasizes the urgent need for international intervention to safeguard civilians and ensure the unhindered flow of humanitarian assistance.

3.4.2 Destruction of the Agricultural Sector in Gaza

The Gaza Strip's 36,410 hectares of land includes 15,053 hectares (42%) dedicated to cropland. The agricultural sector, a vital source of food and income for many Gazans, has also been severely impacted due to restrictions on land access, water scarcity, and damage from military operations. Israeli attacks have resulted in substantial damage to its agricultural land and assets, impacting cropland, greenhouses, wells, and other food production components.

Recent assessment utilizing satellite data from the Food and Agriculture Organization of the United Nations (FAO) and the United Nations Satellite Centre (UNOSAT) indicated a significant increase in damage to farmland and agricultural infrastructure in the Gaza Strip. As of September 1, 2024, 68% of cropland in Gaza—equivalent to 10,183 hectares (ha)—has been reported damaged. Specifically, damage levels include 71.1% of orchards and other trees, 67.1% of field crops, and 58.5% of vegetables. Other agricultural infrastructures, such as home barns, broiler farms, and sheep farms, have been severely damaged, especially in the governorate of Khan Younis (FAO/UN 2024). FAO Deputy Director-General Beth Bechdol emphasized the gravity of the situation, noting "The scale of destruction to agricultural land in the Gaza Strip is unprecedented. This poses significant concerns for the potential of food production both now and in the future, as food aid alone is insufficient to meet the daily needs of Gaza's population. The loss of agricultural land compounds the already imminent risk of famine across the Gaza Strip" (FAO 2024). A recent study identified five distinct types of urban destruction, namely aerial bombardment, artillery shelling, bulldozing, mine blasting, and infrastructure sabotage that have collectively inflicted extensive damage on vital infrastructure in Gaza, including agricultural lands (Holail et al. 2024).

The destruction of the agricultural sector in Gaza primarily affects the availability pillar of food security, which refers to the physical presence of sufficient quantities of food through local production, imports, or aid. When agricultural lands, infrastructure, and resources are destroyed, local food production declines significantly, reducing the overall availability of food in the region. This destruction also indirectly impacts other pillars of food security. The reduced local production increases reliance on imports, which often drives up food prices, making it harder for households already struggling with poverty to afford food. Additionally, damage to agricultural production limits the availability of diverse and nutritious foods, negatively affecting the population's diet quality and nutritional outcomes. The repeated destruction of agricultural assets creates long-term instability in food production, undermining the ability of this region to maintain consistent food availability over time.

3.4.3 Damage in the Marine Environment and Fishing Industry

Restrictions imposed by Israeli authorities have largely dictated the fishing zone off the coast of Gaza. At its most permissive, the fishing zone off the coast of Gaza has historically fluctuated between three and six nautical miles, rarely extending further. This starkly contrasts with the 20-nautical-mile zone designated under the Oslo Accords, which allowed for broader fishing rights. However, Israeli-imposed restrictions have consistently limited these provisions, reducing the fishing zone to a size far smaller than originally agreed upon. These limitations severely impact Gaza's fishing industry, forcing fishers into polluted and overfished areas near the shore, reducing their livelihoods, and contributing to food insecurity in the region (Hussein et al. 2022; Qafisheh 2024). These severe limitations on access to marine resources have significantly worsened food security challenges in the Gaza Strip. Moreover, the adverse effects of climate change further intensify water insecurity in the Strip. Rising sea temperatures have disrupted the marine ecosystem, directly impacting fishing communities that rely heavily on the sea for their livelihoods. The warming waters have altered fish migration patterns and reduced fish stocks, a situation compounded by the declining quality of coastal waters due to pollution and untreated sewage discharge. These factors have collectively led to significantly smaller catches, depriving fishing communities of their primary income and diminishing a vital source of protein for the population (UNEP 2020; PAX 2023). This interplay between water insecurity, climate change, and environmental degradation underscores the urgent need for sustainable solutions to support both marine ecosystems and the communities that depend on them.

The recent Israeli assault has severely impacted the already fragile fishing sector of Gaza, compounding the food insecurity of the region. Alongside the physical destruction of fishing boats, equipment, and storage facilities, the attacks have introduced the added threat of marine pollution, with potential contamination adversely affecting fish growth, quality, and safety within the restricted fishing zones of Gaza. Fishermen who attempt to operate in these dangerous conditions face life-threatening risks from naval fire and harsh enforcement (Gordon and Haddad 2024; UNEP 2024). This combination of destruction and contamination has drastically reduced the availability of fish, a vital source of protein for the population of Gaza, while depriving thousands of families dependent on the fishing industry of their livelihoods. These developments further deepen the reliance of Gaza on humanitarian aid and exacerbate the already critical food crisis in the region.

3.4.4 Damage in Water Resources and Its Impact on Food Security

The recent Israeli assault on Gaza has devastated its already fragile water resources, exacerbating food insecurity and humanitarian crisis. Israeli attacks have systematically targeted water infrastructure, including desalination plants, wells, reservoirs, and pumping stations, crippling water supply systems in Gaza. Newly constructed facilities, such as those funded by the UAE, have not been spared, and damaged sanitation systems have further contaminated water sources. By March 2024, the daily water supply had plummeted by 94%, leaving the population in Gaza with just 4.74 L per person per day—far below the emergency minimum of 15 L recommended by the WHO (ESCWA 2023b; Abdul Samad et al. 2024; Cole 2024).

The destruction of water infrastructure has had severe repercussions for Gaza's agricultural sector. Farmers, already struggling with limited access to irrigation and rising water salinity, face near-total water scarcity, resulting in widespread crop failures and a dramatic decline in food production. Marine pollution caused by damaged sanitation systems further threatens fisheries, an essential food source, and livelihood for thousands of Gazans. These combined factors have led to soaring food prices, pushing basic nutrition out of reach for many families and deepening the dependence on aid (Middle East Institute 2024; UNEP 2024).

Water insecurity extends beyond agriculture, directly impacting household food preparation and sanitation. The lack of clean water for cooking and hygiene has increased foodborne illnesses and malnutrition, particularly among children and internally displaced persons. Diseases such as diarrhea, scabies, and hepatitis are spreading due to unsanitary conditions, with the destruction of water, sanitation and hygiene (WASH) infrastructure compounding these risks. A recent study revealed that malnutrition, combined with infectious diseases linked to poor water quality, has created a feedback loop that worsens food insecurity and weakens immunity (Zinszer and Abuzerr 2024). The systematic destruction of water resources is not merely collateral damage but a calculated strategy to weaponize water and undermine food security in Gaza. Reports, including Oxfam's "water war crimes" (Abdul Samad et al. 2024), highlight how these actions violate international humanitarian law, constituting potential war crimes and crimes against humanity.

By crippling agricultural production, restricting access to food and clean water, and obstructing humanitarian relief, these Israeli actions have fueled the unprecedented levels of famine and hunger that have plagued Gaza throughout the duration of the war. The dire situation highlights the urgent need for international intervention to rebuild food production facilities, restore access to water, and address the escalating food security crisis in the region.

3.4.5 Population-Based Reports on Famine and Hunger and the Role of NGOs

In a recent research report by Faris et al. (2024), a May–July 2024 survey of 1209 Gaza households revealed a catastrophic post-October 7th, 2023, food insecurity and hunger crisis. Using the Household Food Security Survey Module (HFSSM), Household Food Insecurity Access Scale (HFIAS), and Household Hunger Scale (HHS), this study found that 100% of households experienced food insecurity (HFSSM), with 97.7% severely food insecure (HFIAS) and 88% experiencing hunger (HHS), including 48% experiencing severe hunger—levels far exceeding those before this war. The average household member lost approximately 10.5 kg, dramatically increasing underweight prevalence while decreasing overweight and obese individuals, showcasing widespread malnutrition.

In another recent preprint, the authors investigated the psychometric properties of the Food Insecurity Experience Scale (FIES) among 534 Arabic-speaking adults in Gaza amid the Israeli war on the Gaza Strip (Fekih-romdhane et al. 2024). The study used the FIES because it focuses on lived experiences of food insecurity, making it particularly relevant to Gaza's acute, artificial hunger crisis. A concerningly high prevalence of food insecurity (50.2% of the sample, with 31.3% experiencing moderate to severe insecurity) was reported, which were lower than the findings reported by Faris et al. (2024).

NGOs are crucial in minimizing poverty and famine by providing humanitarian assistance, monitoring the situation on the ground, and collaborating with governments and other agencies. NGOs utilize early warning systems to gather accurate information about growing crises and mobilize resources to address urgent needs. They also raise awareness about impending famines, focusing on humanitarian emergencies to facilitate timely interventions (Stephenson 2020; Nassar et al. 2023).

The Israeli attacks on the Gaza Strip have drastically worsened the pre-existing humanitarian crisis in Gaza, triggering a substantial response from numerous NGOs and international agencies focused on alleviating famine and widespread food insecurity. While efforts concentrate on immediate needs such as food aid distribution (ready-to-eat meals, staples, nutritional supplements), water and sanitation (infrastructure repair, purification tablets, hygiene promotion), healthcare (malnutrition treatment, vaccinations, addressing famine-related illnesses), shelter and essential household items, and cash transfers, access and effectiveness remain severely hampered by significant challenges. These challenges include access restrictions caused by the blockade and military operations, which severely limit the entry of personnel and supplies; the volatile security situation, which poses significant risks to humanitarian workers; persistent funding shortfalls that impede aid efforts; and difficulties in achieving effective coordination among the numerous agencies involved, compounded by a large number of actors and bureaucratic hurdles.

The United Nations Relief and Works Agency for Palestine Refugees in the Near East (UNRWA) plays a critical role in supporting Palestinian refugees, particularly in Gaza. UNRWA provides essential services, including education, healthcare, food assistance, and social services, which form a lifeline for millions of Palestinians. However, the blockade imposed by Israel, along with repeated attacks on UNRWA facilities, has severely hampered the ability of this agency to operate effectively. These restrictions exacerbate food insecurity and poverty, as many Palestinians rely on various services of UNRWA for survival. The deliberate targeting and weakening of UNRWA not only undermine immediate humanitarian responses but also threaten long-term development and stability in the region (Hamad et al. 2024).

With the capacity of UNRWA constrained, vulnerable populations face worsening conditions, including heightened rates of malnutrition and disease. The failure of the international community to safeguard the operations of UNRWA and address the broader impacts of the blockade perpetuates a cycle of dependency and despair. Addressing these challenges requires a concerted effort to lift restrictions, ensure the protection of humanitarian agencies, and provide sustainable solutions to the underlying causes of food insecurity in Palestine.

3.5 Implications of the War on Gaza for SDG 2 Targets

The war in Gaza has profoundly undermined progress toward achieving SDG 2 targets. Each of the targets has been severely affected, as outlined below:

Target 2.1: End Hunger and Ensure Access to Food

The war and subsequent blockade have significantly reduced access to food for the people of Gaza. With over 80% of the population reliant on humanitarian aid before these recent escalations, the destruction of food supply chains and limited entry of aid have pushed many into starvation. Civilians face acute hunger, with reports highlighting widespread food shortages and increased reliance on insufficient aid supplies (IPC 2024b; WFP 2024).

Target 2.2: End Malnutrition, Especially Among Children

The destruction of health services and the collapse of water and sanitation infrastructure have exacerbated malnutrition among children. Rising cases of wasting and stunting highlight the war's devastating impact on child health. A significant proportion of children in the Gaza Strip are now at risk of severe hunger-related illnesses (Global Nutrition Cluster 2024; Nasreddine and Jomaa 2024).

Target 2.3: Double Agricultural Productivity and Incomes of Small-Scale Food Producers

The destruction of farmland, greenhouses, and irrigation systems has decimated the agricultural sector in Gaza. The blockade restricts access to essential farming inputs, leaving farmers unable to produce food or sustain their livelihoods. Small-scale food producers, who formed the backbone of local food security, are unable to function, crippling the food system in the region (FAO/UN 2024; Hassoun et al. 2024).

Target 2.4: Ensure Sustainable Food Production Systems

The war has rendered sustainable food production impossible in Gaza. Widespread environmental degradation, destruction of agricultural infrastructure, and overexploitation of limited resources have eliminated the possibility of sustainable farming practices. Long-term damage to food systems in the Gaza Strip poses a significant barrier to recovery (UNEP 2024).

Target 2.5: Maintain the Genetic Diversity of Seeds, Plants, and Animals

The destruction of agricultural facilities and restrictions on imports have severely impacted efforts to preserve genetic diversity in Gaza. Limited access to seeds and agricultural resources hampers the ability to maintain diverse crops and livestock, further diminishing food security (Buheji and Hasan 2024; Uyeda 2024).

Target 2.a: Increase Investment in Rural Infrastructure and Agricultural Research

The war has obstructed efforts to invest in the agricultural sector in Gaza. Infrastructure projects, including irrigation and farming support systems, have been destroyed, leaving the agricultural sector with minimal capacity for recovery. The blockade has further limited international funding and technical support, worsening the situation (FAO/UN 2024; UNCTAD 2024).

Target 2.b: Correct and Prevent Trade Restrictions and Distortions in World Agricultural Markets

The blockade on Gaza exemplifies the distortion of trade and access to agricultural markets. The inability to export goods or import necessary inputs has crippled the economy and food systems in Gaza, deepening dependency on international aid (Euro-Mediterranean Human Rights Monitor 2021; Devereux 2024).

Target 2.c: Adopt Measures to Ensure the Proper Functioning of Food Commodity Markets

Food markets in Gaza have collapsed due to the blockade and destruction of infrastructure. Supply chain disruptions and inflated food prices have left households unable to afford basic nutrition, undermining the stability and functionality of local markets (WFP Palestine Food Security Analysis 2024).

In summary, the war has reversed progress on all SDG 2 targets in Gaza. The international community must urgently address these challenges to mitigate the catastrophic impact on food security and rebuild pathways toward achieving SDG 2.

3.6 Recommendations, Mitigating Measures, and Conclusion

Addressing the acute food insecurity and impending famine in Gaza require a multifaceted approach that combines immediate humanitarian interventions with long-term strategies to rebuild and strengthen food systems. The following measures are critical:

1. Immediate humanitarian aid: Expand the delivery of emergency food assistance and nutritional support, focusing on vulnerable groups such as children, women, and displaced persons. This necessitates expanding partnerships with organizations such as WFP and UNICEF to ensure the swift deployment of resources.
2. Restoration of WASH infrastructure: Prioritize the repair of water and sanitation systems to reduce disease prevalence and improve food utilization. Investment in desalination plants and wastewater treatment is essential to address the water scarcity crisis in Gaza.
3. Agricultural recovery and support: Implement programs to rehabilitate damaged farmlands, greenhouses, and irrigation systems. Provide farmers with essential inputs such as seeds, fertilizers, and equipment to restart local food production.
4. Unimpeded humanitarian access: Advocate for lifting blockades and establishing secure humanitarian corridors to facilitate the entry of food, medical supplies, and reconstruction materials.
5. Capacity building and resilience programs: Empower local communities by introducing capacity-building initiatives that promote sustainable agriculture, enhance livelihoods, and reduce dependency on aid.
6. Strengthening international advocacy: Mobilize the global community to pressure political actors to end the blockade, cease military actions, and uphold international humanitarian law.
7. Support for UNRWA and NGOs: Ensure adequate funding and protection for UNRWA and other NGOs to continue their critical work in providing essential services to Palestinians.
8. Monitoring and accountability: Establish mechanisms to monitor the delivery of aid and hold parties accountable for actions that obstruct humanitarian efforts or violate international laws.

In conclusion, the humanitarian crisis in Gaza underscores the devastating interplay of conflict, political instability, and food insecurity. Decades of systemic challenges, compounded by a prolonged blockade, intensified by recent hostilities, have pushed the region to the brink of famine. Vulnerable groups, particularly children and displaced families, are disproportionately affected, facing acute malnutrition, rampant disease, and pervasive insecurity.

To address this dire situation, immediate international intervention is essential to alleviate suffering and prevent further loss of life. However, lasting solutions must extend beyond short-term relief. Achieving SDG 2 requires comprehensive and targeted efforts. These efforts must focus on rebuilding sustainable agricultural and food systems in Gaza, ensuring access to adequate resources, and lifting the crippling blockade that has perpetuated deprivation and dependency. Additionally, addressing the six pillars of food security, namely availability, access, utilization, stability, agency, and sustainability, is critical to creating a robust and resilient food system.

The international community must urgently prioritize the protection and strengthening of institutions like UNRWA, which provide critical lifelines to millions of Palestinians. Concurrently, there must be a concerted effort to prevent future conflicts and hold accountable those responsible for policies that undermine the stability of the region. Only through coordinated, sustained global efforts can the catastrophic conditions in Gaza be addressed, ensuring a future where its people can achieve food security, resilience, and dignity, aligning with the broader vision of SDG 2.

References

Abdul Samad L, Butcher M, Khalidi B (2024) Water war crimes: how Israel has weaponised water in its military campaign in Gaza. Oxfam Int 6

Abu-Ras W, Mohamed RA (2018) Child poverty and youth unemployment in Palestine. Poverty Public Policy 10:354–370. https://doi.org/10.1002/pop4.227

Al Jazeera (2024) Israel kills three World Central Kitchen aid workers as it pounds Gaza. https://www.aljazeera.com/news/2024/11/30/israel-kills-world-central-kitchen-aid-workers-in-gaza. Accessed 13 Dec 2024

Alnabih A, Alnabeh N, Aljeesh Y, Aldabbour B (2024) Food insecurity and weight loss during wartime: a mixed-design study from the Gaza Strip. J Health Popul Nutr 43:222

Alshwaikh J (2024) The racial role of mathematicians and mathematics in maintaining the Israeli military occupation. Educ Stud Math 116:391–402. https://doi.org/10.1007/s10649-023-10255-8

Arora NK, Mishra I (2022) Current scenario and future directions for sustainable development goal 2: a roadmap to zero hunger. Environ Sustain 52(5):129–133. https://doi.org/10.1007/S42398-022-00235-8

B'TSELEM (2024a) Manufacturing famine: Israel is committing the war crime of starvation in the Gaza Strip

B'TSELEM (2024b) Israel is starving Gaza. https://www.btselem.org/gaza_strip/20240108_israel_is_starving_gaza. Accessed 13 Dec 2024

Béné C, Devereux S (2023) Resilience, food security and food systems: setting the scene. Palgrave Macmillan, Cham

Brück T, d'Errico M (2019) Food security and violent conflict: introduction to the special issue. World Dev 117:167–171. https://doi.org/10.1016/J.WORLDDEV.2019.01.007

Brück T, d'Errico M, Pietrelli R (2019) The effects of violent conflict on household resilience and food security: evidence from the 2014 Gaza conflict. World Dev 119:203–223. https://doi.org/10.1016/J.WORLDDEV.2018.05.008

Buheji M, Hasan A (2024) Beyond famine and chaos—case of Gaza. Int J Manag. https://doi.org/10.17605/OSF.IO/B6SPQ

Butt KM, Butt AA (2016) Blockade on Gaza strip: a living hell on earth. J Polit Stud 23:157–182

Checchi F, Ververs M, Jamaluddine Z (2024) Wartime food availability in the Gaza strip, October 2023 to August 2024: a retrospective analysis. medRxiv 2023–2024

Clapp J, Moseley WG, Burlingame B, Termine P (2022) Viewpoint: the case for a six-dimensional food security framework. Food Policy 106:102164. https://doi.org/10.1016/J.FOODPOL.2021.102164

Cole J (2024) Weaponizing a water crisis: the destruction of water supplies in Gaza. New Lines Inst:1–7

de Waal A (2024a) Hunger in global war economies: understanding the decline and return of famines. Disasters e12661. https://doi.org/10.1111/disa.12661

de Waal A (2024b) Famine in Gaza: an example of the global humanitarian crisis. Am J Clin Nutr 119:1383–1385. https://doi.org/10.1016/j.ajcnut.2024.04.015

Devereux S (2024) Was there a famine in Gaza in 2024?

Doocy S, Walton S, West KP (2023) Famine: causes, consequences and responses. In: Caballero B (ed) Encyclopedia of human nutrition, vol 1–4, 4th edn. Elsevier, pp 411–419

ESCWA (2023a) Social and economic situation of Palestinian women and girls: July 2020–June 2022

ESCWA (2023b) War on Gaza: weaponizing access to water, energy and food

Euro-Mediterranean Human Rights Monitor (2021) Suffocation and isolation: 15 years of Israeli blockade on Gaza

FAO (2024) Gaza: Geospatial data shows intensifying damage to cropland. https://www.fao.org/newsroom/detail/gaza-geospatial-data-shows-intensifying-damage-to-cropland/en. Accessed 13 Dec 2024

FAO, IFAD, WHO, UNICEF W (2024) Food security and nutrition in the World financing to end hunger, in all its forms

FAO/UN (2024) Agricultural Damage Assessment in the Gaza Strip from October 7th 2023 to September 1st 2024

Faris M, Abutair AS, Elfarra RM, et al (2024) Catastrophic hunger in Gaza: unprecedented levels of hunger post-October 7th. A real population-based study from the Gaza Strip. https://www.medrxiv.org/content/10.1101/2024.08.21.24312386v1

Fekih-romdhane F, Jebreen K, Swaitti T, et al (2024) Dying of starvation if not from bombs: assessing measurement properties of the food insecurity experiences scale (FIES) in Gaza's civilian population experiencing the world's worst hunger crisis. Int J Equity Health in review

Global Nutrition Cluster (2024) Nutrition vulnerability and situation analysis/Gaza

Gordon N, Haddad M (2024) The road to famine in Gaza. New York Rev Books 1:1–12

Guo H (2024) SDG 2: zero hunger 2. In: Big earth data in support of the sustainable development goals (2022)—the belt and road. Sustainable development goals series

Hamad BA, Jamaluddine Z, Aly S et al (2024) UNRWA's work is at risk again. Lancet 404:2157. https://doi.org/10.1016/S0140-6736(24)02476-0

Hassoun A, Al-Muhannadi K, Hassan HF et al (2024) From acute food insecurity to famine: how the 2023/2024 war on Gaza has dramatically set back sustainable development goal 2 to end hunger. Front Sustain Food Syst 8:1402150. https://doi.org/10.3389/fsufs.2024.1402150

HLPE (2020) Impacts of COVID-19 on food security and nutrition: developing effective policy responses to address the hunger and malnutrition pandemic. Rome

Holail S, Saleh T, Xiao X et al (2024) Time-series satellite remote sensing reveals gradually increasing war damage in the Gaza strip. Natl Sci Rev 11:nwae304. https://doi.org/10.1093/NSR/NWAE304

Horino M, Bahar L, Al-Jadba G et al (2020) Dietary inadequacy, micronutrient deficiencies, and approaches to preventing poor nutrition in the Gaza strip. Food Nutr Bull 41:503–511. https://doi.org/10.1177/0379572120967819

Human Rights Watch (2023) Israel: starvation used as weapon of war in Gaza

Hussein DMA, Lucchetti MC, Zaqoot HA et al (2022) Status of fisheries in Gaza strip: past trends and challenges. Int J Euro-Mediterranean Stud 15:179–216

Insight Insecurity (2024) Flash analysis report: over five months of attacks on food security in Gaza

IPC (2024a) Gaza strip: acute food insecurity situation for September–October 2024 and projection for November 2024–April 2025

IPC (2024b) IPC famine review committee alert Gaza strip

Iriqat D (2024) International law: Israel's fig leaf or a route to Palestinian statehood? Polit Geogr 115:103208. https://doi.org/10.1016/j.polgeo.2024.103208

Iriqat D, Alousi R, Aldahdouh TZ et al (2025) Educide amid conflict: the struggle of the Palestinian education system. Qual Educ All 2:81–99. https://doi.org/10.1108/QEA-10-2024-0120

Juaidi A, Montoya FG, Ibrik IH, Manzano-Agugliaro F (2016) An overview of renewable energy potential in Palestine. Renew Sust Energ Rev 65:943–960. https://doi.org/10.1016/j.rser.2016.07.052

Mahomed S (2024) International humanitarian laws: applicable to all or a privilege for some? SAJBL 17:22–26

Manikas I, Ali BM, Sundarakani B (2023) A systematic literature review of indicators measuring food security. Agric Food Secur 12:1–31. https://doi.org/10.1186/s40066-023-00415-7

Martin-Shields CP, Stojetz W (2019) Food security and conflict: empirical challenges and future opportunities for research and policy making on food security and conflict. World Dev 119:150–164. https://doi.org/10.1016/j.worlddev.2018.07.011

Médecins Sans Frontières (2024) Gaza: Life in a death trip

Middle East Institute (2024) The devastating impact of Gaza's acute and prolonged water crisis. https://www.mei.edu/publications/devastating-impact-gazas-acute-and-prolonged-water-crisis?utm_source=chatgpt.com. Accessed 13 Dec 2024

Mockshell J, Nielsen Ritter T (2024) Applying the six-dimensional food security framework to examine a fresh fruit and vegetable program implemented by self-help groups during the COVID-19 lockdown in India. World Dev 175:106486. https://doi.org/10.1016/J.WORLDDEV.2023.106486

Nasreddine L, Jomaa L (2024) Safeguarding the SDG promise to end hunger and leave no one behind: the plight of children in the Gaza Strip. Lancet Glob Heal 12:e1748–e1749. https://doi.org/10.1016/S2214-109X(24)00367-X

Nassar S, Tóth ZN, Vasa L (2023) Economic empowerment as a result of achieving SDGs with resource access: a comparative research between Gaza Strip and Hungary. J Int Stud 16:9–33. https://doi.org/10.14254/2071-8330.2023/16-2/1

OHCHR (2024) UN experts condemn 'flour massacre', urge Israel to end campaign of starvation in Gaza. https://www.ohchr.org/en/press-releases/2024/03/un-experts-condemn-flour-massacre-urge-israel-end-campaign-starvation-gaza. Accessed 6 Mar 2024

PAX (2023) Uninhabitable? The reverberating public health and environmental risks from the war in Gaza

Qafisheh MM (2024) From Constantinople to Oslo to Gaza: developments of Palestine's maritime legislation under colonization. Mar Policy 160:105954. https://doi.org/10.1016/j.marpol.2023.105954

Radi SM, El-Sayed NA, Nofal LM, Abdeen ZA (2013) Ongoing deterioration of the nutritional status of Palestinian preschool children in Gaza under the Israeli siege. East Mediterr Heal J 19:234–241. https://doi.org/10.26719/2013.19.3.234

Ridao-Cano C, Rother F, Sanchez-Reaza J (2017) Enhancing job opportunities report

Roy S (2016) The Gaza strip: the political economy of De-development, 3rd edn. Institute for Palestine Studies, Washington, DC

Samuel MT (2024) The Israel-Hamas war one year later: mass violence and Palestinian dispossession. Middle East Policy 31:3–16. https://doi.org/10.1111/mepo.12792

Shomar B, Rovira J (2023) Human health risks associated with the consumption of groundwater in the Gaza Strip. Heliyon 9:e21989. https://doi.org/10.1016/J.HELIYON.2023.E21989

Smith RJ, Isleem M (2017) Farming the front line: Gaza's activist farmers in the No Go Zones. City 21:448–465. https://doi.org/10.1080/13604813.2017.1331566

Sporchia F, Antonelli M, Aguilar-Martínez A et al (2024) Zero hunger: future challenges and the way forward towards the achievement of sustainable development goal 2. Sustain Earth Rev 7:1–6. https://doi.org/10.1186/S42055-024-00078-7

Stephenson M (2020) NGOs and humanitarian assistance. In: International encyclopedia of civil society. Springer, Cham, pp 1–7

Sultany N (2024) A threshold crossed: on genocidal intent and the duty to prevent genocide in Palestine. J Genocide Res. https://doi.org/10.1080/14623528.2024.2351261

Think Global Health (2024) Gaza's food crisis began long before the Israel-Hamas conflict. https://www.thinkglobalhealth.org/article/gazas-food-crisis-began-long-israel-hamas-conflict?utm_source=chatgpt.com. Accessed 13 Dec 2024

UNCT (2012) Gaza in 2020 a liveable place? A report by the United Nations country team in the occupied Palestinian territory

UNCTAD (2024) Preliminary assessment of the economic impact of the destruction in Gaza and prospects for economic recovery

UNEP (2020) State of environment and outlook report for the occupied Palestinian territory

UNEP (2024) Environmental impact of the conflict in Gaza: preliminary assessment of environmental impacts. Nairobi

United Nations in Palestine (2022) Food and nutrition fact sheet 2022

United Nations Sustainable Development (2015) Goal 2: zero hunger. https://www.un.org/sustainabledevelopment/hunger/?utm_source=chatgpt.com. Accessed 12 Dec 2024

Uyeda RL (2024) Even as agriculture in Gaza nears extinction farmers vow to plant their lands. https://prismreports.org/2024/10/22/farmers-vow-to-plant-lands-in-gaza/. Accessed 13 Dec 2024

WFP (2023) Gaza faces widespread hunger as food systems collapse, warns WFP. https://www.wfp.org/news/gaza-faces-widespread-hunger-food-systems-collapse-warns-wfp. Accessed 13 Dec 2024

WFP (2024) Gaza: urgent action needed as hunger soars to critical levels. https://www.wfp.org/stories/gaza-urgent-action-needed-hunger-soars-critical-levels?utm_source=chatgpt.com. Accessed 13 Dec 2024

WFP Palestine Food Security Analysis (2024) Market monitor—Gaza

Whyte J (2024) A "tragic humanitarian crisis": Israel's Weaponization of starvation and the question of intent. J Genocide Res. https://doi.org/10.1080/14623528.2024.2339637

Yang YT (2024) Protecting humanitarian aid workers: an urgent call. Lancet 403:1847. https://doi.org/10.1016/S0140-6736(24)00854-7

Zinszer K, Abuzerr S (2024) Water, Sanitation, and Hygiene insecurity and infectious disease outbreaks among internally displaced populations in Gaza: implications of conflict-driven displacement on public health. J Water Sanit Hyg Dev 14:1182–1192. https://doi.org/10.2166/washdev.2024.361

4

Gaza's Health Emergency: Unprecedented Humanitarian Impact of the Recent War on Public Health

Yara Ashour, Abdallah AbuJlambo, and Abdo Hassoun

Abstract

The recent war (October 2023–January 2025) on Gaza has triggered an unparalleled health crisis, resulting in a devastating humanitarian toll and extensive public health repercussions. This chapter explores the multifaceted impact of continuous military actions and hostilities on Gaza's healthcare system, infrastructure, and population health. The relentless bombardment and ground operations have caused huge loss of life and devastated medical facilities, severely restricting access to essential healthcare services. Moreover, the destruction of critical infrastructure, including hospitals and medical personnel and facilities, and the lack of fuel have led to acute shortages of medical supplies and workers, further exacerbating the public health crisis. The chapter provides a detailed analysis of the health effects of this assault on the civilian population in Gaza, highlighting the dramatic increase in malnutrition, infectious diseases, and mental health disorders. By drawing on recent data and reports from government and nongovernmental agencies, as well as recent scientific journal publications, the chapter underscores the urgent need for comprehensive international humanitarian intervention and sustainable solutions to address the public health disaster in Gaza. This work aims to contribute to the global discourse on war-related health emergencies, advocating for enhanced support mechanisms and policy changes to mitigate the impact of such crises on vulnerable populations.

Keywords

Child health · Disease outbreaks · Health access · Healthcare infrastructure · Humanitarian aid · Infectious diseases · Malnutrition · Maternal health · Mental health · Public health

Y. Ashour (✉)
Ministry of Health, Gaza, Palestine

Al-Quds University, Jerusalem, Palestine
e-mail: yara.ashour98@gmail.com

A. AbuJlambo
Ministry of Health, Gaza, Palestine

Department of ICU, Al-Shifa Hospital, Gaza, Palestine
e-mail: abdallah.abojlambo@gmail.com

A. Hassoun
Sustainable AgriFoodtech Innovation & Research (SAFIR), Arras, France

Faculty of Agricultural Engineering, University of Aleppo, Aleppo, Syria
e-mail: a.hassoun@saf-ir.com

A. Hassoun (ed.), *War on Gaza*, Sustainable Development Goals Series,
https://doi.org/10.1007/978-3-031-88500-6_4

4.1 Introduction

The World Health Organization (WHO) defines health as a "state of complete physical, mental, and social well-being and not merely the absence of disease or infirmity" (WHO 1948). Health and well-being are among the 17 Sustainable Development Goals (SDGs) established by the United Nations, specifically addressed in SDG 3 "Good Health and Well-Being." This goal emphasizes ensuring healthy lives and promoting well-being for all ages as a fundamental aspect of sustainable development (Guégan et al. 2018; UN 2024e). Globally, SDG 3 has nine Targets (Fig. 4.1) that should be achieved by 2030.

Compared to other SDGs, SDG 3 intersects critically with various aspects of life, as health underpins overall human development and well-being, so that progress in its implementation contributes to progress towards other goals, and action on other goals in turn contributes to attaining SDG 3 (Guégan et al. 2018; Smith et al.

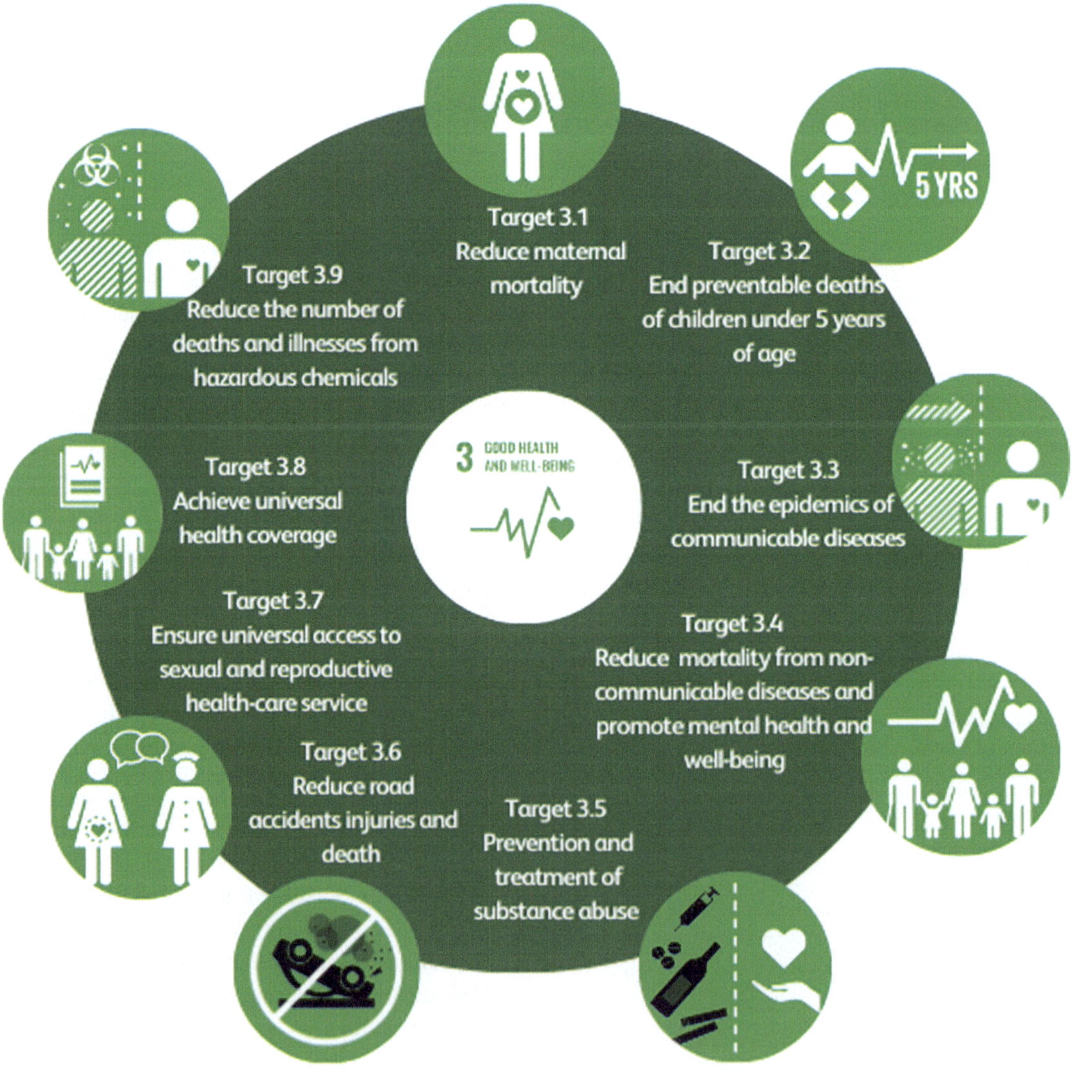

Fig. 4.1 Sustainable Development Goal 3 (SDG 3) and its nine targets

2023). While some countries, such as India and Brazil, have made notable progress toward achieving this goal—India has significantly reduced child mortality rates and improved maternal health (PIB 2022) and Brazil has advanced in reducing maternal and child mortality and in expanding healthcare access for its large population (da Silva Bueno et al. 2023)—others continue to face considerable challenges. Key obstacles to progress include limited resources, inadequate funding and infrastructure, persistent inequalities and disparities, the impacts of epidemics and pandemics, insufficient data and evidence, and the outbreak of wars and conflicts (Guégan et al. 2018; Sherin 2024).

The Gaza Strip, home to over 2.2 million people, has been subjected to repeated cycles of conflict and violence since the early 2000s. In June 2007, following the military takeover of Gaza by Hamas, the Israeli authorities significantly intensified existing movement restrictions, virtually isolating the Gaza Strip from the rest of the occupied Palestinian territory (oPt) and the world. This land, sea, and air blockade has significantly exacerbated previous restrictions, limiting the number and specified categories of people and goods allowed in and out through the Israeli-controlled crossings (UNICEF 2022). Years of socioeconomic decline, conflict, and closure have left the health system across Gaza lacking adequate physical infrastructure and training opportunities. Facilities are overstretched, and service is frequently interrupted by power cuts. These challenges further exacerbate the risks faced by an already vulnerable population (Blanchet et al. 2024; UNRWA 2022; WHO 2024d).

Healthcare provision in Gaza faces severe challenges, including restricted access to specialized services, construction limitations, and barriers to importing medical supplies and equipment. Patient and health staff movement is also heavily restricted. Before the recent crisis, Gaza's health system struggled with a fiscal crisis, leading to chronic shortages of health workers, particularly in primary care and subspecialties, and essential medicines, with 45% of the Essential Medicines List out of stock in 2022. Additionally, 60% of essential laboratory items were reported to have less than a 1-month supply. Limited specialized services and significant capacity gaps further impede the provision of adequate treatment. Government systems accounted for 45% of healthcare expenditures, while households contributed 39%, predominantly through out-of-pocket expenses (WHO 2024d). Despite these challenges, Gaza's healthcare system showcased relative strengths before October 2023, offering advanced surgical care, conducting research, and managing antimicrobial resistance, among other capabilities (MSF 2024a).

While Gaza has experienced previous escalations, the recent war is unprecedented in its scale of death, destruction, and human suffering, with long-lasting repercussions (WHO 2024d). This war has profoundly impacted Gaza's public health system. Hospitals and clinics, already under-resourced due to the blockade, have been frequently targeted or have suffered collateral damage during military operations (OCHA 2024c; Shellah 2024). The Gaza Strip has experienced an unparalleled humanitarian crisis characterized by high mortality and morbidity rates, displacement, massive degradation of the health system, and disruption of public health and other vital infrastructure (Abed Alah 2024; OCHA 2024a). Evidence shows that 90% of Gaza population (approximately 1.9 million) has been subjected to displacement orders, to make-shift shelters or overcrowded areas with limited access to basic necessities (Kerbaj et al. 2024; ReliefWeb 2024c).

The massive degradation of the health system in the Gaza is result from extensive bombardment, widespread destruction, escalating health needs, restrictions on the entry of fuel, goods, medical supplies and medical personnel, as well as significant limitations on patient's movement, including the injured. Health facilities are operating below optimal levels, having served patients far beyond their capacities with reduced staffing, which necessitated prioritizing services for emergency and critical life-saving care only (OCHA 2024a; Perugini and Gordon 2024).

4.2 Rationale for Chapter Focus and Methodology

4.2.1 Undermining Progress Toward SDG 3

The war on Gaza represents a significant barrier to achieving SDG 3 in the region. The destruction of healthcare facilities, combined with severe shortages of medical staff and supplies, directly undermines efforts to achieve SDG3 Targets such as reducing mortality rates, combating diseases, promoting mental health, and ensuring universal access to health service. The collapse of public health systems due to war not only halts progress but, in many ways, reverses gains made in previous years (UNRWA 2022). Without stability and sufficient resources, the prospect of realizing SDG 3 in Gaza remains bleak.

4.2.2 Importance of Addressing the Public Health Crisis in Gaza

Addressing the public health crisis in Gaza is not only a local imperative but also a global one. As the international community strives to achieve SDGs, the situation in Gaza highlights the deep inequities that persist in conflict zones. The Israeli military has been accused of deliberately destroying Gaza's health system through constant attacks on hospitals, ambulances, and doctors, including by air strikes, detentions, and denial of medical equipment (Irfan et al. 2024; Middle East Eye 2024). The Gaza Strip has experienced an unprecedented humanitarian crisis, marked by soaring levels of civilian mortality and morbidity, forced displacements, destruction of essential infrastructure, and targeting of the healthcare system. Between October 7, 2023, and December 31, 2024, the health authorities reported 45,541 fatalities, 108,338 injuries, and more than 10,000 reported missing in the Gaza Strip (OCHA 2024e). Tackling this issue is critical to ensuring that no one is left behind in the global effort to meet SDG 3.

4.2.3 Addressing the Sources of Collected Data

The data and insights presented in this chapter are drawn from a variety of sources, including reports by international organizations such as the WHO, the United Nations Office for the Coordination of Humanitarian Affairs (OCHA), and Médecins Sans Frontières (MSF). Additionally, local and national health authorities in Gaza such as Ministry of Health reports, as well as peer-reviewed papers, have provided essential data on the impact of the war on the region's health system. These sources ensure a comprehensive understanding of the recent public health crisis in Gaza.

4.3 Destruction of Healthcare Infrastructure

4.3.1 Impact of the War on Health Facilities

Hospitals in Gaza face severe shortages of electricity, medication, medical equipment, food, and personnel due to Israeli restrictions on aid movement, further compounded by significant damage to healthcare infrastructure (WHO 2024d). As of 25 September 2024, 17 of 36 hospitals (47%) were only partially functioning, with most facilities operations (3 in North Gaza, 7 in Gaza, 3 in Deir Al Balah, 4 in Khan Younis, 0 in Rafah) severely compromised due to airstrikes and military ground operations (WHO 2024c). Healthcare services across the Gaza Strip were collapsing, particularly in northern Gaza, where health services have been largely destroyed, and southern Gaza's healthcare system was on the brink of collapse (OCHA 2024b). The Ministry of Health has made appeals for the establishment of field hospitals, yet the health system continues to deteriorate. Nasser Medical Complex, once the second-largest referral hospital in Gaza and the last main hospital in southern Gaza, became exposed to many bombardments and siege throughout the war. MSF calls on all parties to ensure safe access to care and avoid the evacuation of Nasser hospital, which would endanger hundreds of patients (MSF 2024b).

Al-Shifa Hospital, the largest hospital in Gaza, was besieged for 14 days in April 2024, leading to widespread destruction, denial of medical access, and mass casualties. During the siege, attempts by humanitarian organizations to evacuate patients were denied, and dozens of dismembered bodies were found on hospital grounds. Israeli forces detained and interrogated healthcare workers, leaving the hospital in ruins (WHO 2024f). Al-Ahli and Kamal Adwan Hospitals, which remain functional but severely under-resourced, now struggle to accommodate patients, with minimal resources available. As of September 2024, at least 350,000 Palestinians in the north are served by as few as 200 hospital beds, with Kamal Adwan and al-Ahli hospitals operating at just 30% and 70% capacity, respectively, and lacking many of the specialist services (WHO 2024d). The destruction of Nasser, Al-Amal, and Al-Khair Hospitals, alongside damage to medical warehouses, has further crippled Gaza's ability to meet the overwhelming healthcare needs. Already in November 2023, Al-Nasr Children's Hospital, a critical facility in Gaza City, suffered extensive damage during an invasion and was forcibly evacuated. Tragically, five infants in the intensive care unit (ICU) incubators lost their lives after being left behind. These infants required respirators to survive and could not be safely moved during the evacuation (CNN 2023; IPS 2024a). Kamal Adwan Hospital has suffered multiple Israeli military bombardments, ultimately rendering it nonoperational in December 2024. This has placed the lives of approximately 75,000 Palestinians remaining in the area at significant risk, according to the WHO (UN 2024d). On 27 December 2024, the Israeli forces arbitrarily detained human rights defender Dr. Hussam Abu Safiya, the director of Kamal Adwan Hospital, who was the director of the last remaining hospital in northern Gaza (Mahase 2024). Until the time of writing, his fate remained uncertain, with reports indicating that he had experienced torture in Israeli detention facilities.

The healthcare system in Gaza has collapsed under the weight of the war, leaving over 2.2 million people without adequate medical care. This devastation severely undermines SDG 3, by obstructing access to essential health services and reversing progress toward universal health coverage.

4.3.2 Health Workforce Challenges

As per September 2024, the Israeli onslaught on Gaza resulted in the deaths of 1151 Palestinians working in the health sector, including 165 doctors, 260 nurses, 300 management and support personnel, 184 health associate professionals, 76 pharmacists, and 12 other healthcare workers (Middle East Eye 2024). Due to the overwhelming pressure on the health system, life-saving amputations are often prioritized over limb reconstruction, and premature discharges occur due to the critical shortage of hospital beds. Wound infections are also prevalent due to limited resources and overcrowded conditions. Tragically, approximately 10 children become amputees every day, with more than 3500 children now living with amputations in Gaza (WHO 2024e).

Healthcare workers (HCWs) in conflict zones are accustomed to working under extreme pressure. However, the critical situation in Gaza has pushed these limits. HCWs have worked relentlessly, often for many days without a break. The sheer volume of casualties and the severity of injuries in this conflict have surpassed any previous war, setting the stage for a heightened risk of mental health issues, particularly posttraumatic stress disorder (PTSD). The detrimental mental health effects of war on HCWs can endure for years, often persisting long after this war has ended (Abed Alah 2024; ReliefWeb 2024b).

The recent war on Gaza presents unique challenges for HCWs, significantly impacting their mental health. Attacks on medical facilities and personnel, combined with severe shortages of supplies and staff, create a climate of fear and uncertainty. HCWs must operate under dire conditions, often performing procedures without anesthesia, and face emotional trauma from losing colleagues and witnessing patient suffering. Overcrowding in hospitals due to displaced individuals further complicates their ability to provide care and exacerbates feelings of burnout and inadequacy (Abed Alah 2024). Additionally, the threat of arrest adds to the stress faced by HCWs, undermining their capacity to deliver care effectively. In less than a year, more than 300 healthcare workers were detained by Israeli forces, and over 1000 were killed during the war (Middle East Eye 2024). Personal worries for their families and the absence of basic needs contribute to their physical and mental exhaustion. Despite the critical role HCWs play, their struggles remain underreported and unsupported, leading to a cumulative mental health crisis marked by high rates of PTSD and related issues (Abed Alah 2024).

4.4 Public Health Crisis and Disease Burden

4.4.1 Infectious Disease Outbreaks

Target 3 of SDG 3 aims to "end the epidemics of AIDS, tuberculosis, malaria and neglected tropical diseases, and combat hepatitis, waterborne diseases and other communicable diseases" (WHO 2024g). However, the destruction of infrastructure, coupled with poor living conditions in Gaza, has led to a rise in waterborne and communicable diseases, posing a major setback to the global efforts outlined in this SDG.

Conflicts often lead to infectious disease outbreaks due to displacement, overcrowding, healthcare system collapse, damaged water and sanitation infrastructure, and malnutrition, causing death rates that may rival or exceed those from direct violence (Goniewicz et al. 2021; Kearney et al. 2024). The crisis in Gaza has created an ideal environment for infectious disease outbreaks (Kearney et al. 2024). The WHO has warned of the high risk of infectious diseases in shelters for internally displaced persons (IDPs) in Gaza (Reuters 2023a). The lack of control measures, basic hygiene items, and limited access to healthcare has made containing these outbreaks increasingly difficult, exacerbating public health challenges and making it difficult for families to protect themselves against communicable diseases (Hussein et al. 2024; ReliefWeb 2024d). WHO reports have noted outbreaks of acute respiratory infections, scabies, lice, diarrhea, skin rashes, chickenpox, and jaundice linked to hepatitis (WHO 2024d). As of July 7, 2024, there were 577,000 cases of acute watery diarrhea and 107,000 acute jaundice syndromes (OCHA 2024d).

Hepatitis A, which causes liver inflammation and is transmitted through contaminated food and water or direct contact with an infected person, has surged. UNRWA shelters and clinics have recorded 40,000 cases of hepatitis A, a sharp increase compared to just 85 cases in the same period before the war (UN 2024c). WHO data shows that, on average, 500 people are sharing a single toilet, and over 2000 people must use one shower, further increasing the risk of disease transmission (UN 2024a).

Respiratory infections are also rampant in Gaza. By 7 July 2024, approximately 995,000 cases of acute respiratory infections had been reported (OCHA 2024a). In the oPt, respiratory diseases account for 70% of deaths caused by infectious diseases, with a mortality rate of 17.0 per 100,000 population, making it the sixth most common cause of death (WHO 2024g). Handwashing with soap is one of the most effective ways to prevent the spread of WASH-related diseases, such as diarrhea, respiratory infections, and scabies and other skin infections. This simple practice can reduce diarrheal diseases by up to 40% and respiratory infections by about 20%. It can protect approximately one in three children who suffer from diarrhea and prevent the spread of germs to food, drinks, and surfaces (ReliefWeb 2024d). However, access to soap and clean water is often impossible for many due to severe shortages and blockades, leaving vulnerable populations without this essential preventive measure.

In August 2024, a 10-month-old baby in Gaza became partially paralyzed from polio, the first confirmed case in 25 years. This paralysis is likely permanent, as there are no treatments for polio (UN 2024b). Healthcare workers in Gaza have long warned of a potential polio outbreak, and this case confirms their fears. Prior to the war, Gaza's children were largely vaccinated against polio, but vaccination campaigns have since been disrupted. The WHO estimates that for every confirmed case of polio-related paralysis, hundreds of additional individuals are likely infected but remain asymptomatic (WHO 2024a).

Malnutrition is a contributor for developing infectious diseases especially for children, as hunger and thirst weaken their bodies and immune systems, making them more susceptible to illnesses. WHO report finds that at least 90% of children under 5 are affected by one or more infectious disease (WHO 2024b). Cases of diarrhea in children under 5 years of age rose from 48,000 to 71,000 in just 1 week in December 2023, equivalent to 3200 new cases of diarrhea per day. In comparison, an average of 2000 cases of diarrhea in children under 5 was reported per month before October 7. This recent climb represents a staggering increase of about 2000% and a 23-fold increase compared with the 2022 baseline (UNICEF 2024). According to reports, more than 50,000 children require treatment for acute malnutrition in 2024 (OCHA 2024d). Both malnutrition and infectious diseases are deadly and should be entirely preventable in children under the age of 5 (Target 2.3).

4.4.2 Noncommunicable Diseases

Target 3.4 aims to reduce mortality from noncommunicable diseases (NCDs), yet they remain the leading cause of death among Palestinians. Cardiovascular diseases top the list accounting for 30.6% of deaths recorded; cancer was the second leading cause of death, with 14.0% of deaths; complications of diabetes came in the fourth rank with a proportion of 8.0% (Albelbeisi et al. 2022). According to the latest WHO report on NCDs from September 2024, more than 2000 people are diagnosed with cancer each year, including 122 children. There are 45,000 patients living with cardiovascular diseases. More than 650,000 people have raised blood pressure, over 60,000 people with raised blood glucose, and over 1500 patients in need of kidney dialysis to maintain life (WHO 2024g).

Patients with NCDs rely on a consistent supply of medications to manage their conditions effectively. The recent war has disrupted the pharmaceutical supply chain, leading to shortages of essential drugs and exacerbating health risks for those requiring continuous medication. Moreover, the destruction of infrastructure, shortage of medical personnel, and prioritizing treating war injuries over chronic illnesses in Gaza's emergency departments have further impeded the delivery of essential services and raised the toll of preventable fatalities (Ashour et al. 2024; Mahmoud and Abuzerr 2023).

Gaza lacks basic NCD treatments, such as cancer care, which makes it highly dependent on patient referrals outside Gaza. Patients face significant challenges in accessing care for NCDs because of bureaucracy, high patient volumes, long waiting periods, and issues with the referral system. These challenges include delays and restrictions in receiving permission from Israeli authorities and lack of coordination with referral facilities (UN 2022). From October 2023 to September 2024, 5012 patients were evacuated abroad seeking medical treatment, representing 35% of the 14,192 requested cases (WHO 2024g).

The destruction of Gaza's only cancer treatment facility, the Turkish Friendship Hospital, by the Israeli army, coupled with border closures, severe medication shortages, fuel depletion, and the blockade on essential medical equipment such as X-ray machines and medical radioisotopes, has left 10,000 cancer patients in Gaza without access to life-saving treatment (Reuters 2023b). National reports have already warned from imminent death for those patients (PCHR 2024). Neglecting them all from enjoying their health rights falls within the acts of genocide committed by the occupation against civilians in the Gaza Strip. For cancer patients requiring urgent travel for treatment, the Ministry of Health prepares an approved list of individuals. This list is shared with countries willing to receive the patients and is published online. Each group of patients is assigned a specific travel date for evacuation (BBC 2024). But whether a patient is ultimately allowed to leave depends on the discretion of the Palestinian border authorities. This bureaucratic delay exacerbates the urgent need for care, adding another indirect consequence of this war.

4.4.3 Maternal and Child Health

Target 3.1 aims to reduce the global maternal mortality ratio, while Target 3.7 emphasizes universal access to sexual and reproductive healthcare services, including family planning, information, and education. However, the war on Gaza has severely undermined both of these targets, creating dire consequences for maternal and child health.

Pregnancy is a crucial and sensitive period for both mother and fetus, requiring specialized care. Pregnant women are especially vulnerable to high levels of stress and mental health disorders due to hormonal changes (Davis and Narayan 2020). However, the war on Gaza has stripped them of access to essential maternal care services. More than 500,000 women of reproductive age currently lack access to crucial services such as antenatal care, postnatal care, family planning, and the management of sexually transmitted infections. Among these women, 50,000 are pregnant, with 5500 expected to give birth each month, and an estimated 70,000 babies have been born since the recent war began (Muhammad and Rizek 2024; Ramadan and Abouzid 2024; WHO 2024c).

The recent war on Gaza has not only destroyed vital health infrastructure but has also severely limited access to care for pregnant women (Elnakib et al. 2024). Frequent displacement, malnutrition, lack of clean water, and the traumatic loss of loved ones significantly amplify the risks faced by expectant mothers (Keasley et al. 2017; SCI 2024). Cesarean sections are frequently performed without anesthesia or painkillers due to medical shortages, putting both mothers and babies at risk of complications (UN 2024a). The destruction of roads and infrastructure has created additional barriers, forcing pregnant women to navigate dangerous areas and delayed access to care, contributing to maternal and newborn deaths (DWB 2024). Postbirth conditions are no better. Many mothers return to unsanitary and overcrowded shelters, exposing both themselves and their newborns to infections, malnutrition, and other health risks. The lack of nutritious food and clean water, combined with elevated stress levels, has made breastfeeding difficult for many women, leading to further malnutrition among infants. This is particularly devastating given the absence of affordable alternatives such as milk formula (DWB 2024).

The long-term impacts on children born in such conditions are profound. Studies show that high levels of maternal stress, poor nutrition, and trauma during pregnancy and early childhood lead to adverse developmental outcomes. The violence and deprivation experienced by mothers and children in Gaza will likely result in a generational cycle of poor health, impairing children's cognitive and physical development (IPPF 2024). The war has decimated maternal healthcare in Gaza, with maternal mortality on the rise, directly undermining Target 3.1. Miscarriages have increased by 300%, and many women are experiencing premature deliveries due to stress, malnutrition, and lack of access to essential prenatal care (IPPF 2024). Service providers report widespread cases of anemia and malnutrition among pregnant women, many of whom lack access to essential prenatal vitamins and supplements. Studies indicate that 50% of pregnant women in Gaza suffer from anemia (Elnakib et al. 2024). These adverse outcomes increase maternal mortality and result in long-term health complications for the next generation (IPPF 2024).

The collapse of the healthcare system has also severely impacted family planning and sexual health services, directly hindering progress toward Target 3.7. The lack of access to family planning services, reproductive health information, feminine hygiene products, and emergency obstetric care exacerbates the vulnerabilities faced by women of reproductive age in Gaza and elevates the risks of preventable infections, including reproductive and urinary tract infections (Sabet et al. 2024). Furthermore, the physical and emotional stress endured by these women is likely to contribute to future reproductive health complications, including high-risk pregnancies and poor birth outcomes (IPPF 2024).

Children are among the most vulnerable victims of the war. As per September 2024, 11,000 children have been killed, including 710 infants under 1 year of age. Many were born during the war and killed in its midst (Middle East Eye 2024). Moreover, over 10,000 people are missing, and 17,000 children remain separated from their families or unaccompanied, leaving them in need of urgent mental health and psychosocial support (OCHA 2024d). The collapse of pediatric care, coupled with attacks on health facilities, has left many children without access to basic health services. As a result, children in Gaza face severe malnutrition, with nearly 3000 children at risk of dying due to food shortages and the lack of functional healthcare centers. The disruption of vaccination programs has significantly increased the risk of preventable diseases among children under 5, including polio, with the first reported case in Gaza on August 25, 2024 (UN 2024f).

The direct and indirect effects of the war will have long-lasting impacts on Gaza's children. Short-term consequences include delays in motor and cognitive development, while long-term effects are likely to result in permanent health and developmental impairments. The war has created conditions that will make it nearly impossible for children in Gaza to achieve the milestones of good health and well-being, as envisioned in SDG 3.

4.4.4 Mental Health

Target 4 of SDG 3 aims to reduce by one-third premature mortality from NCDs by 2030, with a strong focus on prevention and treatment, including the promotion of mental health and well-being. Mental health is integral to the success of SDG 3, as unaddressed mental health challenges, particularly in crisis-affected populations, hinder the achievement of global health and well-being goals. Prior to October 2023, Gaza already had significant mental health and psychosocial support (MHPSS) needs due to decades of conflicts, occupation, and economic deprivation which contributed to widespread depression, PTSD, and other mental health problems (Buheji 2024; Rockowitz et al. 2024). A study by the International Security and Development Center (ISDC) in 2023 found that 70% of Gazans screened positive for depression (ISDC 2023).

Gaza's population has experienced unprecedented number of violence, displacement, and loss. These traumatic events have significantly increased the already high levels of anxiety, depression, and other mental health conditions (Aldabbour et al. 2024; Kienzler et al. 2024). Destruction of the healthcare system has left Gaza unable to meet even basic mental health needs, with treatment capacities severely constrained since the region's only specialized treatment center stopped functioning in November 2023 (WHO 2024d). According to a 2024 WHO report, over 485,000 people in Gaza were estimated to be living with mental health disorders (WHO 2024g).

Children of Gaza, who had lived through six phases of conflict since 2008, inherited trauma from their caregivers' experiences and memories. As of 2023, approximately 816,000 children required MHPSS due to their exposure to traumatic events (ReliefWeb 2024a; Taha et al. 2024). The war has severely impacted the children of Gaza, leaving nearly all of the 1.2 million children in need of MHPSS (ReliefWeb 2024a).

The mental health of women in Gaza has significantly worsened, with a notable increase in cases of physical and verbal gender-based violence driven by escalating economic and emotional stress within households. Women, often tasked with the primary responsibility of meeting family needs, face heightened challenges due to shortages of water and food, coupled with limited access to sexual and reproductive health services. Many are also forced to manage their menstrual health with inadequate resources, further undermining their dignity and well-being (Sabet et al. 2024).

Gazan men have faced identity and self-esteem crises as they are unable to fulfill their traditional roles as providers and protectors due to the high levels of unemployment (80% by June 2024, compared to 46% in 2023) (ILO 2024). Social expectations and norms often prevent men from seeking help for their mental health needs, further complicating their ability to cope. Older individuals and those with disabilities have been disproportionately affected by the collapse of healthcare services. The war has left many without access to essential care, including assistive devices and social support, exacerbating feelings of isolation and helplessness. Newly disabled individuals due to the war are struggling to adapt to their reality, often facing traumatic flashbacks to the event of their injury, with limited or no access to mental health support (Mostoller and Mickelson 2024).

4.4.5 Environmental Risk

Target 3.9 seeks to reduce deaths and illnesses caused by hazardous chemicals and air, water, and soil pollution by 2030. However, the war is expected to leave Gaza with enduring environmental damage, jeopardizing the health and well-being of its population for generations. Pollution of air, water, and soil is a significant threat to global health, as demonstrated by "The Lancet Commission on Pollution and Health," which documented pollution as the leading environmental cause of disease and premature death worldwide, accounting for an estimated 16% of all deaths worldwide and three times more than from AIDS, malaria, and tuberculosis together (Landrigan et al. 2018).

Conflict-generated pollution is further exacerbated by the destruction in Gaza. After October 7, 2023, Gaza has experienced unprecedented levels of environmental degradation. Military operations have generated massive quantities of debris. Conflict debris differs from normal construction and demolition waste in that it contains harmful substances including asbestos, heavy metals, fire contaminants, unexploded ordnance (UXOs), explosive residue, household chemicals, and other hazardous substances. This debris is more than 13 times that generated by all previous conflicts in Gaza since 2008, with 107 kg of debris per square meter of land (Buheji and Al-Muhannadi 2023; Mahmoud 2024; UNEP 2024). Exposure to asbestos, from the debris in refugee camps and other areas, is a major concern, as inhalation can cause lung cancer, mesothelioma, and other severe health conditions (UNEP 2024). The war's destruction of agricultural land further threatens food security, as contamination of soils with heavy metals from military munitions will persist for decades, leading to widespread health risks through food webs and water supplies (Mahmoud 2024; UNEP 2024). Additionally, the shortage of cooking gas during this period led to a shift toward less clean energy sources, such as firewood and open-air waste burning. This transition posed serious air pollution and increases the risk of respiratory diseases among the Gaza population (UNEP 2024).

Moreover, the discharge of approximately 60,000 m^3 of untreated wastewater and sewage per day, particularly into the Mediterranean Sea, has significantly contaminated water sources. This has greatly increased the risk of pathogen exposure, further escalating the likelihood of waterborne diseases. The destruction of Gaza's tunnel system and military actions has also disrupted water supplies, compounding the public health crisis (Dardona et al. 2024; UNEP 2024). This contamination poses a threat to Gaza's water sources and marine ecosystems and carries the risk of spreading environmental damage beyond Gaza's borders, potentially affecting neighboring areas (Shaheen et al. 2024). The contamination will likely have long-term effects, resulting in chronic diseases, including respiratory, cardiovascular issues, and cancer, especially among vulnerable populations like children.

4.4.6 Palestinians in Detention

Between October 7, 2023, and July 2024, Israel arrested over 14,000 Palestinians in Gaza, the West Bank, and East Jerusalem, marking a 130%

increase in detainee numbers since the war began (TRT 2024; UN 2024f). Among that number were 4000 Palestinians arrested in Gaza, including journalists, human rights defenders, medical staff, patients, United Nations staff, and relatives of suspects (UN 2024f). Reports indicate harsh detention conditions, with limited access to basic necessities, insufficient food, and inadequate medical care. Detainees faced severe hygiene issues, untreated injuries, and, in some cases, amputations due to neglect (UN 2024f). Such conditions contradict Target 3.8, which emphasizes universal health coverage and access to essential health services.

Sexual and gender-based violence, affecting both men and women, was reported in facilities like Negev Prison and Sde Teiman camp, with violence used as a tool of punishment and intimidation (OHCHR 2024a; UN 2024f). Children also suffered under similar conditions, raising concerns about long-term mental and physical impacts (UN 2024f). By mid-July 2024, at least 53 detainees had died in Israeli custody, 44 of them from Gaza, largely due to dire conditions at facilities like Sde Teiman camp (OHCHR 2024b). Detainees endured reduced food and water rations, restricted sanitation, and medical neglect, leading to preventable deaths and disease outbreaks like scabies (UN 2024f). These practices have prompted calls for accountability, with reports of torture, deprivation of rights, and inhumane treatment drawing international condemnation (UN 2024e; OHCHR 2024a).

4.5 Call for International Action: Rebuilding Gaza's Healthcare System

The immediate cessation of the military assault on Gaza is essential, followed by urgent and coordinated international action to rebuild its devastated healthcare system. Governments, humanitarian organizations, and advocacy groups must work together to restore health services, rebuild critical infrastructure, and address the root issues of the ongoing health crisis, focusing on both immediate relief and long-term recovery. This collective effort must focus on several key areas to ensure the long-term sustainability of Gaza's healthcare system.

4.5.1 Permanent Ceasefire and Unrestricted Humanitarian Aid

A permanent ceasefire is a crucial first step in stabilizing the region. Without an immediate end to hostilities, efforts to restore healthcare services and rebuild infrastructure cannot proceed effectively. Unrestricted access to essential humanitarian aid, including medical supplies, food, water, and fuel, must be guaranteed. This requires the removal of barriers that limit the movement of goods and humanitarian aid, such as blockades and checkpoints (Hassan et al. 2024; Oxfam 2024).

4.5.2 Rebuilding Healthcare Infrastructure

Hospitals, clinics, and public health facilities—including vital electrical, sanitary, and medical logistical systems—should be the main targets of reconstruction efforts. These institutions are crucial for handling trauma care, providing basic health services, and providing the requirements for patients with chronic diseases. In order to respond to future health emergencies, reconstruction efforts must also prioritize improving surveillance, emergency preparedness, and disease prevention. The global community must commit to building a sustainable and resilient healthcare system in Gaza. This will involve the long-term allocation of resources to support disease prevention, healthcare infrastructure, and the development of evidence-based health programs (Beiraghdar et al. 2023; Blanchet et al. 2024). The is not just to address the recent crisis but to build capacity within the healthcare system to withstand future challenges, including conflicts, natural disasters, and public health emergencies.

4.5.3 Supporting Healthcare Workers and Addressing Workforce Shortages

The sheer number of patients, the scarcity of resources, and the hazardous working conditions in Gaza are overwhelming healthcare professionals. It is essential to support these workers by offering them financial assistance, psychosocial support, and training. International efforts should also alleviate the shortage of medical personnel by allowing more medical professionals from neighboring areas and international organizations to enter Gaza and the oPt and by guaranteeing local health workers' free movement throughout Gaza (Blanchet et al. 2024; Mohammed et al. 2024).

4.5.4 Addressing Food, Water, Environmental Challenges, and Mental Health Crises

Food insecurity and water shortages, exacerbated by the destruction of infrastructure and long-term environmental pollution, present significant challenges. Long-term recovery will require addressing these issues to ensure a healthy population. Programs aimed at food security, nutrition, and clean water access need to be expanded. Furthermore, the war has triggered a severe mental health crisis in Gaza, underscoring the need for international support to expand mental health services, including trauma counseling and long-term psychiatric care (Hassoun et al. 2024; Kienzler et al. 2024).

The WHO Director-General, in a report to the 77th General Assembly, recommended several actions to improve health conditions in Gaza and the oPt, including (a) an immediate ceasefire and the release of hostages; (b) unrestricted access to medical and humanitarian aid; (c) free access to health services for patients and workers across the region; (d) safe passage for essential goods like medicine, food, and fuel, along with the protection of humanitarian operations; and (e) full restoration of Gaza's health system, ensuring the functionality of hospitals and primary healthcare centers, and the resumption of essential public health measures (WHO 2024e).

4.6 Conclusion

The war on Gaza has severely set back the progress toward achieving SDG 3, which aims to ensure good health and well-being for all. The destruction of health infrastructure, increased poverty, and restricted access to medical care have severely impacted health outcomes. Rebuilding medical infrastructure is essential, but the effort should focus beyond the infrastructure. Only 10–20% of health outcomes are related to health care, with broader social, environmental, economic, and political conditions playing crucial roles (IPS 2024b). Therefore, to restore health and well-being in Gaza, efforts must focus on addressing the following underlying determinants:

- **Social factors**, such as education disruption, food insecurity, loss of loved ones, and broken social networks.
- **Environmental challenges**, including limited access to clean water, inadequate sanitation and hygiene, and the impacts of overcrowding.
- **Economic recovery**, to combat job loss, destroyed economic sectors, and child labor.
- **Political factors and safety** that are essential for health and security.

A public health approach that integrates these factors into recovery efforts is essential for long-term resilience. SDG 3 is central to Gaza's recovery and future development. Reaffirming the importance of this goal highlights the need for sustained global support to ensure health and well-being in Gaza. By providing global support and fostering international collaboration, Gaza can move toward achieving SDG 3, ensuring that health and well-being become a cornerstone of its recovery and sustainable development, despite the challenges posed by the recent unprecedented war.

References

Abed Alah M (2024) Echoes of conflict: the enduring mental health struggle of Gaza's healthcare workers. Confl Heal 18(1):1–6. https://doi.org/10.1186/S13031-024-00577-6/METRICS

Albelbeisi AH, Albelbeisi A, El Bilbeisi AH, El Afifi A (2022) Do patients with major non-communicable diseases receive advice on health behaviors from healthcare professionals in the Gaza strip, Palestine? Ethiop J Health Sci 32(1):45. https://doi.org/10.4314/EJHS.V32I1.6

Aldabbour B, Abuabada A, Lahlouh A, Halimy M, Elamassie S, Sammour AAK, Skaik A, Nadarajah S (2024) Psychological impacts of the Gaza war on Palestinian young adults: a cross-sectional study of depression, anxiety, stress, and PTSD symptoms. BMC Psychol 12(1):1–12. https://doi.org/10.1186/S40359-024-02188-5/TABLES/4

Ashour Y, Jlambo A, Abuzerr S (2024) Patients in Gaza with chronic conditions need urgent interventions. Lancet 403(10439):1847–1848. https://doi.org/10.1016/S0140-6736(24)00705-0

BBC (2024) Cancer patients stopped from leaving Gaza for treatment. https://www.bbc.com/news/world-middle-east-68552314

Beiraghdar F, Momeni J, Hosseini E, Panahi Y, Negah SS (2023) Health crisis in Gaza: the urgent need for international action. Iran J Public Health 52(12):2478. https://doi.org/10.18502/IJPH.V52I12.14309

Blanchet K, Najem M, Shadid L, Fehmi RA, Al Hammouri F, Saed G, Saleh K, Nasser M, Moukaddam N, Polonsky J, Lattouf O (2024) Rebuilding the health sector in Gaza: alternative humanitarian voices. Confl Heal 18(1). https://doi.org/10.1186/S13031-024-00599-0

Buheji M (2024) Addressing human factors in Gaza: challenges and solutions for post-war recovery. Int J Manag (IJM) 15(3):35–45. https://doi.org/10.17605/OSF.IO/RXB5V

Buheji M, Al-Muhannadi K (2023) Mitigating risks of environmental impacts on Gaza-review of precautions & solutions post (2023 war). Int J Adv Res Eng Technol (IJARET) 14(7):15–47. https://doi.org/10.17605/OSF.IO/2JMCP

CNN (2023) Al-Nasr hospital: infants found dead and decomposing in evacuated Gaza ICU. Here's what we know. https://edition.cnn.com/2023/12/08/middleeast/babies-al-nasr-gaza-hospital-what-we-know-intl/index.html

da Silva Bueno LA, Ribeiro MAL, Martins CB, de Abreu LC, de Almeida AD, Correa JA (2023) Describing trends in maternal mortality in the State of São Paulo, Brazil, from 2009 to 2019. Healthcare 11(18):2522. https://doi.org/10.3390/HEALTHCARE11182522

Dardona Z, Amane M, Dardona A, Boussaa S (2024) Health and environmental impacts of Gaza conflict (2023–2024): A review. One Heal Bulletin. https://doi.org/10.4103/OHBL.OHBL_42_24

Davis EP, Narayan AJ (2020) Pregnancy as a period of risk, adaptation, and resilience for mothers and infants. Dev Psychopathol 32(5):1625. https://doi.org/10.1017/S0954579420001121

DWB (2024) Gaza: Maternal and child health suffer under a decimated system | Doctors Without Borders—USA. https://www.doctorswithoutborders.org/latest/gaza-maternal-and-child-health-suffer-under-decimated-system

Elnakib S, Fair M, Mayrhofer E, Afifi M, Jamaluddine Z (2024) Pregnant women in Gaza require urgent protection. Lancet 403(10423):244. https://doi.org/10.1016/S0140-6736(23)02835-0

Goniewicz K, Burkle FM, Horne S, Borowska-stefańska M, Wiśniewski S, Khorram-manesh A (2021) The influence of war and conflict on infectious disease: A rapid review of historical lessons we have yet to learn. Sustainability 13(19):10783. https://doi.org/10.3390/SU131910783

Guégan JF, Suzán G, Kati-Coulibaly S, Bonpamgue DN, Moatti JP (2018) Sustainable development goal #3, "health and well-being", and the need for more integrative thinking. Veterinaria Mexico OA 5(2). https://doi.org/10.21753/vmoa.5.2.443

Hassan F, London L, Manjra S (2024) The global health community must call for an immediate ceasefire and unrestricted humanitarian aid in Gaza. BMJ 385. https://doi.org/10.1136/BMJ.Q782

Hassoun A, Al-Muhannadi K, Hassan HF, Hamad A, Khwaldia K, Buheji M, Al Jawaldeh A (2024) From acute food insecurity to famine: how the 2023/2024 war on Gaza has dramatically set back sustainable development goal 2 to end hunger. Front Sustain Food Syst 8:1402150. https://doi.org/10.3389/FSUFS.2024.1402150/BIBTEX

Hussein S, Ahmed SK, Qurbani K, Fareeq A, Essa RA (2024) Infectious diseases threat amidst the war in Gaza. J Med Surg Publ Heal 2:100067. https://doi.org/10.1016/J.GLMEDI.2024.100067

ILO (2024) War propels unemployment to close to 80 per cent and shrinks GDP by 83.5 per cent in the Gaza Strip | International Labour Organization. https://www.ilo.org/resource/news/war-propels-unemployment-close-80-cent-and-shrinks-gdp-835-cent-gaza-strip

IPPF (2024) Press release: Gaza nine months on, pregnant women carry the burden of conflict. https://www.ippf.org/media-center/press-release-gaza-nine-months-pregnant-women-carry-burden-conflict

IPS (2024a) Al-Nasr Children's hospital | Documenting the targeting and destruction of the health sector in the Gaza strip. Institute of Palestine Studies. https://gazahcsector.palestine-studies.org/index.php/en/node/529

IPS (2024b) A public health approach to rebuilding health in Gaza—A contribution from the public health subgroup to the Amman conference (February 7, 2024) | Documenting the targeting and destruction of the health sector in the Gaza strip. Institute of Palestine Studies. https://gazahcsector.palestine-studies.org/en/node/381

Irfan B, Lulu I, Hamawy A, Shammala AA, Kullab S, Fawaz M, Sammour AAK, Khawaja H, Alshaer N, Abu-Sittah G, Alser O, Perlmutter M, Al-Shembari A, Hickey J, Nayfeh T, Tahir M, Sultan MJ, Nasser E, Wajahath M et al (2024) Combating infections under siege: healthcare challenges amidst the military assault in Gaza. World Med Health Policy. https://doi.org/10.1002/WMH3.642

ISDC (2023) Mental health in the West Bank and Gaza. https://isdc.org/publications/mental-health-in-the-west-bank-and-gaza/

Kearney JE, Thiel N, El-Taher A, Akhter S, Townes DA, Trehan I, Pottinger PS (2024) Conflicts in Gaza and around the world create a perfect storm for infectious disease outbreaks. PLOS. Glob Public Health 4(2). https://doi.org/10.1371/JOURNAL.PGPH.0002927

Keasley J, Blickwedel J, Quenby S (2017) Adverse effects of exposure to armed conflict on pregnancy: a systematic review. BMJ Glob Health 2(4). https://doi.org/10.1136/BMJGH-2017-000377

Kerbaj M, Traboulsi J, Maasri Z, Toukan H (2024) Gaza in plain sight: witnessing in solidarity. J Vis Cult 22(3):1–19. https://doi.org/10.1177/14704129241227847/ASSET/IMAGES/LARGE/10.1177_14704129241227847-FIG15.JPEG

Kienzler H, Daniel G, Hammoudeh W, Nashashibi R, Abu-Jamei Y, Giacaman R (2024) Unbearable suffering: mental health consequences of the October 2023 Israeli military assault on the Gaza Strip. BMJ Glob Health 9(9):14835. https://doi.org/10.1136/BMJGH-2023-014835

Landrigan PJ, Fuller R, Acosta NJR, Adeyi O, Arnold R et al (2018) The lancet commission on pollution and health. Lancet 391(10119):462–512. https://doi.org/10.1016/S0140-6736(17)32345-0

Mahase E (2024) Gaza: Israeli raids leave Kamal Adwan Hospital "out of service" with head doctor detained. BMJ 387:q2885. https://doi.org/10.1136/BMJ.Q2885

Mahmoud K (2024) The impact of Gaza war on environmental security. Serb J Eng Manag 9(1):1–5. https://doi.org/10.5937/SJEM2401001M

Mahmoud H, Abuzerr S (2023) State of the healthcare system in Gaza during the Israel–Hamas war. Lancet 402(10419):2294. https://doi.org/10.1016/S0140-6736(23)02634-X

Middle East Eye (2024) Nearly 1,000 Palestinian health workers killed by Israeli forces in Gaza named. https://www.middleeasteye.net/news/nearly-1000-palestinian-health-workers-killed-israeli-forces-gaza-named

Mohammed F, Elgailani USA, Ibrahim Ali SY, Mohamed RFA, Su Yin ET, Bravo-Vasquez ML (2024) Defending the right to health in Gaza: a call to action by health workers. Confl Heal 18(1):1–4. https://doi.org/10.1186/S13031-024-00613-5/METRICS

Mostoller AM, Mickelson KD (2024) Masculinity and mental well-being: the role of stigma attached to help-seeking among men. Sex Roles 90(3):353–362. https://doi.org/10.1007/S11199-024-01457-2/METRICS

MSF (2024a) Gaza before 7 October. https://www.msf.org/gaza-7-october

MSF (2024b) Nasser hospital must be protected. https://www.msf.org/nasser-hospital-must-be-protected

Muhammad A, Rizek J (2024) Caring for women in an active war zone. J Emerg Nurs 50(6):722–726. https://doi.org/10.1016/J.JEN.2024.08.002

OCHA (2024a) Health conditions in the occupied Palestinian territory, including east Jerusalem Report by the Director-General. https://www.ochaopt.org/page/gaza-strip-critical-

OCHA (2024b) Hostilities in the Gaza Strip and Israel | Flash Update #147 [EN/AR/HE]. https://www.unocha.org/publications/report/occupied-palestinian-territory/hostilities-gaza-strip-and-israel-flash-update-147-enarhe

OCHA (2024c) Occupied Palestinian territory. https://www.unocha.org/occupied-palestinian-territory

OCHA (2024d) Reported impact snapshot | Gaza Strip (25 September 2024) | United Nations Office for the Coordination of Humanitarian Affairs—occupied Palestinian territory. https://www.ochaopt.org/content/reported-impact-snapshot-gaza-strip-25-september-2024

OCHA (2024e) Reported impact snapshot | Gaza Strip (31 December 2024) | United Nations Office for the Coordination of Humanitarian Affairs—occupied Palestinian territory. https://www.ochaopt.org/content/reported-impact-snapshot-gaza-strip-31-december-2024

OHCHR (2024a) Israel's escalating use of torture against Palestinians in custody a preventable crime against humanity: UN experts | OHCHR. https://www.ohchr.org/en/press-releases/2024/08/israels-escalating-use-torture-against-palestinians-custody-preventable

OHCHR (2024b) UN report: Palestinian detainees held arbitrarily and secretly, subjected to torture and mistreatment | OHCHR. https://www.ohchr.org/en/press-releases/2024/07/un-report-palestinian-detainees-held-arbitrarily-and-secretly-subjected

Oxfam (2024) How to help Gaza | Make your voice heard. https://www.oxfam.org.uk/oxfam-in-action/current-emergencies/how-to-help-gaza/

PCHR (2024) "Women with cancer in Gaza face imminent death risk amid ongoing genocide" | Palestinian Centre for Human Rights. https://pchrgaza.org/women-with-cancer-in-gaza-face-imminent-death-risk-amid-ongoing-genocide/

Perugini N, Gordon N (2024) Medical Lawfare: the Nakba and Israel's attacks on Palestinian healthcare. J Palest Stud 53(1):68–91. https://doi.org/10.1080/0377919X.2024.2330366

PIB (2022) India achieves significant landmarks in reduction of Child Mortality. https://pib.gov.in/Pressreleaseshare.aspx?PRID=1861710

Ramadan A, Abouzid M (2024) Resilience beyond borders: Gaza's women and their silent battles. Women Birth J Aust Coll Midwives 37(4). https://doi.org/10.1016/J.WOMBI.2024.101623

ReliefWeb (2024a) Humanitarian action for children 2024—State of Palestine, Revision 3 (June 2024)—occupied Palestinian territory. https://reliefweb.int/report/occupied-palestinian-territory/humanitarian-action-children-2024-state-palestine-revision-3-june-2024

ReliefWeb (2024b) Resilience amidst chaos: 100 days of UNRWA's health response to the humanitarian crisis in the Gaza Strip, 7 October 2023 to 14 January 2024—occupied Palestinian territory. https://reliefweb.int/report/occupied-palestinian-territory/resilience-amidst-chaos-100-days-unrwas-health-response-humanitarian-crisis-gaza-strip-7-october-2023-14-january-2024

ReliefWeb (2024c) UNICEF State of Palestine humanitarian situation report No. 30–1 to 31 August 2024—occupied Palestinian territory. https://reliefweb.int/report/occupied-palestinian-territory/unicef-state-palestine-humanitarian-situation-report-no-30-1-31-august-2024

ReliefWeb (2024d) Urgent hygiene crisis: combatting diarrhea and skin infections in Gaza—occupied Palestinian territory. https://reliefweb.int/report/occupied-palestinian-territory/urgent-hygiene-crisis-combatting-diarrhea-and-skin-infections-gaza

Reuters (2023a) Disease could be bigger killer than bombs in Gaza - WHO | Reuters. https://www.reuters.com/world/middle-east/more-people-risk-death-disease-than-bombings-gaza-who-2023-11-28/

Reuters (2023b) Gaza's only cancer hospital goes out of service, health officials say. https://www.reuters.com/world/middle-east/gazas-only-cancer-hospital-goes-out-service-health-officials-2023-11-01/

Rockowitz S, Bayoumi R, Parr N, Awad A, Altawil M, Elmusharaf K (2024) Fostering cultural resilience: assessing the success of a locally engaged and adapted mental health intervention in Gaza. Front Public Health 12. https://doi.org/10.3389/FPUBH.2024.1390211

Sabet C, Nguyen D, Nada SA, Abuzerr S, Taha AM (2024) Women and girls in Gaza face increasingly dire physical and mental health challenges. BMJ 384. https://doi.org/10.1136/BMJ.Q625

SCI (2024) Women self-inducing labour and facing life-threatening complications in pregnancy after nine months of Gaza conflict | Save the Children International. https://www.savethechildren.net/news/women-self-inducing-labour-and-facing-life-threatening-complications-pregnancy-after-nine

Shaheen A, Dajani R, Zinszer K, Ashour Y, Abuzerr S (2024) The war on the Gaza Strip and its consequences on global warming. Front Human Dynam 6:1463902. https://doi.org/10.3389/FHUMD.2024.1463902/BIBTEX

Shellah D (2024) War on Gaza: the impossible duty to care for the critically ill. Intensive Care Med 50(2):311–313. https://doi.org/10.1007/S00134-023-07309-Z/METRICS

Sherin A (2024) Challenges and solutions for developing countries in promoting global health and achieving SDG 3. Khyber Med Univ J 16(3):179–181. https://doi.org/10.35845/KMUJ.2024.23761

Smith TB, Vacca R, Mantegazza L, Capua I (2023) Discovering new pathways toward integration between health and sustainable development goals with natural language processing and network science. Glob Health 19(1). https://doi.org/10.1186/S12992-023-00943-8

Taha AM, Sabet C, Nada SA, Abuzerr S, Nguyen D (2024) Addressing the mental health crisis among children in Gaza. Lancet Psychiatry 11(4):249–250. https://doi.org/10.1016/S2215-0366(24)00036-1

TRT (2024) Palestinian detainees in Israeli jails increased 130% after October 7. https://www.trtworld.com/middle-east/palestinian-detainees-in-israeli-jails-increased-130-after-october-7-17811425

UN (2022). United Nations common country analysis for the occupied Palestinian territory. https://palestine.un.org/en/200011-united-nations-common-country-analysis-occupied-palestinian-territory

UN (2024a). Gaza crisis: Babies being born 'into hell' amid desperate aid shortages. https://news.un.org/en/story/2024/01/1145677

UN (2024b) Gaza: First polio case confirmed in war-shattered enclave | UN News. https://news.un.org/en/story/2024/08/1153486

UN (2024c) Gaza: 'Frightening increase' in Hepatitis A cases | UN News. https://news.un.org/en/story/2024/08/1152791

UN (2024d) Last hospital in northern Gaza out of service following raid | UN News. https://news.un.org/en/story/2024/12/1158596

UN (2024e) The sustainable development goals report. https://unstats.un.org/sdgs/report/2024/The-Sustainable-Development-Goals-Report-2024.pdf

UN (2024f) Treatment of detainees and hostages and attacks on medical facilities and personnel (7 October 2023 to August 2024)—Third Report of the Independent International Commission of Inquiry on the Occupied Palestinian Territory and Israel—Question of Palestine. https://www.un.org/unispal/document/report-of-the-independent-international-commission-of-inquiry-on-the-occupied-palestinian-territory-including-east-jerusalem-and-israel-11sep24/

UNEP (2024) Environmental impact of the conflict in Gaza: preliminary assessment of environmental impacts | UNEP—UN Environment Programme. https://www.unep.org/resources/report/environmental-impact-conflict-gaza-preliminary-assessment-environmental-impacts

UNICEF (2022) The Gaza Strip | The humanitarian impact of 15 years of blockade—June 2022 | UNICEF Middle East and North Africa. https://www.unicef.org/mena/documents/gaza-strip-humanitarian-impact-15-years-blockade-june-2022

UNICEF (2024). Intensifying conflict, malnutrition and disease in the Gaza Strip creates a deadly cycle that threatens over 1.1 million children. https://www.unicef.org/press-releases/intensifying-conflict-malnutrition-and-disease-gaza-strip-creates-deadly-cycle

UNRWA (2022). Gaza_15 years of blockade. https://www.unrwa.org/gaza15-years-blockade

WHO (1948) Summary report on proceedings, minutes and final acts of the International Health Conference held in New York from 19 June to 22 July 1946. https://iris.who.int/handle/10665/85573

WHO (2024a) Around 560 000 children vaccinated in first round of polio campaign in Gaza. https://www.who.int/news/item/13-09-2024-around-560-000-children-vaccinated-in-first-round-of-polio--campaign-in-gaza

WHO (2024b) Children's lives threatened by rising malnutrition in the Gaza Strip. https://www.who.int/news/item/19-02-2024-children-s-lives-threatened-by-rising-malnutrition-in-the-gaza-strip

WHO (2024c) oPt emergency situation update. Issue 45. https://www.emro.who.int/images/stories/Sitrep_45b.pdf

WHO (2024d) Public Health Situation Analysis (PHSA) on Hostilities in the occupied Palestinian territory (oPt). https://www.un.org/unispal/wp-content/uploads/2024/05/WHO-PHSA-oPt-020524-FINAL.pdf

WHO (2024e) SEVENTY-SEVENTH WORLD HEALTH ASSEMBLY A77/12 Provisional agenda item 14.1 WHO's work in health emergencies Health conditions in the occupied Palestinian territory, including east Jerusalem. https://doi.org/10.1016/S0140-6736(19)30934-1

WHO (2024f) Six months of war leave Al-Shifa hospital in ruins, WHO mission reports. https://www.who.int/news/item/06-04-2024-six-months-of-war-leave-al-shifa-hospital-in-ruins%2D%2Dwho-mission-reports

WHO (2024g) Targets of sustainable development goal 3. https://www.who.int/europe/about-us/our-work/sustainable-development-goals/targets-of-sustainable-development-goal-3

5

'Waiting for a Better Time Is a Waste of Time': Motivations of Learning Amid the Genocide in Gaza in Light of Sustainable Development Goal 4

Tahani Aldahdouh, Khalid Dader, Alaa Aldahdouh, Rawia Awadallah, and Imed Romdhani

Abstract

This study, set against the backdrop of the recent genocide in Gaza (October 2023–January 2025), investigates the motivations driving university students, graduates and academics to independently enrol in an initiative offering free access to onlinecourses through platforms such as Coursera, edX and DataCamp. This study adopted self-determination theory as the conceptual framework. A convergent mixed-methods design was employed, and all participants in the initiative were invited to complete a brief online questionnaire incorporating the situational motivation scale and open-ended questions. A total of 503 responses were collected: 138 participants completed one or more courses, 78 indicated ongoing enrolments, 92 did not complete any courses and 195 did not specify their status. The results of the quantitative analysis highlighted that the participants were mainly driven by autonomous internal motivation, and the thematic analysis of the qualitative data further highlighted this motivation, identifying six key categories: career-related, war-

The original version of this chapter was previously published without open access. A correction to this chapter is available at https://doi.org/10.1007/978-3-031-88500-6_19

We dedicate this work to Gaza—its people, students, teachers, and all who contributed to this study. We do not know whether the study participants are still here to read this acknowledgement or if they have fallen victim to the brutal machinery of the Israeli occupation. Yet, we are certain that this work will stand as part of the body of knowledge, reinforcing the Palestinian academic presence in global research communities, where such voices are too often marginalized.

T. Aldahdouh (✉)
Faculty of Education and Culture, Tampere University, Tampere, Finland
e-mail: tahani.aldahdouh@tuni.fi

K. Dader
Faculty of Management and Business, Tampere University, Tampere, Finland
e-mail: khalid.dader@tuni.fi

A. Aldahdouh
Independent Researcher, Tampere, Finland
e-mail: AlaaAlDahdouh@gmail.com

R. Awadallah
Faculty of Social Sciences, Department of Archivistics, Library and Information Science, Oslomet University, Oslo, Norway

Faculty of Information Technology, Islamic University of Gaza, Gaza, Palestine
e-mail: raaav6942@oslomet.no

I. Romdhani
School of Computing Engineering and the Built Environment, Edinburgh Napier University, Edinburgh, Scotland, UK
e-mail: I.Romdhani@napier.ac.uk

A. Hassoun (ed.), *War on Gaza*, Sustainable Development Goals Series,
https://doi.org/10.1007/978-3-031-88500-6_5

related, belief-related, personality-related, social and affectional motivations. War-related and belief-related motivations set this study apart from prior work in the field. The findings provide critical insights into how educational initiatives can support learners in extreme circumstances in alignment with Sustainable Development Goal 4. Implications for education policy and practice in conflict zones, along with avenues for future research, are discussed.

Keywords

Gaza · Genocide · Learning · Motivation · Palestine · Resistance · SDG 4 · Self-determination theory · Sustainable development goals

5.1 Introduction

In a place where schools and universities serve as shelters for the displaced and schoolbooks fuel the fires that cook food for the hungry, the very act of learning becomes a distant luxury that is beyond reach. What does it mean to hunger for knowledge in a place where even bread is scarce? How do you learn when there are no schools, no universities, no books and no safety to gather? In such spaces, the pursuit of education is transformed into an act of defiance, as remarkable acts of learning persist. This study explores the motivations behind Gaza students and academics joining the 'Free DataCamp, Edx and Coursera Memberships' initiative, launched in July 2024 by Dr. Imed Romdhani (one of this chapter's co-authors) and partners. The programme offered free access to online learning platforms, enabling participants to earn certificates despite the destruction of their educational infrastructure. What is it that motivates these learners to persist, despite the actualised scholasticide or educide and constant threats to their lives?

In this chapter, we explore the motivations behind these learners' perseverance, framing their efforts within the context of Sustainable Development Goal 4 (SDG 4), which calls for inclusive and equitable quality education for all. By centring the experiences and actions of Gaza's learners, we highlight the immense challenges they face and the defiance they demonstrate in their continued pursuits of education. While their actions are inspiring, we do not seek to promote or romanticise their resilience in a way that mitigates justice or the necessary action, particularly in line with SDG 4. Instead, we aim to highlight the profound hardships and sacrifices they endure just to access what is fundamentally their right to access (Brück et al. 2019). The persistence of these learners must not be understood as a celebration in any way but as a reminder of the injustices that force them to fight for their education in the first place. This study is guided by the following research question:

- What types of motivations drive learners to enrol in and complete free online courses during the genocide in Gaza?

5.2 Contextual and Theoretical Background

This section provides an overview of the education system in Gaza, both before and after October 7, 2023. Even prior to recent escalations, the education sector in Palestine faced continuous challenges and disruptions due to geopolitical conditions, Israeli restrictions, blockades and engineered immobility (Fassetta et al. 2017). The education system in Gaza, primarily governed by the Ministry of Education and the United Nations Relief and Works Agency (UNRWA), has operated under severe limitations, including frequent school overcrowding and a lack of infrastructure due to constraints on construction materials and limited resources. This section outlines the structure and challenges of the education system before October 2023 and examines the extensive impact of recent escalations, framing how these compounded issues have led to the systematic breakdown, or 'scholasticide' (Dader et al. 2024), of Gaza's educational framework.

5.2.1 Education System in Gaza Before October 7

The political ramifications of the Gaza Strip have exerted a heavy toll on the education sector. Israel imposed a blockade on Gaza in 2007 that has remained in place through multiple destructive aggressions from 2008 to 2021 (Nijim 2022), in addition to the genocidal attacks that have continued for more than a year at the time of writing this manuscript. These events have debilitated economic development, exacerbated poverty and restricted access to essential services, all of which have severely impacted education (Abed Alah 2024). The Israeli blockade limited access to electricity and construction materials and hindered rebuilding efforts after military aggression (GCPEA 2022; UNICEF 2022). The destruction and re-destruction of universities and campus infrastructures was framed by Jebril (2023) as a process of 'de-development' of higher education, in which rebuilding faces further destructive bombardment and resource scarcity.

The blockade also stifles intellectual growth by restricting travel and cultural exchange (Fassetta et al. 2017; Norwegian Refugee Council 2021). High unemployment rates have increased poverty, leading many students to drop out because they struggle to cover educational costs (OCHA 2018). The blockade's effects on teachers have also been severe, with restricted funding leading to low salaries, heavy workloads and limited professional development, resulting in dissatisfaction and turnover (Ramahi 2015). Poor living conditions further strain teachers, depleting their time for self-development and harming their well-being (Aldahdouh et al. 2024) and competencies (Alenius et al. 2019). Students show low engagement and diminished clarity regarding their future professions (Korhonen et al. 2024). As of 2018, around 20,000 students were not awarded their higher education certificates due to financial challenges and accumulated fees (OCHA 2018). Furthermore, the COVID-19 pandemic exacerbated these challenges, as inadequate infrastructure and a lack of preparedness for online education hindered learning further, affecting students' mental health and digital literacy (UNICEF 2022).

5.2.2 Scholasticide to Gaza's Education System Since October 7

In addition to the aforementioned difficulties, the recent Israeli genocidal assault has kept the education sector in a state of paralysis since October 2023 (Dader et al. 2024). According to the United Nations' Human Rights Office of the High Commissioner (OHCHR), at least 10,490 students, including 651 higher education students, 261 teachers and 95 university professors have been killed. Even more have been injured—at least 16,700 students, including 1317 higher education students and 756 teachers (Al-Mqadma et al. 2024; OHCHR 2024; ReliefWeb 2024). Israel has destroyed more than 85.8% (UNOSAT 2024) of all educational facilities across the Gaza Strip, including several archives and 13 public libraries that have been either partially or completely obliterated (Al-Mqadma et al. 2024). All universities in Gaza have now been destroyed, leading to a range of severe consequences, including the erasure of culture, memory and knowledge and the prevention of around 88,000 students from accessing higher education. Palestinian scholar Nabulsi introduced the term 'scholasticide' to describe the systematic destruction of education in various forms—students, teachers, universities, archives, cultural centres and indigenous knowledge systems—with a long history in Palestine (Ahmad and Vulliamy 2009; Glück 2024; Spectre 2024).

The attacks on infrastructure have also impacted internet connectivity, causing prolonged outages and compromised telecommunications (Access Now 2023). Although partial restoration began in January 2024, services remain inconsistent (Al Arabiya 2024). These disruptions pose substantial barriers to students' access to online learning, further exacerbating the scholasticide landscape that Israel has intentionally and pointedly engineered in Gaza (Iriqat et al. 2025).

5.2.3 Sustainable Development Goal 4 in Times of Crisis

In the face of this targeted destruction, the principles enshrined in SDG 4 become both a moral compass and a stark contrast to the lived realities of Gaza's learners. SDG 4 seeks to ensure inclusive, equitable and quality education while promoting lifelong learning opportunities for all individuals. This goal covers a wide range of educational aspects to advance quality and accessible education globally, particularly in areas and populations where educational opportunities have historically been limited or underfunded (Benavot et al. 2024; Klees 2024). For example, Target 4.1 aims to provide all children with free high-quality primary and secondary education, while Target 4.2 covers the preparation of children for school life. Equal access to technical, vocational and higher education (Target 4.3) addresses the need for accessible skills training and employment readiness, while Target 4.5 highlights inclusivity and gender equality to eliminate educational disparities and promote equal opportunities. Other targets focus on improving literacy and numeracy (Target 4.6) and fostering education for sustainable development, global citizenship and peace (Target 4.7).

In crisis regions, specific SDG 4 targets are essential for maintaining access to education and providing safe, inclusive environments for learning, as crises often disrupt traditional education structures and create obstacles to regular attendance, such as those we have highlighted in Gaza. The urgent adoption of flexible education delivery models that employ online learning platforms and temporary schooling setups is required, which is the crux of this sustainable goal. Teachers in these regions require additional support, resources and training to cope with the complexities of delivering education amid crises (McKinney and Farrar 2023; Menashy and Zakharia 2023). The inclusivity and safety of educational environments must be guaranteed in such contexts, given the serious risks that students face. Local partnerships with community-based organisations and leaders also play a vital role in strengthening the resilience of educational systems and addressing specific needs within crisis settings. The incorporation of mental health and psychosocial support into educational settings is also important (McKinney and Farrar 2023; Menashy and Zakharia 2023).

The context of Gaza exemplifies these broader and even more specific challenges in an extreme form, where the recent indiscriminate genocidal war has decimated the educational infrastructure and uprooted the foundations of the student landscape in Gaza from the classroom. The deliberate targeting of schools and educational institutes and the systemic deprivation of resources represent not just an obstacle but an outright assault on education—that is, scholasticide. Schools that have not been bombed have all been used as shelters for displaced populations.

5.2.4 Attempts to Achieve SDG 4 Amid the Collapse of the Education Sector

Even within this grim reality, grassroots initiatives and efforts highlight the undeterred resolve of communities and organisations striving to maintain access to education. Most of these initiatives came to light amid the current genocidal violence, and the majority of such attempts, if not all, have relied on online materials, distance teaching and short-term projects.

The following section examines the *diverse initiatives* designed to sustain educational continuity, including efforts to support higher education and professional learning, and focuses on how universities in Gaza have navigated these tumultuous times to provide opportunities for students (Birzeit University 2024; The Emergency Committee of Universities in Gaza 2024).

5.2.4.1 West Bank Universities' Support

West Bank universities have taken a leading role in supporting Gaza students' access to education during the recent genocidal war. By opening their courses to Gaza students through online platforms, these institutions have enabled approximately 5000 students to continue their studies remotely over two academic semesters since March 2024, overcoming significant barriers to in-person attendance (Ministry of Higher Education & Scientific Research 2023). To further support educational continuity, West Bank universities have implemented specialised campaigns and partnerships addressing students' financial, technological and psychological needs. For example, Birzeit University launched the 'Campaign for Gaza 2024', which provides scholarships to alleviate financial burdens and resources like laptops and internet access to bridge the digital divide. Birzeit University also integrated mental health resources to support students' psychological well-being, acknowledging the emotional strain of the war (Birzeit University 2024). Following suit, An-Najah National University, in collaboration with the Mediterranean Universities Union (UNIMED), implemented a scholarship programme to aid Gaza students in accessing remote education. This initiative combined financial support with infrastructure enhancements critical for digital learning (UNIMED 2024). Other universities in the West Bank have followed in the footsteps of Birzeit and accommodated good numbers of students digitally within their capacities.

5.2.4.2 Gaza Universities' Coping Attempts

Gaza universities collectively initiated an emergency committee to seek international support and coordinate recovery efforts with global academic partners. This initiative aimed to rebuild academic facilities and help reintegrate students and educators into academic environments. The committee developed a priority framework outlining immediate and long-term needs, including digital resources, infrastructure repair and reliable internet access for remote education. This framework has not only addressed the current educational disruptions but also provided a resilience model for crisis-prone settings (The Emergency Committee of Universities in Gaza 2024). Through this framework, Gaza universities have implemented flexible learning solutions, enabling online education through platforms such as Moodle. Online course offerings have gradually expanded, accompanied by financial aid for economically impacted students. Recognising connectivity challenges, some universities have provided internet subsidies and established safe access points to facilitate online learning (The Emergency Committee of Universities in Gaza 2024).

International initiatives like Academic Solidarity with Palestine and Academics for Palestine address faculty and staff shortages in Gaza universities to sustain academic progress. These programmes enable remote academic instruction and ensure continuity in education by organising subgroups based on academic disciplines. Each subgroup includes representatives from Gaza's university faculties, who identify immediate staffing needs in specific subjects or courses. Volunteer academics then connect with these representatives to arrange teaching opportunities (Academic Solidarity with Palestine 2024; Academics for Palestine 2024). Similarly, Academics4Gaza focuses on strengthening Gaza-based faculty by providing a support network for academics to continue research and teaching during the recent war (Academics4Gaza 2024).

5.2.5 Motivation to Learn Amid the War

While the preceding sections have highlighted the structural and institutional efforts to restore education in Gaza, the heart and soul of such initiatives lie in the individuals they serve. The motivation and resilience of Gaza's higher education students who persist in their pursuit of learning amid unimaginable odds epitomise the essence of SDG 4—ensuring inclusive, equitable and quality education for all.

Motivation serves as a fundamental force in sustaining learning, especially in conflict-ridden regions where students face significant psychological, infrastructural and societal barriers. Previous research has demonstrated that motivated learners often exhibit remarkable resilience and defiance as their intrinsic drive to pursue education has the power to counterbalance the severe disruptions of war and violence (HuffPost 2017; Sheykina et al. 2024). Motivated students have found innovative ways to sustain learning amid crises, leveraging online education platforms, as seen in conflict zones such as Syria and Ukraine (Sheykina et al. 2024; Tozan 2023). Furthermore, studies on global responses to the Ukraine war highlight the importance of emotional well-being in maintaining student motivation and academic persistence in war-torn contexts (Raccanello et al. 2024). Related research from Ethiopia and other conflict-ridden nations emphasises the necessity of strong governance and adaptive educational systems to minimise the long-term impact of violence on learning outcomes (Assefa et al. 2023).

In the absence of a motivation theory specifically tailored to education in war- and conflict-affected settings, this study uses self-determination theory (SDT) to examine the motivations of students and academics in Gaza who persist in engaging with online learning platforms (e.g. edX and Coursera). SDT, developed by Ryan and Deci (2000), offers a comprehensive framework for understanding human behaviour and motivation, particularly in educational contexts. SDT posits that there are three basic psychological needs—autonomy, competence and relatedness—that are essential for fostering intrinsic motivation and individual well-being (Ryan and Deci 2017). *Autonomy* refers to the need to self-regulate experiences and to feel that one's actions are self-chosen; *competence* involves the need to achieve self-referenced success in tasks and to exert meaningful effects on one's environment; and *relatedness* pertains to the need to feel valued and socially connected to others. The degree of satisfying those needs results in three primary forms of motivation: *amotivation*, *intrinsic motivation* and *extrinsic motivation*. *Amotivation* occurs when individuals lack a clear reason for engaging in an activity. *Intrinsic motivation* refers to engagement in activities for their inherent enjoyment or interest. *Extrinsic motivation* involves engaging in activities for external reasons, such as rewards or obligations. SDT further categorises extrinsic motivation into several types based on the degree of autonomy involved. *External regulation* is the least autonomous form of extrinsic motivation, in which individuals are forced to act solely to meet external demands or avoid punishment. *Introjected regulation* refers to engagement in activity driven by internal pressures such as guilt, anxiety or pride. *Identified regulation* is a more autonomous form of extrinsic motivation; individuals with identified regulation motivation strive to grasp the value and importance of engaging in the activity. *Integrated regulation* is the most autonomous form of extrinsic motivation, in which the activity becomes fully aligned with an individual's personal values, identity and beliefs (Ryan and Deci 2000).

In Gaza, where educational infrastructure has been severely disrupted, online learning may offer students a way to regain autonomy, demonstrate competence and connect with a broader academic community. A recent study by Taka (2023) explored how learners in post-genocide Rwanda made sense of their educational journey and their learning motivations. Their study revealed that both intrinsic and extrinsic motivations were driving students' learning. They also found that intrinsic motivations were largely overlooked in the education in emergency context (Taka 2023).

Dozens of empirical studies have shown the essential role of motivation in students' learning and thus the promotion of SDG 4 (Cachia et al. 2018; Gorges 2015; Vansteenkiste et al. 2004; Wei et al. 2024). SDG 4 emphasises quality education, which includes effective learning outcomes, and research evidence has shown that the intrinsic motivations of learners are strongly correlated with improved academic performance and persistence (Vansteenkiste et al. 2004), higher learning engagement (Wei et al. 2024) and lower dropout rates (López-Angulo et al. 2024). Moreover, lifelong learning, a core component of SDG 4, depends heavily on learners' motivations, as indicated in a study by Gorges (2015).

5.3 Methods

5.3.1 Research Context

This study explores the motivations behind the participation and commitment of university students and academics in Gaza in the initiative titled 'Free DataCamp, Edx, and Coursera Memberships to Palestinian Students and Academics'. This initiative was launched in July 2024 by Dr. Imed Romdhani in collaboration and partnership with Scottish Charity Scot Aid (https://www.scotaid.org.uk), UNDP (https://www.undp.org), DataCamp (https://www.datacamp.com), Edx (https://www.edx.org), Protected Consulting Tunisia (https://protected-consulting.com) and a group of academics from Gaza. This initiative aims to help students and academics in Gaza continue studying and earn valuable verified academic and professional certificates. The initiative grants participants unlimited access for at least 3 months, up to what the completion of the chosen course requires. The initiative has been welcomed by many students and academics in Gaza. At the time of writing this chapter, 12,686 requests have been filed to participate in those courses. Table 5.1 shows some of the initiative statistics as of April 10, 2025.

The initiative has attracted learners from both hard and soft disciplines, with a notable interest from technical fields of study. For instance, Fig. 5.1 shows the distribution of skills and areas of interest among learners in Gaza on the edX platform during the period from July to April 2025.

5.3.2 Research Design and Data Collection

This study employed a convergent mixed-methods design to corroborate and validate findings, allowing qualitative insights to illustrate and expand on the quantitative results (Creswell and Plano Clark 2018, p. 125). Data were collected through a self-reported online questionnaire administered to all participants of the initiative via the LimeSurvey platform, featuring both close- and open-ended questions in two primary sections.

The first section gathered background information, while the second explored the participants' experiences in relation to the initiative. This section included questions on enrolment in and completion of online courses and motivations for participation. The questionnaire was structured with conditional logic to adapt questions based on each participant's enrolment and course completion status.

- Participants who had not completed any courses were asked to specify why.
- Participants who had completed one or more courses or were still actively studying were prompted to provide open-ended responses on their motivations. They were asked to indicate their motivations by answering 16 items from the situational motivation scale (SIMS), rated on a 7-point Likert scale (1 = does not correspond at all to 7 = corresponds exactly).

Table 5.1 Statistics for the Free DataCamp, Edx and Coursera Memberships to Palestinian Students and Academics Initiative from June to April 2025

	Coursera	Edx	DataCamp
Number of enrolled students on each platform	310	2000	2359
Number of verified certificates	655	2434	12
Number of hours	6611	17,086	3054
Number of courses	1230	8745	4197

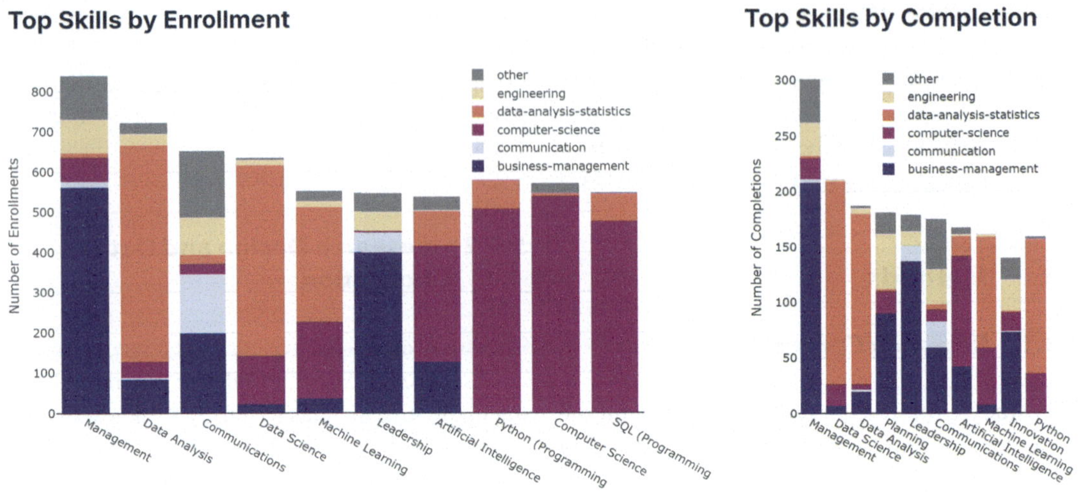

Fig. 5.1 Distribution of skills and fields of study among learners on the edX platform. Data generated from the edX dashboard

The SIMS, grounded in self-determination theory, assesses four types of motivation along the self-determination continuum: amotivation (four items, e.g. 'There may be good reasons to take these courses, but personally, I don't see them'), external regulation (four items, e.g. 'Because I am required to take these courses'), identified regulation (four items, e.g. 'Because I believe these courses are beneficial for my personal development') and intrinsic motivation (four items, e.g. 'Because I enjoy exploring new topics through these courses'). Previous research has demonstrated the strong internal consistency and construct validity of the SIMS (Guay et al. 2000). For this study, the SIMS was adapted to fit the context of online learning within the initiative's courses. The first author translated the SIMS items into Arabic, and the translation was then reviewed and verified for accuracy by two additional native Arabic-speaking researchers (see Appendix 1).

5.3.3 Participants

In collaboration with the initiative's founder (the last author), we sent the online questionnaire to all participants registered in the initiative by the end of August 2024. Ethical guidelines were explained on the first page, and informed consent was given as a precondition to start the questionnaire.

A total of 756 responses were received, of which 503 were complete and usable. The majority of the participants were female (59.8%, n = 301). Ages ranged from 18 to 57 years (M = 25.05, SD = 6.762). Most participants were university students (56%, n = 281), followed by unemployed graduates (25%, n = 127), employed graduates (12%, n = 60) and university teachers (7%, n = 35).

Based on Biglan's (1973) taxonomy of academic disciplines, the majority belonged to hard disciplines (62%, n = 313), followed by soft disciplines (24%, n = 119) and mixed disciplines (14%, n = 71). Most participants (61%, n = 308) reported enrolling in one or more online courses, while the remaining (39%, n = 195) had not yet enrolled. Among those who had enrolled, 138 participants (45%) reported completing one or more courses, 92 participants (30%) had not completed any courses and 78 participants (25%) were still in the process of studying.

5.3.4 Analysis

Quantitative data were analysed using SPSS (version 29.0) with descriptive statistics (mean and standard deviation) and factor analysis. The qualitative data were analysed using inductive thematic analysis in Atlas.ti (version 24.1), following Braun and Clarke's (2006) guidelines. Initially, the data were pseudonymised and reviewed multiple times to gain a comprehensive understanding. They were then systematically coded to identify features relevant to the research questions. The coded data were examined for patterns, with similar codes grouped into themes and subthemes. In a second iteration, themes and subthemes were revisited to validate their relevance and clarify distinctions, ensuring comprehensive and clearly defined categories.

5.4 Results

5.4.1 Quantitative Results

As the current study adapted the SIMS instrument, it was first necessary to examine the underlying structure of the data, especially since the instrument had not previously been examined in the Arabic language or in the Palestinian higher education context. Using data from 216 participants who completed the SIMS, we performed exploratory factor analysis using the maximum likelihood extraction method with Promax rotation on all SIMS items. The results of sampling adequacy were acceptable and indicated that the data were appropriate for factor analysis: Kaiser–Meyer–Olkin (KMO) was 0.864, which is above the cut-off value of 0.50 (Yong and Pearce 2013), while the result of Bartlett's test of sphericity was significant ($X^2 = 1553.533$, $p < 0.001$). As recommended by Tabachnick and Fidell (2007), a cut-off value of 0.32 was used as the minimum loading of an item on any factor. To determine the number of items to be retained, eigenvalues greater than 1 were looked for, as suggested by Tabachnick and Fidell (2007). Based on the inspection of eigenvalues, cumulative variance and scree plot, three factors were extracted, explaining 52.8% of variance. The communality of items and loading on their corresponding factors are presented in Table 5.2.

Contrary to expectations, exploratory factor analysis revealed that intrinsic motivation and identified regulation emerged as a single factor, while external motivation and amotivation formed distinct, separate factors. Reliability for the resulting SIMS dimensions was assessed using Cronbach's alpha, with all three dimensions demonstrating acceptable reliability values. Distribution properties for each factor were evaluated, and the results indicated normality, as skewness and kurtosis values were within acceptable ranges: skewness <2 and kurtosis <7 (Kim 2013). A mean score was then calculated for the items within each factor (see Table 5.2).

As expected, the participants predominantly endorsed the merged intrinsic and identified regulation motivation factor, with an average score of 5.93 on a 1–7 scale. This was followed by external motivation, with an average score of 3.13, and amotivation, with the lowest average score of 1.98.

To further investigate the high merged intrinsic and identified regulation motivation score and to understand the specific characteristics driving this motivation, we conducted a qualitative analysis. The qualitative findings presented in the following section provide deeper insights into these motivational features.

5.4.2 Qualitative Results

The thematic analysis identified several key motivations driving students' and academics' engagement in the online learning initiative, structured here by frequency of occurrence in the data. These motivations encompassed career-related, war-related, belief-related, academic-related, personality-related, social and affectional motivations. Notably, war-related motivations emerged as a defining theme, capturing the unique impact of the recent war on participants' engagement in learning.

5.4.2.1 Career-Related Motivations

Career-related motivations were the most frequently cited reason for participating in online learning. Many participants viewed the initiative as an opportunity for *skills development* and a way to improve their job prospects, either by directly aiding in *job hunting* or enhancing their *CVs* to stand out in a competitive job market (e.g.

Table 5.2 Factor Analysis, Alpha, Means and Standard Deviations of the SIMS Dimensions

	Factor			
	1	2	3	Communality
Intr01	0.77			0.72
Intr02	0.74			0.54
Intr03	0.69			0.52
Intr04	0.77			0.58
Iden01	0.76			0.52
Iden02	0.77			0.57
Iden03	0.51			0.33
Iden04	0.79			0.59
Extr01			0.80	0.68
Extr02			0.43	0.27
Extr03			0.57	0.33
Extr04			0.69	0.47
Amot01		0.62		0.60
Amot02		0.76		0.56
Amot03		0.87		0.69
Amot04		0.72		0.49
Eigenvalues	5.53	2.26	2.03	
Explained variance (%)	31.82	10.45	10.55	
Alpha	0.89	0.83	0.70	
M (*SD*)	5.93 (1.002)	1.98 (1.289)	3.13 (1.396)	

'I want to be more competitive in the job market by acquiring many new skills' [Participant 131]). Some participants, particularly those already engaged in academia, were motivated to *develop their own online teaching* abilities, seeing it as a way to stay current with teaching practices. For others, online learning served as a form of job security, driven by a *fear of job loss* in an uncertain and unstable economic climate.

5.4.2.2 War-Related Motivations

War-related motivations emerged as a profound and distinct category, illustrating how the recent genocide significantly influenced their reasons for engaging in online learning. Many participants saw learning as a form of *resistance*, aiming to counteract the destructive impact of the war on their personal and community growth. Learning became a tool to defy adversity, embodying the notion of 'resisting the enemy' through intellectual empowerment.

I am from the northern part of Gaza. They [the occupation forces] always ask us to move to the south. However, we refuse to leave our houses, so we stand firm even with starvation, bombing and fear. I concentrate on the positive side; I am still in my home, and I have electricity and an internet connection, which allow me to sit at my desk and concentrate only on success (Participant 586).

A future-focused mindset was also evident, with participants driven by *future-focused resistance*: a commitment to building skills and knowledge that could contribute to post-war recovery and resilience (e.g. Participant 274 described, 'building a future from nothing', while Participant 328 elaborated that 'the current situation requires a responsible person, and responsibility includes the future we are going to; I would like to face the future with a university degree'). For some, this was intertwined with *compensating for loss*; by gaining new knowledge and skills, they sought to replace opportunities and stability taken by the war.

Learning also served as a psychological strategy for *distancing from distress*, offering the participants a productive way to shift their focus from the hardships of war to personal growth and self-improvement (e.g. 'Investing our energy in learning will enhance our mental health by feeling accomplished rather than helpless' [Participant 144]). Many described education as a source of hope, driven by the belief that *change begins with learning* (e.g. 'I believe that education is the most powerful weapon to overcome cuffs and borders' [Participant 43]). This perspective was often accompanied by a drive to combat broader societal issues, with participants emphasising the importance of *wrestling ignorance* as a way to rebuild and strengthen their community.

5.4.2.3 Belief-Related Motivations

Belief-related motivations also played a significant role, rooted in personal values and faith. Many participants expressed a strong desire to *make every moment count* (e.g. Participant 259 pointed out, 'The scene in Gaza is foggy. Waiting for a better time is a waste of time. Life will pass away for nothing' [Participant 259] and 'Every day you don't learn something new has no value' [Participant 323]). With the ever-present uncertainty of genocide, participants sought to use their time as meaningfully as possible, turning to education as a constructive and hopeful way to navigate a challenging reality. As Participant 715 indicated, 'The current situation is actually the reason for this. We have become accustomed to filling our time with beneficial activities. Since life is not as it once was, we are now trying to create an educational opportunity'. For others, *faith beliefs* provided a spiritual foundation for learning, encouraging them to persevere, emanating from their vision that they should 'satisfy God' (Participant 688). This sense of purpose was echoed in the participants' determination to *strive with trust*, using education as a means to maintain momentum despite the circumstances (e.g. Participant 586 recited, 'Man shall have nothing but what he strives for'; Quran, Sura An-najm, verse 39).

5.4.2.4 Academic Motivations

Academic motivations were another common theme, with participants focusing on achieving specific educational goals. Many viewed the initiative as a way to facilitate *graduation* and obtain *certifications* that would validate their learning. For some, these certificates were a way to *promote their expertise* within their fields, ensuring that they stayed current and competitive. Participants also expressed a desire for *knowledge renewal*, driven by an intrinsic appreciation for learning (e.g. 'No delay in acquiring knowledge' [Participant 431]) and an ambition to continue *post-graduate studies* in the future.

5.4.2.5 Personality-Related Motivations

Personal growth and self-actualisation were additional motivators for engagement. Participants expressed a deep *passion for learning* and a determination to *grasp opportunities* as they arose (Participant 719: 'Life is passing away even if we are in the midst of war, and this is an opportunity offered to us on a golden dish'). Many were driven by a desire to achieve *personal goals*, find *self-fulfilment* and seek *continuing success*, seeing learning as essential to their identity and self-actualisation ('To make use of my time and to feel accomplished after one year of my life has gone for nothing' [Participant 148]).

5.4.2.6 Social Motivations

Social motivations reflected the participants' commitment to contributing positively to their communities. Many were driven by a sense of responsibility to *serve their society* because 'it is Gaza, it needs us deeply—and now more than ever, it needs its educated students, both men and women, in all scientific and practical fields' (Participant 646). Others wanted to *support their families*, seeing education as a way to strengthen social bonds and provide support in trying times. Some participants also valued the chance to *interact with the international community*, expanding their perspectives and fostering global connections (e.g. 'The current circumstances impose significant restrictions ... making online education an essential avenue for overcoming these barriers and achieving personal growth. These initiatives empower individuals to acquire new skills ... and engage with global learning communities, thereby opening greater professional horizons' [Participant 78]). Others expressed a desire to *be a role model* for others, hoping to inspire resilience and determination through their own engagement in learning.

5.4.2.7 Affectional Motivations

Finally, affectional motivations highlighted the emotional aspects of learning engagement. For some participants, *hope* for a better future was a significant driver, with online learning representing a source of possibility despite adversity ('It is the hope and our positive vision to the future even with all that pain and sorrow our people is passing through. With every woe comes a gift' [Participant 673]). Additionally, a *fear of losing mastery* of previously acquired skills motivated the participants to stay engaged and maintain proficiency. A few participants expressed that learning helped them *avoid feelings of guilt* over missed opportunities, emphasising the emotional complexity intertwined with their pursuit of education. (e.g. 'I don't want to waste any more of my life and regret it later' [Participant 009]).

5.5 Discussion

Using the lens of self-determination theory, this study aimed to understand the motivations driving university students and academics in Gaza to persist with online learning, despite the recent genocide. As part of this investigation, the study validated the Arabic version of the SIMS to quantitatively measure their motivations. Scale validation revealed a three-factor solution in which intrinsic and identified regulation motivations emerged as a single factor, alongside external regulation motivation and amotivation as distinct factors. This outcome could be partially explained by the fact that both intrinsic and identified regulation motivation dimensions share a high level of autonomy. We named the combined factor *autonomous internal motivation*. The Palestinian sample also appeared to perceive actions driven by the pure enjoyment of online learning (intrinsic motivation) and those guided by recognising its value (identified regulation) as fundamentally similar. They saw little difference between finding the learning process fulfilling (intrinsic) and seeing online courses as valuable for achieving other goals, even if the courses themselves were not inherently enjoyable (identified regulation). In both scenarios, the decision to engage in these online courses was viewed as self-initiated, reflecting a sense of personal autonomy.

The qualitative findings uncovered seven main themes of the high *autonomous internal motivation* score reported in the quantitative results: career-related, war-related, belief-related, academic-related, personality-related, social and affectional motivations.

Career-related motivations, at first glance, might suggest an orientation towards external regulation, in which the aspiration for a job acts as an external motivator. However, within the context of this study, we argue that these career-related motivations align more closely with the identified regulations. This interpretation stems from the observation that study participants voluntarily chose to pursue online courses, exercising full autonomy in recognising the value these courses would bring to their lives. For example, participants identified how the courses could help them develop the skills necessary to compete in the labour market. This finding is consistent with existing literature (Li et al. 2022; Remedios and Sewell 2024), which has found a positive association between motivation and college students' employability.

In line with expectations, our findings in relation to academic-related motivations were consistent with previous studies (Cachia et al. 2018; Gorges 2015; Vansteenkiste et al. 2004; Wei et al. 2024). Subthemes such as obtaining certified certificates, enhancing expertise in a specific field, seeking knowledge renewal and pursuing postgraduate studies reflect identified regulation in this study's context. Participants recognised the intrinsic value of engaging in online courses as a means of achieving their academic goals and aspirations.

Personality-related motivations emerged in the students' narratives as seeking passion in learning, grasping opportunities, appreciating knowledge, achieving personal goals, pursuing self-fulfilment and maintaining continuous success. These motivations clearly reflect both intrinsic and identified regulation motivations. A substantial body of research has demonstrated that intrinsic motivation serves as the driving force behind learning, leading to a deeper understanding and better achievement outcomes (Cachia et al. 2018; Gorges 2015; Vansteenkiste et al. 2004; Wei et al. 2024). The theme of grasping opportunities stands out as a particularly interesting result. The participants demonstrated a proactive mindset, viewing these free online courses as unique opportunities that might not arise again. This perspective highlights their self-starter attitudes in addition to their willingness to take initiative, persevere and act decisively to bring about meaningful change in their lives.

The literature has also recognised social motivations, in which students engage in online learning for social reasons, such as serving their community, supporting their family, interacting with the international community or becoming a role model (Urdan and Maehr 1995). In line with Senko et al. (2023), while SDT does not explicitly recognise social motivations, we interpreted the drive to serve others as a socially oriented identified motive. Similar to personally oriented identified motives, it focuses on how achieving a goal can fulfil a personally significant desire—in this case, the desire to benefit others.

Affectional motivations refer to the drive to perform actions influenced by feelings of pressure, such as avoiding guilt or anxiety or seeking ego enhancement. These motivations align with introjected motivation. Some participants in our study felt obligated to complete courses to avoid feelings of regret for not seizing the opportunity to learn. They also expressed a sense of obligation to preserve the knowledge and skills they had already acquired, which aligns with the concept of mastery-avoidance goal orientation (Van Yperen et al. 2009). Although mastery-avoidance goals are often viewed as non-optimal motivations in terms of their outcomes, research suggests they may be less harmful and even beneficial for older adults during certain life phases (Senko and Freund 2015). Moreover, the role of hope is noteworthy in this context. As Peterson et al. (2006, p. 1100) described, 'People who are high in hope have both the "will" and the "ways" to achieve their goals'. This suggests that hope can serve as a critical factor in driving individuals to overcome challenges and pursue their learning aspirations.

War-related motivations may seem unusual within the context of Western literature, particularly as many scholars emphasise the importance of providing comfortable and engaging environments for learning based on research highlighting their positive impact on learning outcomes (Baeten et al. 2013). However, our findings challenge this prevailing view, demonstrating that the hardships of war can, paradoxically, encourage learners to pursue education under identified regulation motivation. This result, while counterintuitive, should not be entirely surprising. Studies from Palestine have documented similar findings across multiple contexts (Korhonen et al. 2024; Pherali and Turner 2018; Smith and Scott 2023). For instance, education in Palestine is often perceived as a necessity for survival (Shakhshir 2011), a means to achieve empowerment and agency (Korhonen et al. 2024), and a tool intrinsically linked to the struggle for independence and the preservation of national identity. It serves as a source of strength, enabling Palestinians to maintain their cultural heritage and assert their sovereignty (Smith and Scott 2023).

However, war-related motivations should not be used to justify crimes against Palestinians or the failure of international institutions to protect the educational process in Gaza in any way. Although it was not reported in the results of this study, the participants' narratives were overwhelmed by the obstacles they faced in securing the basic requirements to engage in online learning (Abusamra 2024), such as electricity and internet access—needs often taken for granted in other contexts. According to Maslow's hierarchy of needs, for a learner to reach the top of the hierarchy (e.g. engaging in fruitful higher-order thinking and learning), the bottom levels of the hierarchy must be satisfied (e.g. food, safety, tools of online learning such as electricity, internet access and computers; Abed Alah 2024). Given that this study was conducted during the recent genocide in Gaza, significant questions remain: Will war-related motivations persist after the recent military assault ends? If so, for how long? Additionally, although we classified war-related motivations as identified regulation, which is typically associated with positive outcomes, further research is required to track the long-term learning outcomes of students driven by such motivations.

Belief-related motivation is seen as integrative motivation in which participants view online learning as integral to their identity and beliefs. This motivation can be divided into two subthemes: religious motivations and time perspective motivations. Previous studies have linked student motivations to religious beliefs, showing that these beliefs guide meaningful learning by providing a sense of purpose in life (Michaels et al. 2021) and fostering hope (Mvududu and Larocque 2008). In this study, set within an Islamic context, students reported that the ultimate reward for their hard work lay in the hereafter. They expressed a belief that they should strive with the trust of God because 'God will never let the rewards of those who do good go to waste' (Quran, Surah Hud, verse 115).

Although time perspective motivation has been given less attention in the literature, research suggests that perceptions of time—encompassing past, present and future experiences—significantly influence cognition, tendencies and behaviours (Baikeli et al. 2021). Time perspective theory, for instance, identifies five perspectives of time, two of which appear particularly relevant in the war context, as reflected in participants' narratives. The present fatalistic perspective reflects the belief that one's fate is determined by uncontrollable forces. In the context of war, this belief seems to push participants to seek solutions: 'I want to find solutions and make this time—over which we have no control or ability to change or escape—an investment in myself', Participant 09 clarified. The future time perspective is characterised by planning and working towards future goals, with a strong focus on achievement, self-discipline and resisting the temptation to waste time. As Participant 754 explained, 'Continuing to learn and making use of time are essential for creating better opportunities for a post-war future'. This result should be interpreted with caution, as it is unclear whether these beliefs are temporary or shaped by the extraordinary circumstances of living in a war zone. Further studies are needed to examine the persistence of these beliefs and their impact on learning outcomes. It is noteworthy that while these beliefs were prevalent among the participants in this study, they are rarely associated with learning motivations in the existing literature. More studies are needed to explore the underlying drivers of these beliefs in the Palestinian context.

From a holistic view of motivation, the findings revealed that participants' motivations were *hierarchical*. Rather than citing a single reason, the participants identified several motivations that built on each other to achieve an ultimate goal. As Participant 463 explained, 'It is an attempt to make use of the free time in my day to develop my skills, enabling me to qualify for scholarships'. Participant 528 shared that they were 'striving and taking the necessary steps to secure a job as soon as possible to support my family'. This result echoes the literature on goal systems, which argues that motivations are often structured hierarchically, and achieving one goal depends on the attainment of another, more fundamental goal (Shaddy and Fishbach 2018). Research has also shown that people are more successful in pursuing long-term goals when they focus on specific, challenging and concrete goals (Höchli et al. 2018).

5.6 Implications of the Study

The findings of this study have several implications for both theory and practice. Theoretically, the study provides validation for the Arabic version of the SIMS, developed based on self-determination theory, within the Palestinian context. This suggests that the scale is effective in explaining learning motivation in war contexts. From this study's perspective, participants who had lost almost everything found that online learning allowed them to regain a sense of control (autonomy) over their personal and professional development by choosing courses that aligned with their interests and aspirations. It also offered opportunities to demonstrate competence by acquiring and applying new skills, completing courses and earning certifications, despite the extreme challenges they faced (competence). Furthermore, connecting with a global academic community through these platforms provided participants with a sense of belonging and solidarity as they interacted with peers and instructors worldwide, breaking the isolation imposed by war (relatedness). While some motivations identified through qualitative data were not explicitly recognised by SDT, our attempt to explain these motivations under the framework of SDT may provide a reasonable mapping, although further research is needed for deeper exploration. Investigating how autonomy, competence and relatedness are satisfied in the unique contexts of wars and conflict zones remains a fertile area for research.

In practice, the *juxtaposition* between the ideal of universal educational equity and the reality of systemic deprivation compels us to interrogate how SDG 4 can be reimagined and operationalised in contexts of profound crisis. The findings highlight the need for specialised strategies from both local and international actors. Protecting education remains both a legal and moral responsibility, supported by international humanitarian laws that affirm the right to education in emergencies (Klees 2024; McKinney and Farrar 2023). One effective strategy could involve creating *crisis-responsive infrastructure*, such as mobile or temporary classrooms, to offer safe and consistent learning environments in war-affected areas. Implementing safety protocols specifically designed for conflict zones is also critical for protecting students and educators. Global partnerships for funding and resource mobilisation are especially vital (Benavot et al. 2024; Menashy and Zakharia 2023). Additionally, the use of digital tools and information and communication technology has demonstrated its value in maintaining educational continuity during emergencies. By enabling access to quality learning materials, fostering resilience and addressing the unique needs of learners in wartime, these tools offer a flexible and accessible solution to bridge educational gaps and support SDG 4 in conflict zones.

5.7 Conclusion

In this study, we highlighted the critical role of motivation in sustaining educational persistence in conflict-affected contexts, centring Gaza as a case study to illuminate the persistence of learners amid extraordinary circumstances. Grounded in SDT, the findings reveal that intrinsic and identified regulation motivations dominate, driven by personal, social and contextual factors unique to the recent genocide and infrastructural elimination.

The intertwinement of autonomy, competence and relatedness in fostering intrinsic motivation places strong emphasis on the transformative potential of education in rebuilding agency and expanding steadfastness (*sumud*) in such eliminatory environments (Hammad and Tribe 2021; Marie et al. 2018). With these insights, we call for a reimagining of global educational frameworks to properly account for and accommodate the psychological and contextual needs of learners in war and genocidal contexts, positioning motivation as a cornerstone of educational recovery and societal reconstruction.

This study is not without its limitations. First, expanding the scope of disciplines offered in similar initiatives, particularly to include underrepresented fields such as the humanities and social sciences, is fundamental in fostering inclusivity. Additionally, infrastructural barriers, including unreliable electricity, internet connectivity and limited access to digital devices, must be examined in greater scope to advance strategies that ensure equitable access, and this would not be possible without the support of international and state-level initiatives. More importantly, such nuances should be incorporated into SDG 4, ensuring international support and frameworks. Longitudinal studies are needed to track the evolution of learners' motivations and outcomes over a period of time, providing a more comprehensive understanding of long-term sustainability and the impact of educational engagement. Research should also explore the barriers faced by non-participants, who remain underrepresented in the current data. Distinguishing geographic contexts between learners in Gaza and those displaced elsewhere is essential to designing and devising more inclusive and context-sensitive educational frameworks and interventions.

Appendix 1: The Situational Motivation Scale (SIMS)—Arabic Version

النسخة العربية—(SIMS) مقياس الدافعية الموقفية

نرجو أن تشير لدرجة موافقتك ومدى انطباق كل عبارة عليك شخصياً وذلك على المدرج السباعي من 1 (لا ينطبق إطلاقاً) إلى فيما يلي مجموعة من الأسباب والدوافع التي قد تكون وراء مشاركة وتعلم الطلبة بشكل عام عبر منصات التعلم (المفتوح. **7 (ينطبق تماماً).**

Below is a list of reasons and motivations that may be behind students' participation and learning through open learning platforms in general. Please indicate your level of agreement and how each statement applies to you personally, using the seven-point scale from 1 (corresponds not all) to 7 (corresponds exactly).

لقد قمتُ بالتسجيل في المبادرة والمشاركة في المساقات الإلكترونية المطروحة وذلك...

I have enrolled in the initiative and participated in the offered online courses...

الدافع الداخلي (Intrinsic Motivation)

1. لأنني أجد هذه المساقات مثيرة للاهتمام
   .Because I find these courses interesting.
2. لأنني أستمتع باكتشاف مواضيع جديدة من خلال هذه المساقات.
   Because I enjoy exploring new topics through these courses.
3. لأنني أعتقد أن التعلم من خلال هذه المساقات ممتع.
   Because I think learning through these courses is fun.
4. لأنني أشعر بالرضا عند مشاركتي في هذه المساقات.
   Because I feel good when I engage with these courses.

التنظيم المُحدّد (Identified Regulation)

5. لأنني أعتقد أن هذه المساقات مفيدة لتطوري الشخصي.
   Because I believe these courses are beneficial for my personal development.
6. لأنني أعتقد أن هذه المساقات ستساعدني في تحقيق أهدافي.
   Because I think these courses will help me achieve my goals.
7. لأنني اتخذتُ قرارًا شخصيًا بالتسجيل في هذه المساقات.
   Because I made the personal decision to enroll in these courses.
8. لأنني أرى بأن المعرفة والمهارات التي سأكتسبها من هذه المساقات مهمة بالنسبة لي.
   Because I believe the knowledge and skills I gain from these courses are important for me.

التنظيم الخارجي (External Regulation)

9. لأنني مُجبَر على المشاركة في هذه المساقات.
   Because I am compelled to take these courses.
10. لأن المشاركة في هذه المساقات شيء يجب أن أقوم به.
    Because taking these courses is something I have to do.
11. لأنني لا أملك حقًا خيارًا آخر سوى التسجيل في هذه المساقات.
    Because I don't have any choice but to take these courses.
12. لأنني ملزم بالمشاركة في هذه المساقات.
    Because I am required to take these courses.

انعدام الدافع (Amotivation)

13. قد تكون هناك أسباب جيدة للمشاركة في هذه المساقات، لكن شخصيًا لا أرى ذلك.
    There may be good reasons to take these courses, but personally, I don't see them.
14. لست متأكدًا مما إذا كانت هذه المساقات تستحق وقتي.
    I'm not sure if these courses are worth my time.
15. لا أعرف؛ لا أرى الفائدة التي تقدمها لي هذه المساقات.
    I don't know; I don't see what benefits these courses bring me.
16. لقد سجلت في هذه المساقات، لكنني لست متأكدًا مما إذا كان من الجيد الاستمرار.
    I enrolled in these courses, but I'm not sure if it's the right thing to continue.

References

Abed Alah M (2024) Shattered hierarchy: how the Gaza conflict demolished Maslow's pyramid of needs. Curr Psychol 43(21):19562–19568. https://doi.org/10.1007/s12144-024-05784-1

Abusamra A (2024) Brain drain or brain circulation? The impact of Gaza's 7th October war on the displaced academicians and personnel of higher education institutions. J High Educ Theory Pract 24(10):10.33423/jhetp.v24i10.7309

Academic Solidarity with Palestine (2024) Academic Solidarity with Palestine. Retrieved from https://academicsolidaritywithpalestine.org/

Academics for Palestine (2024) Academics for Palestine. Retrieved from https://academicsforpalestine.org/

Academics4Gaza (2024) Introducing Academics4Gaza: empowering academics and students through solidarity initiatives [LinkedIn post]. LinkedIn. Retrieved from https://www.linkedin.com/posts/academics-4gaza_introducing-academics4gaza-empowering-activity-7201

Access Now (2023). Palestine unplugged: how Israel disrupts Gaza's internet. Access Now Retrieved December 25, 2024, from https://www.accessnow.org/publication/palestine-unplugged/

Ahmad A, Vulliamy E (2009) In Gaza, the schools are dying too. The Guardian, 10.01.2009. https://www.theguardian.com/world/2009/jan/10/gaza-schools

Al Arabiya (2024) Internet services gradually returning to Gaza: telecom firm. Retrieved from https://english.alarabiya.net/News/middle-east/2024/01/19/Internet-services-gradually-returning-to-Gaza-Telecom-firm

Aldahdouh TZ, Al-Masri N, Abou-Dagga S, AlDahdouh A (2024) Development of online teaching expertise in fragile and conflict-affected contexts. Front Educ 8:1242285. https://doi.org/10.3389/feduc.2023.1242285

Alenius P, Aldahdouh T, Holubek V, Al-Masri N, El-Holy A, Lindén J, Korhonen V (2019) Examining teaching approaches, academic culture, and self-efficacy beliefs of instructors at a Palestinian university. Int J Teach Learn High Educ 31(3)

Al-Mqadma A, Dittli R, Belotti M (2024) Resilience in the rubble: a needs assessment of higher education in the Gaza strip *report). Swiss Peace Foundation – swisspeace. https://www.swisspeace.ch/assets/publications/Reports/241030_Report_Gaza-Higher-Education-Needs-Assessment_final.pdf

Assefa Y, Tilwani SA, Moges BT (2023) The impact of armed violence on educational institutions, students' educational attainment and the role of actors in governance of the education process. Cogent Educ 10(1):2189503. https://doi.org/10.1080/2331186X.2023.2189503

Baeten M, Dochy F, Struyven K (2013) The effects of different learning environments on students' motivation for learning and their achievement. Br J Educ Psychol 83(3):484–501. https://doi.org/10.1111/j.2044-8279.2012.02076.x

Baikeli R, Li D, Zhu L, Wang Z (2021) The relationship between time perspective and meaning in life across different age stages in adulthood. Personal Individ Differ 174:110668. https://doi.org/10.1016/j.paid.2021.110668

Benavot A, Williams JH, Naidoo J (2024) Should we not be educating for resilience? Leveraging SDG 4 in times of crisis. Int J Educ Dev 104:102929. https://doi.org/10.1016/j.ijedudev.2023.102929

Biglan A (1973) Relationships between subject matter characteristics and the structure and output of university departments. J Appl Psychol 57(3):204–213. https://doi.org/10.1037/h0034699

Birzeit University (2024) Campaign for Gaza 2024. Retrieved from https://www.birzeit.edu/en/campaign-gaza-2024

Braun V, Clarke V (2006) Using thematic analysis in psychology. Qual Res Psychol 3(2):77–101. https://doi.org/10.1191/1478088706qp063oa

Brück T, Di Maio M, Miaari SH (2019) Learning the hard way: the effect of violent conflict on student academic achievement. J Eur Econ Assoc 17(5):1502–1537. https://doi.org/10.1093/jeea/jvy051

Cachia M, Lynam S, Stock R (2018) Academic success: is it just about the grades? High Educ Pedag 3(1):434–439. https://doi.org/10.1080/23752696.2018.1462096

Creswell JW, Plano Clark VL (2018) Designing and conducting mixed methods research (Issue 1). SAGE Publications, Inc

Dader K et al (2024) Topologies of scholasticide in Gaza: education in spaces of elimination. Fennia Int J Geogr

Fassetta G, Imperiale MG, Frimberger K, Attia M, Al-Masri N (2017) Online teacher training in a context of forced immobility: the case of Gaza. Palestine Eur Educ 49(2–3):133–150. https://doi.org/10.1080/10564934.2017.1315538

GCPEA (2022) Measuring the impact of attacks on education in Palestine. Global Coalition to Protect Education from Attack. https://protectingeducation.org/wp-content/uploads/impact_attackeducation_palestine_2022_en.pdf

Glück Z (2024) Introduction—anthropology and the security encounter: toward an abolitionist anthropology in the age of permanent war. Am Anthropol. https://doi.org/10.1111/aman.13977

Gorges J (2015) Warum (nicht) an Weiterbildung teilnehmen? In: Gorges J, Gegenfurtner A, Kuper H (eds) Motivationsforschung im Weiterbildungskontext. Springer Fachmedien, Wiesbaden, pp 9–28. https://doi.org/10.1007/978-3-658-06616-1_2

Guay F, Vallerand RJ, Blanchard C (2000) On the assessment of situational intrinsic and extrinsic motivation: the situational motivation scale (SIMS)

Hammad J, Tribe R (2021) Culturally informed resilience in conflict settings: a literature review of Sumud in the occupied Palestinian territories. Int Rev Psychiatry 33(1–2, 132):–139. https://doi.org/10.1080/09540261.2020.1741259

Höchli B, Brügger A, Messner C (2018) How focusing on superordinate goals motivates broad, long-term goal pursuit: a theoretical perspective. Front Psychol 9:1879. https://doi.org/10.3389/fpsyg.2018.01879

HuffPost (2017) A burning motivation to learn helps students from war zones beat the odds. Retrieved from https://www.huffpost.com/archive/ca/entry/a--burning-motivation-to-learn-helps-students-from-war-zones-beat-the-odds_ca_5cd51372e4b07bc729748b36

Iriqat D, Alousi R, Aldahdouh TZ, AlDahdouh A, Dankar I, Alburai D, Buheji M, Hassoun A (2025) Educide amid conflict: the struggle of the Palestinian education system. Qual Educ All 2(1):81–99. https://doi.org/10.1108/QEA-10-2024-0120

Jebril M (2023) Between construction and destruction: the experience of educationalists at Gaza's universities. Compar J Comp Int Educ 53(6):986–1004

Kim H-Y (2013) Statistical notes for clinical researchers: assessing normal distribution (2) using skewness and kurtosis. Restor Dentist Endodon 38(1):52. https://doi.org/10.5395/rde.2013.38.1.52

Klees SJ (2024) Why SDG 4 and the other SDGs are failing and what needs to be done. Int J Educ Dev 104:102946. https://doi.org/10.1016/j.ijedudev.2023.102946

Korhonen V, Aldahdouh T, Holubek V, Abou-dagga S, Al-Masri N (2024) Student engagement and concerns on studies and future professions: exploratory research in a Palestinian higher education context. Int J Educ Manag. https://doi.org/10.1108/IJEM-03-2023-0132

Li X, Pu R, Phakdeephirot N (2022) The influence of achievement motivation on college students' employability: a chain mediation analysis of self-efficacy and academic performance. Front Psychol 13:972910. https://doi.org/10.3389/fpsyg.2022.972910

López-Angulo Y, Cobo-Rendón R, Sáez-Delgado F, Mella-Norambuena J, Pérez-Villalobos MV, Díaz-Mujica A (2024) Cognitive motivational variables and dropout intention as precursors of university dropout. Front Educ 9:1416183. https://doi.org/10.3389/feduc.2024.1416183

Marie M, Hannigan B, Jones A (2018) Social ecology of resilience and Sumud of Palestinians. Health Interdiscip J Soc Study Heal Ill Med 22(1):20–35. https://doi.org/10.1177/1363459316677624

McKinney SJ, Farrar J (2023) The impact of armed conflict on school education. Educazione Interculturale 21(1). https://doi.org/10.6092/issn.2420-8175/16970

Menashy F, Zakharia Z (2023) Partnerships for education in emergencies: the intersecting promises and challenges of SDG 4 and SDG 17. Int J Educ Dev 103:102934. https://doi.org/10.1016/j.ijedudev.2023.102934

Michaels JL, Petrino J, Pitre-Zampol T (2021) Individual differences in religious motivation influence how people think. J Sci Study Relig 60(1):64–82. https://doi.org/10.1111/jssr.12696

Ministry of Higher Education & Scientific Research (2023) Higher education statistical yearbook for scholastic year 2021/2022. Ramallah, Palestine. https://www.mohe.pna.ps/Portals/0/MOHEResources/Docs/Statistics/StatisticalYearbook2021_2022.pdf?ver=ngGjAGthsul35nVQXBB85g%3d%3d

Mvududu N, Larocque M (2008) Hope, faith, and statistics: an examination of the relationship. Christ High Educ 7(3):171–184. https://doi.org/10.1080/15363750801891069

Nijim M (2022) Genocide in Palestine: Gaza as a case study. Int J Human Right 27(1):165–200. https://doi.org/10.1080/13642987.2022.2065261

Norwegian Refugee Council (2021) Learning on the margins: the evolving nature of educational vulnerability in the occupied Palestinian territory in the time of Covid-19

OCHA (2018) Education undermined by deteriorating humanitarian situation in Gaza. Retrieved from https://www.ochaopt.org/content/education-undermined-deteriorating-humanitarian-situation-gaza

OHCHR (2024) UN experts deeply concerned over 'scholasticide' in Gaza. [Press Release] United Nations Human Rights Office of the High Commissioner 18.04.2024. https://www.ohchr.org/en/press-releases/2024/04/un-experts-deeply-concerned-over-scholasticide-gaza

Peterson SJ, Gerhardt MW, Rode JC (2006) Hope, learning goals, and task performance. Personal Individ Differ 40(6):1099–1109. https://doi.org/10.1016/j.paid.2005.11.005

Pherali T, Turner E (2018) Meanings of education under occupation: the shifting motivations for education in Palestinian refugee camps in the West Bank. Br J Sociol Educ 39(4):567–589. https://doi.org/10.1080/01425692.2017.1375400

Raccanello D et al (2024) Coping and emotions of global higher education students to the Ukraine war worldwide. Sci Rep 14:Article 8561. https://doi.org/10.1038/s41598-024-59009-3

Ramahi H (2015) Education in Palestine: current challenges and emancipatory alternatives. Rosa Luxemburg Stiftung Regional Office Palestine:1–51

ReliefWeb (2024, September 2) Scholasticide: Israel's deliberate and systematic destruction of the Palestinian education system in Gaza - occupied Palestinian territory [Press release]. https://reliefweb.int/report/occupied-palestinian-territory/scholasticide-israels-deliberate-and-systematic-destruction-palestinian-education-system-gaza

Remedios R, Sewell P (2024) Employability and motivation: which motivational theories are most appropriate? High Educ Skill Work Based Learn 14(4):908–919. https://doi.org/10.1108/HESWBL-03-2024-0067

Ryan RM, Deci EL (2000) Intrinsic and extrinsic motivations: classic definitions and new directions. Contemp Educ Psychol 25(1):54–67. https://doi.org/10.1006/ceps.1999.1020

Ryan RM, Deci EL (eds) (2017) Self-determination theory: basic psychological needs in motivation, development, and wellness, Guilford Press. https://doi.org/10.1521/978.14625/28806

Senko C, Freund AM (2015) Are mastery-avoidance achievement goals always detrimental? An adult development perspective. Motiv Emot 39(4):477–488. https://doi.org/10.1007/s11031-015-9474-1

Senko C, Liem GAD, Lerdpornkulrat T, Poondej C (2023) Why do students strive to outperform classmates? Unpacking their reasons for pursuing performance goals. Contemp Educ Psychol 73:102178. https://doi.org/10.1016/j.cedpsych.2023.102178

Shaddy F, Fishbach A (2018) Eyes on the prize: the preference to invest resources in goals over means. J Pers Soc Psychol 115(4):624–637. https://doi.org/10.1037/pspa0000127

Shakhshir, K. (2011). Palestinian education under occupation: successes and challenges. In I. Nasser, L. N. Berlin, & S. Wong (Eds.), Examining education, media, and dialogue under occupation (pp. 3–18). Multilingual Matters. 10.21832/9781847694287-006

Sheykina N, Barannyk M, Nessonova M (2024) Factors influencing students' motivation to learning during martial law in Ukraine. New Educ Rev 76(2):10–25. https://doi.org/10.15804/tner.2024.76.2.02

Smith M, Scott H (2023) Distance education under oppression: the case of Palestinian higher education. Education Sciences 13(7):729. https://doi.org/10.3390/educsci13070729

Spectre (2024) Scholars against the war on Palestine. Spectre J, 23.02.2024. https://spectrejournal.com/scholars-against-the-war-on-palestine-2/ 02.07.2024

Tabachnick BG, Fidell LS (2007) Using multivariate statistics. Pearson Education Inc, Boston

Taka M (2023) When education in emergencies fails: learners' motivations for a second chance education in post-conflict Rwanda. Comp J Comp Int Educ 53(2):217–234. https://doi.org/10.1080/03057925.2021.1886908

The Emergency Committee of Universities in Gaza (2024) Gaza universities emergency committee website. Retrieved December 25, 2024, from https://www.gazauniversities.org/

Tozan O (2023) The impact of the Syrian conflict on the higher education sector in Syria: a systematic review of literature. Int J Educ Res Open 4:100221. https://doi.org/10.1016/j.ijedro.2022.100221

UNICEF (2022) Consolidated emergency report 2022. Retrieved from https://open.unicef.org/sites/transparency/files/2023-05/State%20of%20Palestine%20CER%202022.pdf

UNIMED (2024) Enabling Gaza students to continue university education: UNIMED participates in the initiative launched by the An-Najah National University. Retrieved December 25, 2024, from https://www.uni-med.net/news/enabling-gaza-students-to-continue-university-education-unimed-participates-in-the-initiative-launched-by-the-an-najah-national-university

United Nations (n.d.) Goal 4: quality education. United Nations sustainable development goals. Retrieved December 25, 2024, from https://sdgs.un.org/goals/goal4

UNOSAT (2024) Gaza education system devastated by recent conflict, satellite assessment reveals. UNITAR. https://unitar.org/about/news-stories/press/gaza-education-system-devastated-recent-conflict-satellite-assessment-reveals

Urdan TC, Maehr ML (1995) Beyond a two-goal theory of motivation and achievement: a case for social goals. Rev Educ Res 65(3):213–243. https://doi.org/10.3102/00346543065003213

Van Yperen NW, Elliot AJ, Anseel F (2009) The influence of mastery-avoidance goals on performance improvement. Eur J Soc Psychol 39(6):932–943. https://doi.org/10.1002/ejsp.590

Vansteenkiste M, Simons J, Lens W, Sheldon KM, Deci EL (2004) Motivating learning, performance, and persistence: the synergistic effects of intrinsic goal contents and autonomy-supportive contexts. J Pers Soc Psychol 87(2):246–260. https://doi.org/10.1037/0022-3514.87.2.246

Wei X, Saab N, Admiraal W (2024) What rationale would work? Unfolding the role of learners' attitudes and motivation in predicting learning engagement and perceived learning outcomes in MOOCs. Int J Educ Technol High Educ 21(1):5. https://doi.org/10.1186/s41239-023-00433-2

Yong AG, Pearce S (2013) A Beginner's guide to factor analysis: focusing on exploratory factor analysis. Tutor Quant Method Psychol 9(2):79–94. https://doi.org/10.20982/tqmp.09.2.p079

Extraordinary Women of Gaza: Battling Inequality and the Consequences of War

6

Dunya Ahmed, Mohamed Buheji, and Abdo Hassoun

Abstract

This chapter examines the gendered impact of the recent genocidal war on Gaza, highlighting how these hostilities have severely hindered progress toward achieving Sustainable Development Goal 5 (SDG 5), focused on gender equality. The relentless violence, mass destruction, and killing have exacerbated existing gender inequalities and created new challenges that disproportionately affect women and girls in Gaza. By employing a qualitative research approach, this chapter draws on data from official reports, humanitarian organizations, and peer-reviewed publications to highlight the multifaceted struggles faced by women in Gaza. The war has led to a significant increase in gender-based violence, including sexual assault and discrimination, and has severely restricted access for women to essential services such as healthcare, education, and employment. The findings reveal how this unprecedented war has hindered empowerment and participation for women in society, further entrenching vulnerability and marginalization. The work underscores the critical importance of integrating gender-sensitive strategies in humanitarian interventions and post-conflict reconstruction efforts. Moreover, it advocates for international action to address these gender-specific impacts, emphasizing the need for policies and programs that support women's rights and promote gender equality. By highlighting the severe impact of the war on the lives and livelihoods of Palestinian women, this research aims to contribute to the global discourse on gender equality in conflict zones and propose sustainable solutions to advance SDG 5 in Gaza. The call to action stresses the importance of protecting and empowering women in Gaza, recognizing their leading role in building a more equitable and resilient society.

Keywords

Conflict zones · Gazan · Gender-based violence · Gender equality · Humanitarian impact · Palestinian women · Resilience · SDG 5

D. Ahmed (✉)
Department of Social Science, University of Bahrain & International Institute of Inspiration Economy, Zallaq, Bahrain

M. Buheji
Socioeconomic Institute for Advanced Studies (SIAS), Kigali, Rwanda

Inspiration Economy Institute, Manama, Bahrain

A. Hassoun
Sustainable AgriFoodtech Innovation & Research (SAFIR), Arras, France

Faculty of Agricultural Engineering, University of Aleppo, Aleppo, Syria
e-mail: a.hassoun@saf-ir.com

A. Hassoun (ed.), *War on Gaza*, Sustainable Development Goals Series,
https://doi.org/10.1007/978-3-031-88500-6_6

6.1 Introduction

Throughout history, armed conflicts have disproportionately impacted women and girls, exacerbating pre-existing inequalities and creating new vulnerabilities. In war-torn regions, women often bear the brunt of displacement, food insecurity, and restricted access to essential services such as healthcare and education. Gender-based violence, including sexual assault and exploitation, escalates significantly during conflicts, leaving lasting physical and psychological scars. Addressing these gendered impacts requires a commitment to integrating gender-sensitive approaches into conflict resolution, recovery efforts, and sustainable development initiatives (Meagher et al. 2021; Reynolds 2021; Rubini et al. 2023).

In response to these persistent challenges, Sustainable Development Goal 5 (SDG 5) was established to achieve gender equality and empower all women and girls. SDG 5 emphasizes eliminating discrimination, violence, and harmful practices while ensuring equal opportunities in leadership, education, and access to resources. Achieving this goal is essential for addressing the gendered impacts of conflict and fostering sustainable development in post-conflict societies (Beloskar et al. 2024; Rosamund 2024).

The recent war on Gaza has significantly undermined progress toward all Sustainable Development Goals (SDGs), particularly SDG 5 (Gender Equality). The continuous strain on healthcare, education, and economic systems has severely restricted access to essential services for Gazan women and girls, particularly in areas such as sexual and reproductive health, maternal care, and education. These logistical and material hardships are compounded by psychological burdens, as women grapple with the grief of loss, trauma, and a yearning for stability in an environment marked by violence and deprivation (Bdier and Mahamid 2024; UN Women 2024a, 2024b; UNRWA 2023). In addition to these challenges, the war has escalated gender-based violence, with women in Gaza facing increased risks of sexual assault, discrimination, and exploitation. This violence not only deepens their vulnerability but also erodes their ability to participate in societal and recovery processes. Systemic barriers further marginalize women, excluding them from decision-making spaces critical to rebuilding efforts. Together, these intersecting challenges hinder progress toward achieving gender equality and highlight the urgent need for gender-sensitive approaches in humanitarian aid and recovery planning (Buheji et al. 2024; Muhammad and Rizek 2024; Ramadan and Abouzid 2024; UNFPA 2024).

Despite the profound challenges they face, Gazan women exhibit remarkable resilience and strength, taking on pivotal roles in sustaining their families and communities amid devastation. They adapt to the immense pressures of war by managing scarce resources, supporting displaced families, and assuming responsibilities often overlooked in humanitarian narratives. Amid displacement, violence, and systemic neglect, these women cultivate solidarity, preserve cultural identity, and establish informal networks of care, demonstrating the resilience and leadership crucial for community survival and recovery (Abu Zaiter and Al Khawaja 2024; Buheji et al. 2024; Jalambo et al. 2025).

This chapter explores the complex and multifaceted impact of the recent war (October 2023—January 2025) on Gaza on women, shedding light on how it has exacerbated gender-based inequalities and severely disrupted progress toward SDG 5. This work examines how the war impacts women's livelihoods, societal roles, and community participation, highlighting the urgent need for gender-sensitive strategies. Empowering women is not only a matter of justice but also vital for sustainable development and resilience in Gaza and similar conflict zones. Their leadership and active involvement are crucial to rebuilding equitable and resilient societies amid ongoing challenges.

6.2 Overview of Sustainable Development Goal 5 (SDG 5)

SDG 5—Gender Equality aims to achieve gender equality and empower all women and girls as part of the broader 2030 Agenda for sustainable development. This goal underscores the importance of eliminating all forms of discrimination and violence against women and girls, recognizing their equal rights and opportunities in all aspects of life. SDG 5 sets ambitious targets that include ending harmful practices such as child marriage and female genital mutilation, ensuring universal access to sexual and reproductive health and rights, and promoting full participation of women in decision-making processes across political, economic, and social spheres. Central to SDG 5 is the acknowledgment of unpaid care and domestic work, which disproportionately burdens women and limits their ability to engage in education, employment, and leadership opportunities. The goal also emphasizes the importance of equal access to resources, inheritance rights, and the use of enabling technologies to empower women economically and socially. These targets reflect a holistic approach to dismantling systemic gender inequalities and fostering environments where women and girls can thrive (Beloskar et al. 2024; Eden and Wagstaff 2021; Guaita-Fernández et al. 2024; Khurana and Verma 2023).

However, the pursuit of gender equality is significantly hindered in contexts of war and conflict, where the structures needed to support progress toward SDG 5 are disrupted. Armed conflicts exacerbate pre-existing inequalities, placing women and girls at heightened risk of violence, exploitation, and marginalization. Gender-based violence, including sexual assault and trafficking, often escalates in conflict zones, leaving profound physical and psychological scars on survivors. The breakdown of healthcare systems limits access to essential maternal and reproductive health services, while the destruction of schools interrupts the education of girls, further entrenching cycles of poverty and dependence. Conflicts also amplify systemic barriers that prevent women from participating in decision-making processes, both during and after crises. Traditional gender roles are often reinforced in these contexts, sidelining women from leadership roles in recovery and rebuilding efforts. The lack of inclusion in these processes not only undermines gender equality but also weakens the overall effectiveness of post-conflict reconstruction and governance (Cohen and Karim 2022; Lwamba et al. 2022; Sweileh 2024).

In the context of Gaza, the recent war has illuminated the catastrophic impact of conflict on gender equality. Women and girls are disproportionately affected by displacement, food insecurity, and restricted access to essential services. At the same time, they navigate systemic barriers that prevent them from contributing meaningfully to recovery efforts. Achieving the targets of SDG 5 in such settings requires not only the cessation of violence but also the implementation of gender-sensitive approaches that prioritize rights and empowerment of women in all aspects of recovery and sustainable development. These measures are essential for breaking cycles of inequality and fostering resilience in conflict-affected communities.

6.3 Catastrophic Disruption and Worsening Inequalities

The recent war on Gaza has profoundly disrupted the lives of its inhabitants, inflicting indiscriminate violence on all demographics—men, women, children, the elderly, and the disabled alike. However, for women and girls, the war has compounded pre-existing inequalities, exposing them to unique challenges that undermine progress toward achieving SDG 5.

6.3.1 Escalation of Gender-Based Violence (SDG 5.2: Eliminate All Forms of Violence Against Women and Girls)

Gender-based violence (GBV) against Palestinian women is widespread and exacerbated by the ongoing political conflict, military occupation, and socioeconomic hardships. Due to occupation and systemic Israeli oppression, women in Palestine face various forms of violence, including psychological, economic, physical, and sexual abuse. The situation is particularly acute in the Gaza Strip, where political instability and Israeli military occupation have intensified violence against women and further restricted their access to resources, support, and opportunities, reinforcing systemic gender inequalities (Mahamid et al. 2022; Müller and Tranchant 2017; Veronese et al. 2023). Since the beginning of the recent war on Gaza, a UN expert warned that women are disproportionately bearing the brunt of the conflict, as they constitute a significant proportion of the casualties among vulnerable groups (OHCHR 2023). As of November 2024, killing remains the most direct and horrific manifestation of violence, with the war in Gaza claiming the lives of approximately 43,000 people, the majority of whom are women and children (Samuel 2024). The gendered nature of the genocide is evident (Repo 2024), as women face not only the physical toll of violence but also the psychological and societal consequences of losing their families, homes, and livelihoods. These targeted attacks further entrench systemic gender inequalities, highlighting how women and children are not only collateral victims but also deliberate targets of a broader strategy aimed at dismantling the social fabric of Palestinian communities.

The war has led to a marked increase in GBV, with women and girls in Gaza facing heightened risks of sexual assault, abuse, and exploitation (Marteu 2023; UN Women 2024b). The breakdown of law enforcement and judicial systems has left survivors without avenues for justice, perpetuating cycles of violence and trauma. Overcrowded displacement shelters have exacerbated these risks, as lack of privacy monitoring, counseling, or first aid make women particularly vulnerable to abuse (Wilson Center 2024). These escalations directly undermine SDG 5.2, which aims to eliminate all forms of violence against women and girls in both public and private spheres, including trafficking and sexual exploitation. UN experts have emphasized the availability of "credible allegations" that Palestinian women and girls have been subjected to sexual assaults, including rape, while in Israeli detention (The Guardian 2024). Such conflict-related sexual violence has negative consequences on victims' physical, psychological, and social dimensions of health (Rubini et al. 2023).

Even before the recent war, sexual violence has been systematically used as a weapon against Palestinian women visiting loved ones detained in Israeli prisons (Al Issa and Beck 2021). This violence, encompassing verbal harassment, forced nudity, and physical abuse, aims to intimidate and humiliate women, leveraging cultural notions of honor and modesty to exacerbate psychological trauma and undermine resilience. As detailed in interviews with Palestinian women, experiences of forced nudity, invasive searches, and verbal harassment are pervasive, aiming to exploit cultural associations of honor and shame to inflict psychological trauma (Madar 2023).

In many cultures, stigma and shame associated with sexual violence can follow a woman throughout her life. Worse still, if she becomes pregnant as a result of rape, the child may serve as a constant reminder of the trauma, deeply affecting both the mother and the child over their lifetimes. Such violence often goes underreported due to societal stigmas, fear of retaliation, and a lack of accessible support systems, perpetuating cycles of trauma and marginalization within affected communities. Additionally, the normalization of such violence in conflict settings further undermines women's status, often confining them to roles defined by victimhood rather than agency. This dynamic not only perpetuates enduring gender inequalities but also obstructs paths of women to recovery, empowerment, and meaningful participation in rebuilding their lives and communities (Aroussi 2017; Gaggioli 2014; Sahin and Kula 2018).

6.3.2 Impact on Education and Economic Empowerment (SDG 5.1: End All Forms of Discrimination Against Women and Girls Everywhere)

The recent war has severely disrupted education for thousands of girls in Gaza, with widespread school closures, destruction of educational infrastructure, and pervasive safety concerns making it nearly impossible for many to continue their studies. According to a recent report by Education Cluster, Save the Children, and UNICEF, over 95% of schools sustained some level of damage to their buildings due to the war (Education Cluster/Save the Children/UNICEF 2024; Faculty of Education University of Cambridge/Centre for Lebanese Studies & UNRWA 2024; Iriqat et al. 2025), directly impeding access to education, especially for girls. This disruption not only limits future opportunities for these girls but also perpetuates cycles of poverty, dependency, and gender inequality, as education is a critical pathway to empowerment and societal participation (Boyadjieva and Ilieva-Trichkova 2023). The destruction of educational systems, especially universities, has had a particularly profound impact on women, as the ratio of female university graduates in Palestine significantly exceeds that of male graduates (Wadi 2022), highlighting the critical role education plays in the empowerment of women and social mobility.

In parallel, the economic empowerment of Gazan women has faced devastating setbacks, with their livelihoods particularly vulnerable to the compounded effects of the blockade and repeated cycles of violence. Many women in Gaza rely on home-based businesses or nearby facilities to sustain their income, making the destruction of their homes doubly devastating as it eliminates both shelter and workplace. A UN Women study on the 2021 escalation revealed that 18% of home-based businesses owned by women were destroyed during the hostilities (UN Women 2022). Even before the recent war, Palestinian women had some of the lowest labor force participation rates globally, averaging just 15.5% over the past two decades, with unemployment among women in Gaza reaching a staggering 43% by 2021 (International Monetary Fund 2022). The COVID-19 pandemic further exacerbated this situation, as lockdowns and economic restrictions disproportionately affected women's already limited opportunities for employment and financial independence. While many women globally turned to digital platforms for work and participation during the pandemic, Palestinian women faced the additional challenge of cyberviolence, which deterred their digital engagement and compounded their socioeconomic vulnerabilities (ESCWA 2023).

The war has further eroded women's economic independence, with many losing their livelihoods due to the destruction of agricultural lands, marketplaces, and businesses, as well as the overall economic paralysis. Palestinian women, particularly in Gaza, also experience significant levels of economic violence, including restricted access to financial resources and employment opportunities (Morzk et al. 2025). Conflict and war exacerbate these conditions by intensifying poverty, destroying livelihoods, and reinforcing systemic inequalities, creating a cycle of dependency that further deepens discrimination and economic violence.

The lack of economic stability not only limits the ability of women to support their families or contribute to their communities but also underscores the disproportionate impact of conflict on women and girls. These challenges reflect systemic discrimination, directly violating SDG 5, Target 1, which seeks to eliminate all forms of discrimination against women and girls. The intersection of disrupted education and economic disempowerment highlights broader gender inequalities exacerbated by war, necessitating urgent and targeted interventions to ensure that women and girls in Gaza can reclaim their rights and opportunities.

6.3.3 Heightened Burden of Unpaid Care as Well as Additional Workload and Responsibilities (SDG 5.4: Recognize and Value Unpaid Care and Domestic Work)

Women in Gaza bear the primary responsibility for the physically demanding and time-intensive work required to sustain households in war-torn conditions. Their workload has significantly intensified as they carry heavy water containers, cook over open fires, and care for injured relatives, all while facing the challenges of water shortages, poor water quality, electricity outages, and overcrowded landfills. These issues further increase the burden of unpaid domestic work, which falls disproportionately on women and girls (Ibnouf 2020). The war has compounded these challenges, forcing women to take on additional roles as primary caregivers due to the destruction of infrastructure and the loss of male family members through death, injury, or imprisonment (Buheji et al. 2024; World Bank Group 2023).

In the absence of reliable resources, women must manage scarce supplies of food and water, care for children and the elderly, and provide emotional and physical support to their families. These overwhelming responsibilities further entrench traditional gender roles, leaving women with limited opportunities for education, employment, or leadership (UN Women 2024b). This dynamic severely hinders progress toward achieving SDG 5.4, which emphasizes the recognition and equitable distribution of unpaid care work. Such unpaid care burdens often increase disproportionately for women, further exacerbating gender inequalities and impeding economic and social empowerment.

6.3.4 Severe Disruption to Sexual and Reproductive Health Services (SDG 5.6: Ensure Universal Access to Sexual and Reproductive Health and Rights)

The war and prolonged blockade in Gaza have severely undermined the health and well-being of women and girls, creating a multifaceted crisis with far-reaching consequences. Gaza's healthcare system, already fragile, has been devastated, leaving many facilities destroyed or inoperable. This has drastically reduced access to essential sexual and reproductive health services, including prenatal care, skilled birth attendants, and post-natal support. Pregnant women and new mothers face significant risks, including elevated rates of neonatal mortality, exacerbated by the lack of maternal health services, malnutrition, and the unavailability of vital medical supplies (Elnakib et al. 2024; Ferguson and Desai 2024; Hussein-Sabbah 2023). These challenges echo the experiences of women during the 2008 Israeli attack on Gaza, where the absence of safe spaces and destroyed medical facilities forced many women to give birth in overcrowded shelters or unsafe conditions without skilled medical assistance (Wick and Hassan 2012). Midwives worked under extreme duress with limited resources, often resorting to improvised tools and inadequate sanitation. These dire circumstances led to increases in complications such as premature births and neonatal deaths, highlighting the acute vulnerabilities faced by women during conflict.

Beyond the challenges of pregnancy and childbirth, women in Gaza face additional health risks due to a lack of safe feminine hygiene products such as sanitary pads or tampons, as well as a lack of privacy and sanitation, driven by the scarcity and unaffordability of proper hygiene and sanitation products. These practices expose women to infections and long-term health complications, further burdening the already overwhelmed healthcare system (Ahmed et al. 2024; Irfan et al. 2024; Muhammad and Rizek 2024). Malnutrition and hunger, exacerbated by the blockade and destruction of agricultural resources, have had a severe impact on both women and children. Nutritional deficiencies among pregnant and breastfeeding women contribute to higher risks of complications during pregnancy and childbirth, while children suffer from stunted growth, weakened immunity, and developmental delays due to inadequate diets (Assaf et al. 2023; Global Nutrition Cluster 2024; Gordon and Haddad 2024). The intersection of these issues underscores the urgent need for comprehensive interventions to address the health and nutritional needs of women and children in Gaza.

The lack of reliable access to food, clean water, and basic medical supplies has compounded the humanitarian crisis, leaving women to bear the brunt of managing these challenges. These realities not only impede progress toward achieving SDG 5.6, which advocates for universal access to sexual and reproductive health and rights, but also highlight the broader failures to safeguard the health and dignity of women and children in conflict zones.

6.3.5 Barriers to Leadership and Participation of Women (SDG 5.5: Ensure Women's Full and Effective Participation and Equal Opportunities for Leadership)

The war on Gaza has severely hindered progress toward achieving SDG 5.5 by exacerbating systemic barriers that prevent women from participating in leadership and decision-making roles. Traditional gender norms, intensified by the war, have further marginalized women, excluding them from vital roles in crisis response and community support initiatives. This exclusion not only denies women opportunities to advocate for their needs and perspectives—essential for equitable and effective recovery—but also perpetuates their underrepresentation in leadership and decision-making at all levels (Abu Zaiter and Al Khawaja 2024; Zaiter 2024).

The impact of wars is evident in the sharp decline in participation of women in local governance and leadership roles within communities. As traditional gender roles are reinforced during crises, women face restricted access to platforms where they can voice concerns or influence policies related to recovery and relief efforts (ILO 2022). The recent hostilities have caused extensive destruction of infrastructure, deepened economic instability, and created widespread food insecurity, all of which disproportionately affect women. These compounded challenges leave many women preoccupied with survival and caregiving responsibilities, further sidelining them from public and political life (Wadi 2022).

Additionally, wars and conflicts erode the rights of women and their potential to contribute to rebuilding efforts and shaping the future of communities. The underrepresentation of women in decision-making processes leads to the exclusion of critical perspectives needed for inclusive recovery strategies. This systemic marginalization highlights the urgent need for targeted interventions to address structural barriers that hinder leadership by women and ensure full participation in political, economic, and public domains, in alignment with SDG 5.5.

6.4 Gaza Women's Resilience

Women in Gaza have demonstrated extraordinary resilience in the face of relentless challenges, including protracted conflict and siege, and systemic deprivation. Despite losing homes, livelihoods, and access to essential services, they continue to adapt and persevere, often taking on expanded roles to support their families. Many women have become informal leaders within their neighborhoods, organizing relief efforts, managing scarce resources, and fostering solidarity among those affected by displacement and loss (Abuelaish and Yousufzai 2023; Jalambo et al. 2025). In the economic sphere, despite the destruction of home-based businesses, women have found innovative ways to generate income. From producing goods in small-scale cooperatives to leveraging traditional skills such as sewing and food preparation, their contributions not only sustain their households but also strengthen community cohesion. Psychological resilience is equally evident. Palestinian women have shown immense strength in coping with the trauma of war, often serving as emotional anchors for their families. Community networks and cultural practices, such as storytelling and remembrance rituals, provide women with a way to process grief and maintain a sense of identity amidst chaos. These actions reflect not only their capacity to endure but also their vital role in preserving the social fabric of Gaza (Khouri 2024; Ramadan and Abouzid 2024; Veronese et al. 2021).

6.4.1 Women as Part of Gaza's Internal Front

Women in Gaza have taken on crucial roles in maintaining social stability, health, and resilience amid the severe humanitarian crisis and escalating violence during the war. Despite being disproportionately affected by displacement and deprivation, they remain central to supporting their communities. With over 1.9 million people displaced in Gaza, including nearly 1 million women and girls, many women have become primary caregivers for orphans and widows, managing scarce resources to sustain their families. Their responsibilities extend beyond caregiving, as they navigate extreme shortages of food, water, and essential supplies (ActionAid UK 2024; Kerbaj et al. 2024; Migdad and Buheji 2024). These efforts underscore the resilience and adaptability of Gazan women, whose contributions are essential to preserving the social fabric and ensuring survival under the most challenging conditions. Empowering these women is not only vital for immediate relief but also for the long-term recovery and stability of Gaza.

6.4.2 Role of Gazan Women During Famine and Chaos

The war and siege have pushed Gaza into an unprecedented humanitarian crisis, with food insecurity and famine affecting vast parts of the population. The whole population of Gaza now faces critical food insecurity, with women often tasked with securing sustenance for their families under conditions of restricted movement, disrupted supply chains, and skyrocketing prices (Alnabih et al. 2024; Buheji and Hasan 2024; Hassoun et al. 2024; IPC 2024). Women, as primary caregivers and providers, bear the brunt of these challenges, navigating unimaginable hardships exacerbated by acute shortages of food, clean water, and essential resources. This dire situation disproportionately impacts women and girls, leading to malnutrition, health deterioration, and heightened vulnerability. Pregnant women and mothers with young children are among the most affected, as limited access to nutritious food and essential healthcare puts both maternal and child health at severe risk. The lack of clean water compounds these challenges, forcing families to rely on contaminated sources, which increases the prevalence of waterborne diseases. Power outages further exacerbate the situation by limiting access to basic services such as sanitation, healthcare, and safe childbirth facilities, all of which are critical for the well-being of women during crises (Dardona et al. 2024; Elsawi et al. 2024; Save the Children International 2024; UN Women 2024c). Recent reports highlighted alarming health challenges faced by women and newborns in Gaza. These include babies born underweight and dying due to insufficient medical care, mothers unable to breastfeed because of inadequate nutrition, and critical shortages of essential nutritional supplements for pregnant and breastfeeding women (Ferguson and Desai 2024). The dire situation underscores the urgent need for improved access to food, healthcare resources, and support for maternal and neonatal health.

Despite these overwhelming challenges, Gazan women remain at the forefront of creating stability amidst chaos. Many women have organized small cooperative groups, pooling resources and skills to produce essential items, such as food. These initiatives address immediate survival needs and foster community solidarity and mutual support. The ability of women to mobilize informal networks has been instrumental in sharing scarce resources and providing emotional and practical assistance to those in need. Emotional resilience plays a pivotal role in how women cope with these challenges. Through shared practices such as storytelling, prayer gatherings, and communal activities, women provide one another with emotional support that helps them endure the trauma of war and famine. These acts of solidarity preserve a sense of identity and hope, offering a lifeline in an environment marked by relentless uncertainty and extraordinary challenges (Abu Zaiter and Al Khawaja 2024; Buheji et al. 2024).

The contributions of Gazan women during this crisis underscore their critical role in sustaining their communities. Their resilience and ingenuity are indispensable, not only for addressing immediate challenges but also for fostering long-term recovery and stability. However, these efforts demand greater international support to address the systemic barriers and resource shortages that disproportionately burden women in Gaza. Empowering them with the tools and opportunities to rebuild their lives is essential for ensuring a more sustainable and equitable future in Gaza.

6.4.3 Role of Palestinian Women in Maintaining Family Stability During War

The stability of families is one of the first casualties of war, yet Palestinian women in Gaza have consistently stepped up as the backbone of their households, ensuring survival and cohesion amidst relentless conflict. Often becoming the primary caregivers and providers, women navigate the immense emotional and physical burdens of displacement, loss of income, and the breakdown of community structures. They take on multiple roles, balancing caregiving responsibilities with efforts to secure food, water, and other essential resources for their families (Buheji et al. 2024).

Culturally, women play a vital role in fostering a sense of normalcy for their families, particularly for children who are profoundly affected by the trauma of war. By offering educational and psychological support, they help their children maintain routines and cope with the surrounding chaos. Women also preserve cultural traditions and family rituals that provide a sense of continuity and identity, which are critical for psychological resilience during times of upheaval (Sabet et al. 2024; UN Women 2024b). In addition to their roles within the household, many Gazan women have become vocal advocates on social media platforms, sharing firsthand accounts of the impact of the war on civilians. Their voices bring global attention to the plight of families in Gaza, often mobilizing international support and humanitarian aid (ESCWA 2023).

The resilience and resourcefulness of Palestinian women in Gaza during the recent war highlight their indispensable role in maintaining family stability and community cohesion. Their ability to adapt, persevere, and advocate for their families amidst unimaginable adversity demonstrates not only their strength but also their centrality to the long-term recovery and resilience of Gazan society. However, the immense burdens they bear underscore the urgent need for targeted interventions that address their specific challenges, empower them with resources, and support their critical efforts in ensuring family and community stability during war.

6.4.4 Role of Palestinian Women in Maintaining the Reproductive Power of Society Amid War

Palestinian women have played a crucial role in maintaining fertility rates as a form of resilience, despite the immense challenges they face during the war. In its attack against Palestinian motherhood, Israel has attempted to fragment national identity, dismantle resistant social structures, and control their demographic increase. By targeting the "reproductive power" of Palestinian women through systemic violence, displacement, and restrictions on healthcare, these efforts aim to weaken the social fabric that women so resolutely strive to maintain (Friedrich-Ebert-Stiftung 2024; Jangda et al. 2024; Tayara 2024). Despite these pressures, Palestinian women continue to embody strength and perseverance, using their roles as mothers and caregivers to resist attempts to undermine their communities and identity.

In addition to these efforts, Gazan women have engaged in acts of resilience that extend beyond motherhood. One symbolic yet practical form of resistance is the cutting of their hair, driven by water shortages and unsanitary conditions in overcrowded shelters (Hamamra et al. 2024). Forced to cut their hair as a response to the lack of water and hygiene supplies, this act underscores the dual burden of survival and resistance. Women's hair, traditionally a marker of beauty and femininity, becomes a site of resistance against the weaponization of femininity and the systemic deprivation imposed by the Israeli blockade. Women have resorted to this painful measure to prevent health issues like lice and skin infections in shelters lacking basic hygiene facilities. This act not only reflects the dire conditions under which these women live but also symbolizes their defiance of the constraints imposed on their bodies and identities.

While striving to fulfill cultural and familial expectations regarding childbearing, Palestinian women face compounded challenges due to the lack of access to comprehensive sexual and reproductive health services. This often results in unintended pregnancies and other complications, exacerbating the difficulties of living in an environment already marked by instability and trauma. Reproductive health has been severely compromised by widespread infrastructure damage. Indeed, pregnant women are often forced to give birth in unsafe conditions, lacking adequate medical supplies and care. This situation has contributed to heightened maternal mortality rates and highlights the health risks women endure, including insufficient access to basic hygiene necessities such as menstrual products (Awadallah 2024; Jalambo et al. 2025; OHCHR 2023; UN Women 2024a). These circumstances highlight the urgent need for gender-sensitive policies and interventions that address these issues holistically.

6.4.5 Resilience Through Adaptation in Gaza

Women in Gaza demonstrate extraordinary resilience as they navigate profound loss and the struggle to rebuild their lives. They draw strength from community bonds, cultural traditions, and informal support networks, which serve as pillars of endurance amid relentless challenges. The war has devastated economic opportunities, causing many women to lose their jobs and the small businesses they managed from their homes or neighborhoods. In response, women have adapted by turning to new forms of income generation, primarily in informal sectors such as food preparation, sewing, or crafting. These activities, often carried out in small groups or at home, offer a semblance of stability despite disruptions of the war (Al-Muhannadi and Buheji 2024; Buheji 2024).

Community support is crucial for Gazan women in managing grief and loss. Local networks, often organized by women's groups, provide spaces for sharing experiences, accessing counseling, and receiving essential supplies. These gatherings offer material aid while fostering collective mourning and emotional healing. Culturally, storytelling, remembrance rituals, and other traditions help preserve identity, honor loved ones, and serve as therapeutic acts of resistance. Women also demonstrate resilience through small-scale livelihood initiatives like cooperatives and home-based crafts, which generate income, empower communities, and strengthen collective solidarity (Abu Zaiter and Al Khawaja 2024; Humanitarian Practice Network 2024; UNDP 2024). Through these efforts, Gazan women find ways to cope with the yearning for normalcy and stability, preserving hope and determination amid overwhelming challenges.

6.5 Mitigation Measures and Recommendations for Recovery

To address the long-term impacts of the war and ensure sustainable recovery, a comprehensive and inclusive approach is essential, with particular attention to advancing gender equality as emphasized in SDG 5. Women, who have borne disproportionate hardships throughout the war, must be at the forefront of recovery efforts to ensure inclusivity and resilience. Rebuilding healthcare infrastructure, particularly facilities focused on maternal and reproductive health, is crucial. Women in Gaza have experienced significant barriers to accessing essential services, underscoring the need for targeted interventions. Additionally, psychological support programs should be established to address the trauma endured by women and equip them with the tools to rebuild their emotional resilience (Ahmed et al. 2024; Elnakib et al. 2024). The leadership and participation of women in decision-making processes are key to fostering effective and inclusive recovery. Incorporating the voices of women ensures that recovery efforts reflect diverse needs and priorities, while also enabling women to shape the future. Inclusivity in leadership aligns with broader goals of recovery and sustainable development (Iriqat 2024; UN Women 2022).

Economic empowerment must be a cornerstone of recovery, with initiatives such as microfinance programs, vocational training, and support for small businesses enabling women to regain financial independence (Buheji 2024; Nassar et al. 2023). Addressing systemic barriers like unequal access to credit and markets is essential to fostering economic resilience and ensuring sustainable progress. Community-led initiatives, particularly those focused on storytelling, cultural preservation, and remembrance rituals, are vital for processing trauma and fostering collective resilience. These efforts not only preserve cultural identity but also provide individuals with a sense of belonging and solidarity, which are critical for rebuilding fractured communities (Abu Zaiter and Al Khawaja 2024; Humanitarian Practice Network 2024).

International collaboration is crucial for the success of these efforts. Technical and financial assistance from global organizations can strengthen local recovery programs, ensuring that resources reach those most affected. Partnerships with grassroots organizations can bridge gaps in recovery initiatives while aligning with the broader vision of SDG 5 to promote gender equality and women's empowerment (Abu Zaiter and Al Khawaja 2024; UNDP 2024).

By integrating these measures and fostering women's participation at every level, Gaza can create a recovery framework that is inclusive, sustainable, and resilient. A focus on healthcare, economic empowerment, cultural preservation, and international cooperation will address immediate challenges while laying the foundation for long-term peace and equitable development.

6.6 Implications of This Chapter and Conclusions

This chapter sheds light on the profound impacts of the war on Gaza on women, illustrating how the war has deepened gender inequalities, eroded livelihoods, and hindered women's societal participation. Gazan women have been forced to assume significant roles in community stability, caregiving, and survival, highlighting their resilience under extreme conditions. However, the compounded effects of this war, famine, and restricted resources present a pressing need for gender-sensitive approaches in both humanitarian aid and long-term recovery planning. The implications of these findings underscore that any effective response to the crisis in Gaza must prioritize the unique needs of women and girls while empowering them as leaders in the rebuilding process. Integrating gender equality into post-conflict recovery frameworks aligns with SDG 5 and enhances community resilience and stability. By providing women with resources and opportunities for meaningful participation, recovery efforts can foster sustainable development and help break cycles of poverty, violence, and marginalization. A critical takeaway is that international organizations, policymakers, and humanitarian agencies must actively involve Gazan women in decision-making processes, recognizing their insights as vital to peacebuilding and recovery. Empowering Gazan women to contribute to and lead recovery efforts is essential to achieving long-lasting stability and fostering a future where gender equality is both a goal and a foundational pillar for development in Gaza and beyond.

In conclusion, to achieve meaningful progress, it is imperative to adopt gender-sensitive approaches that prioritize the specific needs of women and amplify their voices in recovery processes. International organizations, policymakers, and humanitarian agencies must actively involve women in decision-making, ensuring their contributions are recognized and supported. Addressing systemic barriers, such as economic discrimination, lack of educational opportunities, and limited healthcare access, is essential for fostering a more equitable and resilient society. Ultimately, empowering women in Gaza is not merely a matter of justice but a prerequisite for sustainable development and peace. By investing in their leadership, resilience, and potential, the international community can help build a future where gender equality is both a goal and a foundation for lasting stability in Gaza and beyond.

References

Abu Zaiter W, Al Khawaja A (2024) Agents of change: The role of Palestine's women-led organisations in crisis

Abuelaish I, Yousufzai S (2023) Resilience under protracted conflict and siege: a new paradigm shift. Med Confl Surviv 39(4):369–388. https://doi.org/10.1080/13623699.2023.2258325

ActionAid UK (2024) Christmas 2024. https://www.actionaid.org.uk/donate/christmas-appeal-24

Ahmed F, Alhajeeli N, Badheeb M, Ghabisha S, Mahyoub B, Salem A (2024) A call for global intervention: the pressing requirement to address the health emergency in Gaza health crisis. Avicenna 2024(1):1–6. https://doi.org/10.5339/avi.2024.5

Al Issa FAR, Beck E (2021) Sexual violence as a war weapon in conflict zones: Palestinian women's experience visiting loved ones in prisons and jails. Affilia Fem Inq Soc Work 36(2):167–181. https://doi.org/10.1177/0886109920978618

Al-Muhannadi K, Buheji M (2024) Value-based resilience—stories from Gaza (during war 2023). Int J Manag 15(1):15–37. https://doi.org/10.17605/OSF.IO/KP2DX

Alnabih A, Alnabeh N, Aljeesh Y, Aldabbour B (2024) Food insecurity and weight loss during wartime: a mixed-design study from the Gaza strip. J Health Popul Nutr 43:222

Aroussi S (2017) Women, peace, and security and the DRC: time to rethink wartime sexual violence as gender-based violence? Polit Gend 13(3):488–515. https://doi.org/10.1017/S1743923X16000489

Assaf EA, Al Sabbah H, Al-Jawadleh A (2023) Analysis of the nutritional status in the Palestinian territory: a review study. Front Nutr 10:1206090. https://doi.org/10.3389/FNUT.2023.1206090/BIBTEX

Awadallah A (2024) The forgotten women and girls in Gaza: a sexual and reproductive health catastrophe. https://www.bond.org.uk/news/2024/02/the-forgotten-women-and-girls-in-gaza-a-sexual-and-reproductive-health-catastrophe/

Bdier D, Mahamid F (2024) Womens' mental health and war catastrophes in Palestine. In: Zangeneh M (ed) Essentials in health and mental health, advances in mental health and addiction. Springer, Cham, pp 177–186. https://doi.org/10.1007/978-3-031-56192-4_11

Beloskar VD, Haldar A, Gupta A (2024) Gender equality and women's empowerment: a bibliometric review of the literature on SDG 5 through the management lens. J Bus Res 172(November 2023):114442. https://doi.org/10.1016/j.jbusres.2023.114442

Boyadjieva P, Ilieva-Trichkova P (2023) Adult education as a pathway to empowerment: challenges and possibilities. In: Holford I, Boyadjieva J, Clancy P, Hefler S, Studená G (eds) Lifelong learning, young adults and the challenges of disadvantage in Europe. Palgrave studies in adult education and lifelong learning. Palgrave Macmillan, Cham, pp 169–191. https://doi.org/10.1007/978-3-031-14109-6_7

Buheji M (2024) Dealing with loss—coping with yearning of lost livelihood—case of Gaza. Int J Soc Sci Res Dev 6(June):200–215

Buheji M, Hasan A (2024) Beyond famine and chaos—case of Gaza. Int J Manag. https://doi.org/10.17605/OSF.IO/B6SPQ

Buheji M, Migdad M, Muqdad SI, Muqdad AI (2024) Palestinian women in the shadow of genocide in Gaza (2024). Int J Soc Sci Res Dev 6(2):157–172

Cohen DK, Karim SM (2022) Does more equality for women mean less war? Rethinking sex and gender inequality and political violence. Int Organ 76(2):414–444. https://doi.org/10.1017/S0020818321000333

Dardona Z, Amane M, Boussaa S (2024) Top five infectious disease outbreaks among displaced populations during the Gaza conflict 2023–2024: a comprehensive review. J Infect Dis Epidemiol 10(9):328. https://doi.org/10.23937/2474-3658/1510328

Eden L, Wagstaff MF (2021) Evidence-based policymaking and the wicked problem of SDG 5 gender equality. J Int Bus Policy 4(1):28–57. https://doi.org/10.1057/s42214-020-00054-w

Education Cluster/Save the Children/UNICEF (2024) Verification of damages to schools based on proximity to damaged sites—Gaza, Occupied Palestinian Territory

Elnakib S, Fair M, Mayrhofer E, Afifi M, Jamaluddine Z (2024) Pregnant women in Gaza require urgent protection. Lancet 403(10423):244. https://doi.org/10.1016/S0140-6736(23)02835-0

Elsawi M, Al N, Hospital Q, Shorrab AA, Nassef M, Subhi A, Giwa BL, Buheji M (2024) Health in the crossfire-analyzing and mitigating the multifaceted health risks of the 2023 war on Gaza. Publ Heal Res 14(1):1–11. https://doi.org/10.5923/j.phr.20241401.01

ESCWA (2023) Social and economic situation of Palestinian women and girls: July 2020–June 2022. In Social and economic situation of Palestinian women and girls: July 2020–June 2022 (Issue June). https://doi.org/10.18356/9789213626771

Faculty of Education University of Cambridge/Centre for Lebanese Studies & UNRWA (2024) Palestinian education under attack in Gaza: restoration, recovery, rights and responsibilities in and through education (Issue September)

Ferguson L, Desai S (2024) Sexual and reproductive health and rights in Palestine—securing spaces to speak out. Sex Reprod Heal Matt 32(1):1–8. https://doi.org/10.1080/26410397.2024.2397956

Friedrich-Ebert-Stiftung (2024) The womb as a weapon in Palestine. https://feminism-mena.fes.de/e/the-womb-as-a-weapon-in-palestine.html

Gaggioli G (2014) Sexual violence in armed conflicts: a violation of international humanitarian law and human rights law. Int Rev Red Cross 96(894):503–538. https://doi.org/10.1017/S1816383115000211

Global Nutrition Cluster (2024) Nutrition vulnerability and situation analysis/Gaza (Issue February)

Gordon N, Haddad M (2024) The road to famine in Gaza. New York Rev Book 1:1–12

Guaita-Fernández P, Martín Martín JM, Ribeiro Navarrete S, Puertas R (2024) Analysis of gender-based violence in the context of the sustainable development goals. Sustain Dev:4946–4958. https://doi.org/10.1002/sd.2940

Hamamra B, Alzaghal N, Mleitat A, Veronese G (2024) Hair cutting as resistance: Gazan women and the failures of global feminism. J Cult Res 00(00):1–15. https://doi.org/10.1080/14797585.2024.2434835

Hassoun A, Al-Muhannadi K, Hassan HF, Hamad A, Khwaldia K, Buheji M, Jawaldeh AA (2024) From acute food insecurity to famine: how the 2023/2024 war on Gaza has dramatically set back sustainable development goal 2 to end hunger. Front Sustain Food Syst 8:1402150. https://doi.org/10.3389/fsufs.2024.1402150

Humanitarian Practice Network (2024) Who will listen to the women of Gaza? https://odihpn.org/publication/who-will-listen-to-the-women-of-gaza/

Hussein-Sabbah S (2023) Pregnant women in Gaza face perilous conditions as maternity services and infrastructure crumble. BMJ (December):1–2. https://doi.org/10.1136/bmj.p2895

Ibnouf FO (2020) War-time care work and peacebuilding in Africa. In: War-time care work and peacebuilding in Africa. Gender, development and social change. Palgrave Macmillan, pp 31–51. https://doi.org/10.1007/978-3-030-26195-5

ILO (2022) Gender equality and women's empowerment in the world of work in fragile, conflict and disaster settings

International Monetary Fund (2022) Palestinian labor market outcomes, dynamics, and policies. In: West Bank and Gaza: selected issues. International Monetary Fund, pp 1–12

IPC (2024) THE GAZA STRIP: Risk of Famine as 495,000 people face catastrophic acute food insecurity (IPC Phase 5) (Issue September). https://www.ipcinfo.org/fileadmin/user_upload/ipcinfo/docs/IPC_Gaza_Strip_Acute_Food_Insecurity_Feb_July2024_Special_Brief.pdf

Irfan B, Lulu I, Hamawy A, Shammala AA, Kullab S, Fawaz M, Sammour AAK, Khawaja H, Alshaer N, Abu-Sittah G, Alser O, Perlmutter M, Al-Shembari A, Hickey J, Nayfeh T, Tahir M, Sultan MJ, Nasser E, Wajahath M et al (2024) Combating infections under siege: healthcare challenges amidst the military assault in Gaza. World Med Heal Pol:1–26. https://doi.org/10.1002/wmh3.642

Iriqat D (2024) Towards an inclusive political system, Palestinian perspectives on the reconstruction of Gaza. In: Antar U, Asseburg M, Shaban O (eds) Palestinian perspectives on the reconstruction of Gaza. Friedrich-Ebert-Stiftung (FES), Stiftung Wissenschaft und Politik (SWP). https://doi.org/10.1007/s12646-009-0001-x

Iriqat D, Alousi R, Aldahdouh TZ, AlDahdouh A, Dankar I, Alburai D, Buheji M, Hassoun A (2025) Educide amid conflict: the struggle of the Palestinian education system. Qual Educ All 2(1):81–99. https://doi.org/10.1108/QEA-10-2024-0120

Jalambo MO, Al-masri N, Somaya M (2025) Resilient Palestinian women facing gender-based conflict in the Gaza strip. In: Yohannes HT, Phipps A, Sitholé T (eds) Cultures of sustainable peace: conflict transformation, gender-based violence, decolonial praxes. Bristol & Blue Ridge Summit: Multilingual Matters, pp 138–157

Jangda HN, Farhan K, Akilimali A (2024) Unheard voices: addressing sexual and reproductive health in Gaza's women. Int J Gynecol Obstet 166(1):139–140. https://doi.org/10.1002/ijgo.15596

Kerbaj M, Traboulsi J, Maasri Z, Toukan H (2024) Gaza in plain sight: witnessing in solidarity. J Vis Cult 22(3):1–19. https://doi.org/10.1177/14704129241227847

Khouri LZ (2024) Gaza's broken lullaby: genocidal unchilding and the impossibility of mothering. Stud Gend Sex 25(3):178–187. https://doi.org/10.1080/15240657.2024.2379204

Khurana, N. V., & Verma, R. K. (2023). SDG 5: addressing gaps and accelerating Progress towards a feminist, intersectional, and transformative vision for change. In R. K. Pachauri, S., Verma (Ed.), Transforming unequal gender relations in India and beyond. Sustainable development goals series (pp. 3–20). Springer Nature Singapore. https://doi.org/10.1007/978-981-99-4086-8

Lwamba E, Shisler S, Ridlehoover W, Kupfer M, Tshabalala N, Nduku P, Langer L, Grant S, Sonnenfeld A, Anda D, Eyers J, Snilstveit B (2022) Strengthening women's empowerment and gender equality in fragile contexts towards peaceful and inclusive societies: a systematic review and meta-analysis. Campbell Syst Rev 18(1):e1214. https://doi.org/10.1002/cl2.1214

Madar R (2023) Beyond male Israeli soldiers, Palestinian women, rape, and war Israeli state sexual violence against Palestinians. Confl Soc 9(1):72–88. https://doi.org/10.3167/arcs.2023.090105

Mahamid F, Veronese G, Bdier D (2022) Gender-based violence experiences among Palestinian women during the COVID-19 pandemic: mental health professionals' perceptions and concerns. Confl Heal 16(1):1–10. https://doi.org/10.1186/s13031-022-00444-2

Marteu, E. (2023). The Israeli-Palestinian conflict and the Gaza war through a gender lens. Strategic sectors | Culture & society, November, 315–318

Meagher K, Attal B, Patel P (2021) Exploring the role of gender and women in the political economy of health in armed conflict: a narrative review. Glob Health 17(1):88. https://doi.org/10.1186/s12992-021-00738-9

Migdad M, Buheji M (2024) Realizing the capacity of the internal front—case of Gaza 2024. Community Pract 21(08):544–565. https://doi.org/10.5281/zenodo.13622097

Morzk MH, Shoaib FT, Abdel-Fatah NA (2025) Socioeconomic and demographic determinants of marital violence against women in Palestine. J Popul Social Stud 33:317–335. https://doi.org/10.25133/JPSSV332025.017

Muhammad A, Rizek J (2024) Caring for Women in an active war zone. J Emerg Nurs 50(6):722–726. https://doi.org/10.1016/j.jen.2024.08.002

Müller C, Tranchant J-P (2017) Tackling violence against women and girls in Gaza. In International institute for environment and development (Issue 2017). https://lordslibrary.parliament.uk/tackling-violence-against-women-and-girls-in-the-uk/

Nassar S, Tóth ZN, Vasa L (2023) Economic empowerment as a result of achieving SDGs with resource access: a comparative research between Gaza strip and Hungary. J Int Stud 16(2):9–33. https://doi.org/10.14254/2071-8330.2023/16-2/1

OHCHR (2023) Women bearing the brunt of Israel-Gaza conflict: UN expert. https://www.ohchr.org/en/press-releases/2023/11/women-bearing-brunt-israel-gaza-conflict-un-expert

Ramadan A, Abouzid M (2024) Resilience beyond borders: Gaza's women and their silent battles. Women Birth 37(4):101623. https://doi.org/10.1016/j.wombi.2024.101623

Repo J (2024) Genocide and the destruction of the means of social reproduction in Gaza. Eur J Polit Gender:1–8. https://doi.org/10.1332/25151088Y2024D000000061

Reynolds AN (2021) Women at work and war: integrating gender and conflict into impact assessment. Impact Assess Proj Apprais 39(3):196–205. https://doi.org/10.1080/14615517.2021.1904375

Rosamund S (2024) Accelerating progress on sustainable development goal 5 (SDG 5) achieving gender equality and empowering women and girls. EPRS Eur Parliament Res Serv 5(September):1–12

Rubini E, Valente M, Trentin M, Facci G, Ragazzoni L, Gino S (2023) Negative consequences of conflict-related sexual violence on survivors: a systematic review of qualitative evidence. Int J Equity Health 22(1):227. https://doi.org/10.1186/s12939-023-02038-7

Sabet C, Nguyen D, Nada SA, Abuzerr S, Taha AM (2024) Women and girls in Gaza face increasingly dire physical and mental health challenges. BMJ 384. https://doi.org/10.1136/BMJ.Q625

Sahin B, Kula SL (2018) What women want before justice: examining justice initiatives to challenge violence against women in the DRC. Int J Trans Justice 12(2):296–313. https://doi.org/10.1093/IJTJ/IJY010

Samuel MT (2024) The Israel-Hamas war one year later: mass violence and Palestinian dispossession. Midd East Pol 31(4):3–16. https://doi.org/10.1111/mepo.12792

Save the Children International (2024) Women self-inducing labour and facing life-threatening complications in pregnancy after nine months of Gaza conflict. https://www.savethechildren.net/news/women-self-inducing-labour-and-facing-life-threatening-complications-pregnancy-after-nine

Sweileh W (2024) Gender equality and women's empowerment in Arab countries: a bibliometric review of the literature on SDG 5. Glob Knowled Memory Comm, ahead-of-print(ahead-of-print). https://doi.org/10.1108/GKMC-01-2024-0010/FULL/XML

Tayara H (2024) Health and humanitarianism: colonial manifestations of violence in the Palestinian body. J Intersect Social Justice

The Guardian (2024) Claims of Israeli sexual assault of Palestinian women are credible, UN panel says | Israel-Gaza war. https://www.theguardian.com/world/2024/feb/22/claims-of-israeli-sexual-assault-of-palestinian-women-are-credible-un-panel-says

UN Women (2022) After the may 2021 escalation: a multi-sectoral gender needs assessment in the Gaza strip (Issue May)

UN Women (2024a) Gender alert: Gaza: a war on women's health (Issue September)

UN Women (2024b) Gender alert: the gendered impact of the crisis in Gaza. In UN Women (Issue January). https://www.un-ilibrary.org/content/books/9789211064704

UN Women (2024c) At least 557,000 women in Gaza are facing severe food insecurity. https://www.unwomen.org/en/news-stories/news/2024/06/at-least-557000-women-in-gaza-are-facing-severe-food-insecurity

UNDP (2024) Impact of the 2023 Gaza war on displaced women in the Gaza strip

UNFPA (2024) "Where are we supposed to go?": Women and girls in northern Gaza endure displacement, hunger and healthcare under attack. https://www.unfpa.org/news/"where-are-we-supposed-go"-women-and-girls-northern-gaza-endure-displacement-hunger-and?utm_source=chatgpt.com

UNRWA (2023) Gaza field office rapid gender analysis: gendered impacts of the October 2023 escalation in Gaza (Issue October).

Veronese G, Sousa C, Cavazzoni F (2021) Survival and resilience among Palestinian women: a qualitative analysis using individual and collective life events calendars. Violence Against Women 27(6–7):900–917. https://doi.org/10.1177/1077801220914406

Veronese G, Mahmid FA, Bdier D (2023) Gender-based violence, subjective quality of life, and mental health outcomes among Palestinian Women: the mediating role of social support and agency. Violence Against Women 29(5):925–948. https://doi.org/10.1177/10778012221099988

Wadi D (2022) Women's empowerment in Palestine: a process in progress. Quad La Medit 34:37–40

Wick L, Hassan S (2012) No safe place for childbirth: women and midwives bearing witness, Gaza 2008–09. Reprod Health Matters 20(40):7–15. https://doi.org/10.1016/S0968-8080(12)40648-6

Wilson Center (2024) Unraveling the multi-faceted impact of armed conflict on women in the Gaza strip.

https://www.wilsoncenter.org/blog-post/unraveling-multi-faceted-impact-armed-conflict-women-gaza-strip?utm_source=chatgpt.com

World Bank Group (2023) West Bank and Gaza country climate and development report (Issue September). https://doi.org/10.1596/40673

Zaiter WA (2024) Women at the heart of Gaza's rebirth: Palestinian perspectives on the reconstruction of Gaza. In: Antar U, Asseburg M, Shaban O (eds) Palestinian perspectives on the reconstruction of Gaza. Friedrich-Ebert-Stiftung (FES), Stiftung Wissenschaft und Politik (SWP)

7 Powerless Under Siege and Devastation: The Struggle for Clean and Affordable Energy in Gaza

Nadia Mahdi, Hasan Abumeteir, Mohammed Migdad, Eman Ashour, Iyad Abuhadrous, and Abdo Hassoun

Abstract

This chapter explores the critical energy challenges faced by Gazans amid the recent war and the resulting destruction, with a special focus on Sustainable Development Goal 7 (SDG 7), which aims to ensure access to affordable, reliable, sustainable, and modern energy for all. The work aims to outline the energy landscape in Gaza, detailing the extensive damage to power plants, electricity grids, and fuel supplies. This war and blockade have severely impacted Gaza's energy infrastructure, leading to chronic electricity shortages and reliance on unsustainable energy sources. It highlights how these disruptions have led to widespread energy poverty, affecting homes, hospitals, schools, and businesses. Key issues examined include the limited availability of affordable and clean energy, the heavy reliance on unreliable electricity sources, and the significant environmental and health impacts resulting from energy shortages. The chapter concludes with recommendations for rebuilding and upgrading Gaza's energy infrastructure to achieve SDG 7. It stresses the importance of international cooperation, investment in renewable energy technologies, and the establishment of a stable political environment to ensure long-term energy sustainability, affordability, and resilience. This comprehensive analysis aims to provide a deeper understanding of the energy crisis in Gaza, offering insights into the urgent actions needed to address these challenges and promote energy security and sustainable development in the region.

N. Mahdi (✉)
Gaza Electricity Distribution Corporation, GEDCO, Gaza, Palestine
e-mail: n.mahdi@saf-ir.com

H. Abumeteir
Gaziantep Islam Science and Technology University, Gaziantep, Turkey
e-mail: hasan.abumeteir@gibtu.edu.tr

M. Migdad
Faculty of Economics and Managerial Sciences, Islamic University of Gaza, Gaza, Palestine
e-mail: mmigdad@iugaza.edu.ps

E. Ashour
Al Azhar University, Gaza, Palestine
e-mail: eman.ashour@azhar.edu.ps

I. Abuhadrous
Palestine Technical College, Deir El-Balah, PTCDB, Gaza, Palestine
e-mail: ihadrous@ptcdb.edu.ps

A. Hassoun
Sustainable AgriFoodtech Innovation & Research (SAFIR), Arras, France

Faculty of Agricultural Engineering, University of Aleppo, Aleppo, Syria
e-mail: a.hassoun@saf-ir.com

A. Hassoun (ed.), *War on Gaza*, Sustainable Development Goals Series,
https://doi.org/10.1007/978-3-031-88500-6_7

Keywords

Electricity · Energy crisis · Energy infrastructure · Energy poverty · Fuel · Renewable energy · SDG 7 · Socioeconomics · Sustainable energy

7.1 Introduction

Sustainable Development Goal 7 (SDG 7) is one of the 17 goals set by the United Nations under the 2030 Agenda for Sustainable Development. Its main aim is to ensure that everyone, no matter where they live, has access to modern, affordable, and sustainable energy by 2030. Energy is crucial in improving lives, from powering economies to enhancing healthcare, education, and overall quality of life. While there has been notable progress over the years, 685 million people are still without access to electricity and 2.1 billion without access to clean cooking in 2022, which negatively impacts their health and the environment. Accordingly, each year, millions of people die prematurely due to health diseases resulting from inefficient cooking methods (IEA et al. 2024; Kumar et al. 2024).

Achieving SDG 7 requires significant advancements in renewable energy technologies, energy efficiency, and policy regulations. Energy systems globally have a profound impact on the environment, contributing to air pollution, climate change, and ecosystem degradation. Transitioning to clean energy sources, such as solar and wind, is essential to reducing carbon emissions and advancing sustainability goals. Attaining SDG 7 has the potential to improve quality of life while fostering a more resilient and low-carbon economy. Efforts should focus on enhancing energy efficiency, promoting investments in clean energy projects, and supporting the adoption of renewable energy solutions across both developed and developing regions. These initiatives aim to ensure access to cleaner energy while driving sustainable economic growth, reducing emissions, and encouraging green technologies (Arora and Mishra 2022; Awosusi et al. 2024).

While the global effort to shift toward sustainable energy is crucial, the Gaza Strip exemplifies the extreme challenges conflict zones face in achieving this goal. Over the past decades, energy shortages have become a critical issue in Palestine, with power cuts of up to 16 h per day severely impacting daily life and essential services (Salem et al. 2021). In Gaza, energy infrastructure has faced significant constraints due to prolonged restrictions and recurring conflicts, leading to chronic electricity shortages. Access to reliable power is limited, disrupting homes, schools, hospitals, and businesses and creating widespread hardship. The lack of consistent electricity has hindered basic services, with critical sectors like healthcare and water systems particularly affected. Gazans have been forced to rely on alternative energy sources that are often inefficient, unsustainable, and harmful to both health and the environment. Efforts to improve the situation through renewable energy initiatives and infrastructure development have faced numerous obstacles, further exacerbating the crisis. These energy challenges highlight broader systemic issues and the need for sustainable, long-term solutions to address energy needs and support the well-being of the population (Fathi Nassar and Yassin Alsadi 2019; Nassar et al. 2024).

The situation in Gaza has worsened significantly during the last war, with energy insecurity reaching critical levels. This has been exacerbated by the cutting off and strict control of access to energy sources, particularly fuel, as well as the widespread destruction of vital infrastructure. Damage to power plants, fuel, and distribution networks has compounded existing challenges, leaving the population with minimal access to electricity. This worsening crisis has deepened the reliance on limited and costly alternatives, such as generators, which are often inaccessible to many families (Amnesty International 2024; Procter 2024).

This chapter explores the profound damage inflicted on Gaza's energy system by years of war and blockade, a situation further exacerbated by the recent war (October 2023–January 2025). It examines the far-reaching effects of these disruptions, which have left homes, hospitals, water facilities, educational institutions, and other institutions struggling with persistent energy shortages. This energy poverty is a critical issue, demonstrating that ensuring reliable access to energy must be a priority for international and civil society organizations seeking to achieve SDG 7, especially in conflict and crisis areas, including the Gaza Strip.

7.2 The Impact of Blockade and War on the Energy Sector in Gaza

Palestine heavily relies on imported energy, with 93.87% of its electricity coming from Israel and 3% from Jordan and Egypt. Local generation accounts for only 4.1% due to limited infrastructure and resources. This dependency leaves Palestine vulnerable to supply disruptions, especially in Gaza and the West Bank (Salah and Abuhelwa 2020; Salah et al. 2021).

For decades, Gaza's energy systems have faced immense strain, primarily due to the long-standing blockade and recurring military conflicts. The blockade has severely restricted access to fuel, equipment, and essential infrastructure maintenance, creating chronic energy shortages and forcing Gaza to rely heavily on external energy sources. Before the most recent war, Gaza's energy landscape relied heavily on two primary sources: 120 MW of electricity imported from Israel and varying output from the Gaza Power Plant (GPP), which fluctuated between 70 and 140 MW depending on fuel availability. GPP frequently operated at just 50% capacity due to the elevated cost of fuel and damage sustained from Israeli airstrikes. An additional 23 MW from Egypt had once supplied the southern region of Gaza, but this supply has been non-operational since 2018 due to security concerns in the Sinai region. Gaza's energy system had been further weakened by persistent attacks dating back to 2006, and by April 2017, political disputes between Gaza's authorities and the Palestinian Authority exacerbated the energy crisis (Nassar and Alsadi 2016; Fathi Nassar and Yassin Alsadi 2019; Ersoy et al. 2022; Fischhendler et al. 2022; Nassar et al. 2024). Before the onset of the recent war, Gaza's power supply was meeting less than 50% of total demand (Fig. 7.1), according to the United Nations Office for the Coordination of Humanitarian Affairs (OCHA).

While electricity generation was at the heart of Gaza's energy production, the broader energy landscape was shaped by other critical resources,

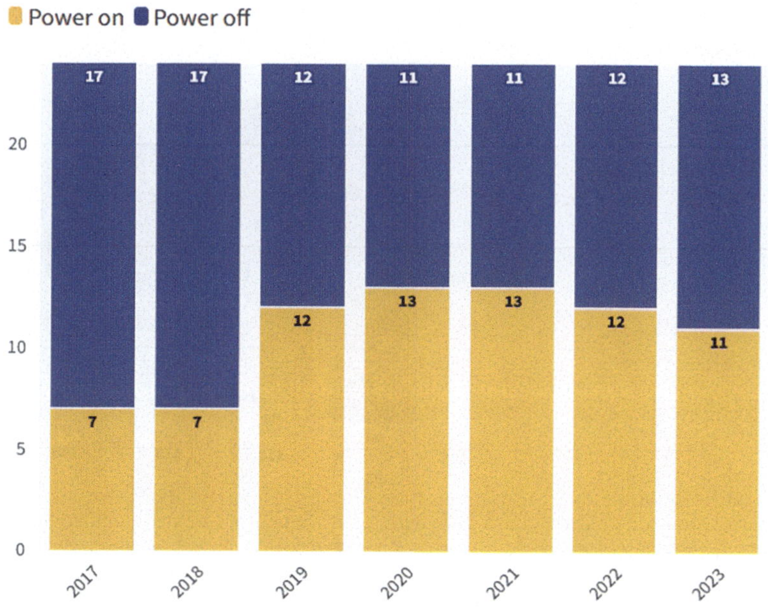

Fig. 7.1 Average daily electricity supply (hours per day) categorized into hours with power (Power on) and hours without power (Power off) (iMEdD Lab 2024; OCHA 2024a)

including diesel, liquefied petroleum gas (LPG), and natural gas, each playing an essential role in meeting the population's energy needs. Diesel fuel powered the GPP and backup generators during frequent outages, ensuring that hospitals and water facilities could continue operating. However, Gaza's dependence on imported diesel made it particularly vulnerable to supply disruptions, often forcing the GPP to reduce capacity or shut down entirely. The high costs associated with diesel further strained Gaza's already fragile economy, making it increasingly difficult to maintain a consistent electricity supply. Similarly, LPG, primarily used for cooking, faced frequent import delays, leading to widespread shortages (Tannira 2021; Juaidi et al. 2022). In contrast, natural gas holds significant long-term potential, particularly from the Gaza Marine gas field. The field is estimated to contain 1.2 trillion cubic feet of reserves to meet Gaza's energy needs for decades. However, the development of this resource has been delayed due to political and logistical challenges, leaving Gaza's energy security in a precarious position amid an escalating crisis (Elkahlout 2024; Rettig and Spanier 2024).

Figure 7.2 illustrates the persistent gap between electricity supply and demand in Gaza from 2017 to 2023, highlighting the insufficiency of contributions from both the GPP and imports from Israel to meet the growing energy needs. The widening gap, particularly in 2023, underscores the worsening energy crisis and its impact on the region's infrastructure and services.

Recurrent conflicts, including the most recent war, have further pushed Gaza's overstretched energy systems to the brink of collapse. The destruction of conventional energy systems and renewable installations has deepened the humanitarian emergency, cutting off access to essential services such as electricity, water, healthcare, and sanitation.

7.2.1 Impact of War on Conventional Energy Systems

The recurring wars in Gaza, notably in 2008, 2014, and the most recent one, which began in October 2023, have compounded the damage caused by the long-standing blockade. Each con-

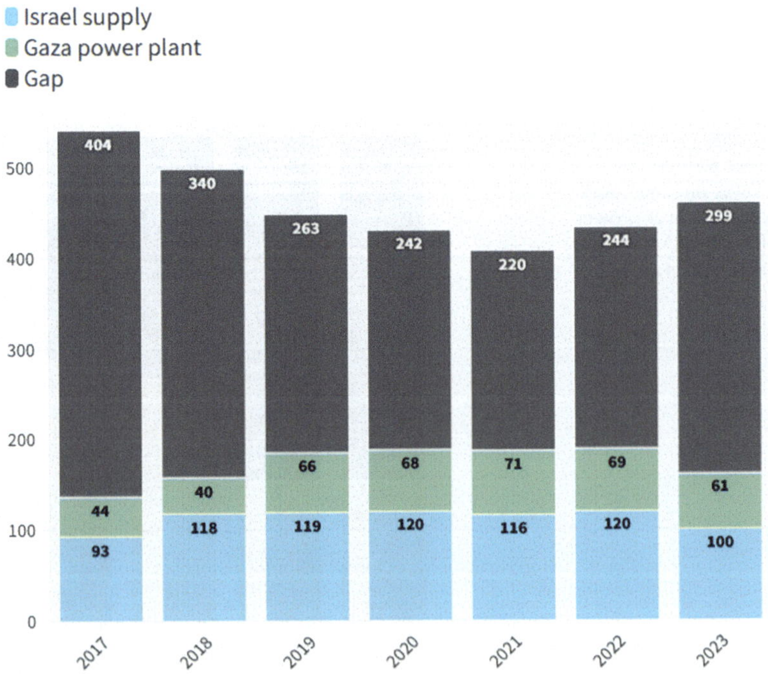

Fig. 7.2 Electricity supply-demand balance in Gaza (2017–2023) (iMEdD Lab 2024; OCHA 2024a)

flict has destroyed key components of Gaza's energy infrastructure. On October 9, 2023, Israeli Defense Minister Yoav Gallant announced a "complete siege" of the enclave: no electricity, no food, no water, and no fuel (Bishara and Elbasri 2024; Gordon and Haddad 2024; Human Rights Watch 2024; Suzuki and Sakai 2024). Since October 11, 2023, Gaza has been plunged into darkness after Israeli authorities cut off the electricity supply, and the fuel reserves for Gaza's sole power plant, the GPP, were depleted. This blackout has further deepened the humanitarian crisis in Gaza, leaving hospitals, schools, and essential services without power, as will be detailed in the following sections.

The electricity sector, in particular, has suffered severe damage, with an initial estimate of US $279 million, primarily affecting the electricity grid distribution networks and off-grid rooftop solar photovoltaic systems installed across public buildings, including schools, hospitals, health facilities, water supply facilities, cultural buildings, and residential buildings. The extent of damage across different sectors, including energy, has escalated since October 12, 2023, as illustrated in Fig. 7.3. By January 21, 2024, over 60% of the energy sector's infrastructure had been damaged or destroyed, significantly impairing Gaza's capacity to sustain essential services and operations (The World Bank 2024a). Gaza Electricity Distribution Company has reported losses of $450 million across the sector over the first 14 months of the recent war on Gaza (Middle East Monitor 2024).

As seen in Fig. 7.4, satellite images vividly capture this dire situation, showing the stark difference in Gaza's lighting before and after the blackout. By mid-October 2023, the territory was in near-complete darkness, symbolizing the loss of energy infrastructure and the worsening conditions for millions of residents.

GPP and the high-voltage connection points that import electricity from Israel have suffered significant damage (Fig. 7.5). Furthermore, the GEDCO has faced extensive losses, including destruction to its warehouses, offices, vehicles, and its critical Supervisory Control and Data Acquisition (SCADA) system (The World Bank 2024b). According to a report by the GEDCO on the first anniversary of the start of the aggression on Gaza, more than 68% of the medium voltage and transmission networks have been destroyed,

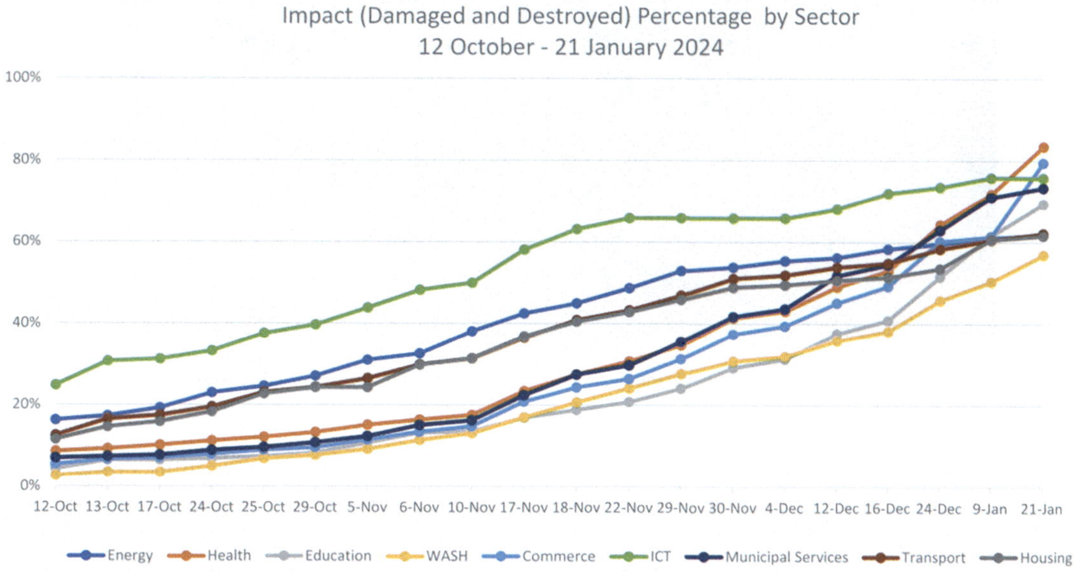

Fig. 7.3 Impact percentage by sector since the beginning of the recent war on Gaza (The World Bank 2024a; Source: World Bank/IPSOS)

Fig. 7.4 Gaza in the dark: on September 14, 2023 (left) and on October 12, 2023 (right) (ESCWA 2023; Wei 2023)

Fig. 7.5 Damage of the electricity distribution network. Source: GEDCO social media website at: https://www.facebook.com/Gedcops

along with 70% of the company's buildings, 90% of its warehouses, and 80% of its vehicles and machinery. In addition, 830 km of medium voltage lines, 2700 km of low voltage lines, and 2105 electrical transformers were completely destroyed. The estimated financial impact of the company's losses in the grid, utility power sales, facilities, and machinery surpassed $450 million (Palestinian Energy and Natural Resources Authority 2024).

The extent of damage caused in the recent war is significantly more catastrophic than reported during previous military hostilities (UNDP 2021), with the scale and severity of destruction to energy facilities underscoring the unprecedented challenges confronting Gazans.

7.2.2 Impact of War on Renewable Energy Infrastructure

Palestine possesses significant potential for renewable energy applications, particularly in solar and wind energy, which are well-suited to the region's high solar radiation levels and geographical conditions. Recent studies highlight that Gaza, despite its constrained resources and geopolitical challenges, has emerged as a fertile ground for renewable energy development. The Gaza Strip's severe energy crisis, stemming from political and logistical barriers to fuel and energy imports, has driven a shift toward decentralized solar energy systems to meet critical needs for homes, hospitals, and water facilities (Salah et al. 2021; Ersoy et al. 2022; Fischhendler et al. 2022).

Before the recent escalation, Gaza had made considerable progress in the renewable energy sector, mainly through the adoption of solar power, which provided over 5% of the region's energy supply. This development was essential in addressing the power shortages caused by the blockade, with solar panels installed in hospitals, schools, and water facilities playing a crucial role in ensuring that essential services like healthcare and water treatment could function uninterrupted. Additionally, solar energy was seen as a critical path to reducing Gaza's dependence on fossil fuels and stabilizing the power supply. Solar energy adoption has increased, with 20% of households in Gaza using solar solutions for their energy needs, reflecting a growing shift toward renewable sources. The number of solar panel installations in Gaza has seen a remarkable increase, growing from only 12 systems in 2012 to an impressive 8760 systems by 2022. However, the war has reversed much of this progress, destroying 67 MW of solar photovoltaic systems and releasing an estimated 1675 kg of lead into the environment (ESCWA 2023; The World Bank 2024a; UNEP 2024).

The outbreak of the recent war in October 2023 has profoundly disrupted progress in renewable energy development. Solar energy systems, which had begun to bridge the gap in energy supply, were among the first casualties of the war. According to the Center for Strategic and International Studies (CSIS), as of November 11, 2023, 17 out of the 29 largest rooftop solar systems had been either completely destroyed or significantly damaged. The remaining solar installations are insufficient to meet the growing demand for emergency services, further deepening Gaza's energy crisis (ESCWA 2023; Fischhendler et al. 2023; UNEP 2024). Furthermore, the destruction of large-scale solar installations has had a profound economic and infrastructural impact. The Gaza Industrial Estate (Fig. 7.6), which relied on solar energy for up to 80% of its electricity needs, was a key example of "finance for climate-friendly investment." The International Finance Corporation (IFC) supported this project, recognizing it as a model for sustainable energy development.

However, as seen in Fig. 7.7, satellite imagery taken on December 16, 2023, reveals that the industrial estate has been almost completely destroyed (UNEP 2024).

Rebuilding Gaza's renewable energy capacity faces significant challenges. The continued blockade restricts access to the equipment and materials necessary for installing and maintaining solar panels. Furthermore, the economic damage inflicted by the war has severely constrained local capacity to invest in renewable energy solutions. Not only has progress in clean energy been halted, but it has also been reversed, deepening Gaza's dependency on external and fragile power sources. The energy crisis in Gaza illustrates the critical need for sustainable energy solutions to be integrated into recovery strategies, ensuring long-term resilience and stability for conflict-affected regions (Fischhendler et al. 2022; Qutaina et al. 2023; Elnaggar 2023; Amnesty International 2024).

Fig. 7.6 Solar photovoltaic panels installed at the Gaza Industrial Estate, part of an IFC-funded project on renewable energy in Gaza, which was recognized as an important example of "Financing for climate friendly investment" (UNEP 2024)

Fig. 7.7 A satellite image taken on December 16, 2023, shows the almost complete destruction of the Gaza Industrial Zone (UNEP 2024)

7.3 Daily Hardships Caused by Energy Insecurity in Gaza

7.3.1 Cooking Using Traditional Methods

Energy insecurity in the Gaza Strip remains a critical issue, characterized by persistent shortages of cooking gas and fuel. These shortages, driven by the blockade imposed by Israel since 2007 and exacerbated by repeated military offensives, have left many residents without reliable access to essential energy resources. Limited supplies of LPG entering Gaza have proven insufficient to meet demand, particularly in Gaza City and northern areas where entry restrictions are most severe. As a result, many households have been forced to adopt traditional cooking methods, using open stoves fueled by firewood, charcoal, or even plastic materials or wood scavenged from destroyed homes or broken trees. This shift places significant burdens on families, as women and children are often tasked with gathering fuel. Baking bread in clay ovens has also become a widespread necessity, requiring daily preparation due to the lack of refrigeration and the inability to store baked goods for extended periods (Abu Riash and Humaid 2023; Alnabih et al. 2024; Amnesty International 2024).

The reliance on these inefficient and labor-intensive methods underscores the urgent need to address the energy challenges in Gaza. This crisis severely hampers progress toward SDG 7, particularly Target 7.1, which aims to ensure universal access to affordable, reliable, and modern energy services.

7.3.2 Access to Water and Sanitation

Severe damage to most water infrastructure in Gaza, coupled with the shutdown of two major pipelines from Israel, has caused a near-collapse of the water sector. Residents face dire shortages of water, with water supply being decreased by 95% and with many residents receiving only 2–9 L per day—significantly below the emergency threshold of 15 L per person, forcing the population to use unsafe water and sanitation facilities (ESCWA 2023; Dardona et al. 2024; Farajallah and Farajallah 2024). Humanitarian organizations depend on municipal wells, powered by diesel generators, to extract water and distribute it via tankers. However, critical fuel shortages significantly hinder these operations. Compounding the issue, many of these wells are contaminated by sewage and seawater infiltration, resulting in high salinity levels. Alarmingly, 97% of Gaza's groundwater fails to meet international standards for safe drinking water (Human Rights Watch 2024; Eshtawi and Abdellatif 2024). Many residents must travel long distances daily to collect small amounts of water, highlighting the physical and mental toll of securing basic needs. Electricity shortages intensify these challenges, particularly for washing clothes.

The severe damage to water infrastructure in Gaza and the reliance on fuel-powered generators for water access highlight the critical connection between energy and essential services. This situation directly impacts SDG 7, particularly Target 7.1. Shortages of fuel hinder water extraction and distribution, worsening the water crisis and forcing the population to rely on unsafe sources.

7.3.3 Cost of Living and Energy Insecurity

The recent war has devastated Gaza's economy, pushing nearly 80% of the population into poverty and driving unemployment rates to among the highest globally, further intensifying the already soaring cost of living (Alagha and Hussein 2024; Buheji and Marouf 2024). Energy scarcity is one of the most critical implications of war, directly affecting the cost of living and access to basic necessities in Gaza.

The destruction of Gaza's agricultural sector has severely worsened food insecurity, leaving approximately 96% of the population (2.15 million people) experiencing acute levels of food insecurity as of September 2024 (IPC 2024). Energy scarcity has played a significant role in this crisis, as agriculture relies on energy for irrigation, harvesting, and food storage. Damaged infrastructure has reduced the ability of this sector to produce and preserve food, while the high cost and scarcity of diesel for irrigation systems have driven production expenses to unsustainable levels.

Energy shortages have also severely impacted food storage capabilities. The lack of electricity for refrigeration leads to significant food spoilage, further driving up food prices (Blouin et al. 2024). Before the recent war, most homes in Gaza relied on refrigerators to store food and water. However, Gazans have to resort to paying for freezing services or purchasing cold water at significantly higher costs, adding to their financial burden. Clean water access has been severely impacted by energy insecurity, with many desalination plants damaged or operating at reduced capacity due to high fuel costs (Ahmed 2023; Hall et al. 2024). A significant portion of Gaza's population is forced to rely on contaminated and saline water from wells, while the cost of desalinated water has risen dramatically, adding to the financial strain on families. The energy needed for water extraction and transportation further increases these expenses.

In response to electricity outages, residents have turned to alternative energy solutions, such as off-grid solar systems, but these have become increasingly unaffordable. The steep costs of such measures highlight the critical connection between energy insecurity and the rising cost of living in Gaza, as residents struggle to secure basic necessities like clean water and electricity (Fischhendler et al. 2023; El-Khozondar et al. 2024).

The severe energy insecurity in Gaza, driven by the war, directly undermines SDG 7. The rising cost and scarcity of energy have significantly increased the financial burden on residents, limiting access to essential needs such as food storage, clean water, and agricultural production. As families struggle to afford alternative energy solutions, the lack of reliable electricity continues to exacerbate poverty and hinder economic recovery, making progress toward sustainable energy access increasingly difficult.

7.3.4 Living Amid Waste and Polluted Water

The war has also severely impacted all five wastewater treatment plants in Gaza, which have ceased operations. Furthermore, the majority of the 65 wastewater pumping stations have ceased operations due to the electricity blockade and fuel shortages, in addition to the damage sustained by the Gaza Central Wastewater Plant (Khalidi 2023; Hall et al. 2024). Fuel supply to the Gaza Strip remains severely limited, falling short of meeting essential humanitarian needs. The fuel shortage has had a profound impact on Gaza's wastewater treatment infrastructure. Without a steady fuel supply, pumping stations cannot transport sewage to treatment facilities, leading to the uncontrolled discharge of untreated wastewater (approximately 60,000 m^3) into streets, open areas, and the Mediterranean Sea (Khalidi 2023; Amnesty International 2024; UNDP/PAPP 2024; UNEP 2024). Due to the porous soil of the Gaza Strip, it heightens the risk of sewage contaminating groundwater. The potential seepage of sewage, which may contain pathogens and chemical pollutants, into the aquifer presents a significant health hazard for residents drawing untreated water directly from wells (Shomar and Rovira 2023; PAX 2024; UNEP 2024). The largest water desalination facility in Gaza, located in Deir al Balah, has the capacity to provide 90 L of clean water per person to 275,000 residents. The plant requires 12 MW of power, consuming approximately 3200 L of diesel per hour (Rettig 2024). However, fuel shortages caused by the war have severely impacted operations, preventing the facility from functioning at full capacity and worsening the water crisis in the region.

The recent war on Gaza has led to a near-total collapse of the solid waste management system, primarily due to the severe fuel shortages that have crippled operations. The inaccessibility of Gaza's two main landfills—Al-Fukhary and Johr Al-Deek—has resulted in the accumulation of nearly half a million tons of solid waste in streets and makeshift shelters, creating hazardous living conditions for displaced populations. Waste collection has been drastically reduced, with 80–90% of waste collection vehicles either destroyed or severely damaged. The lack of fuel has also forced municipalities to halt waste transfer to landfills, leading to the proliferation of unregulated dumping sites that pose serious health risks, such as the spread of vector-borne diseases and contamination of groundwater. Immediate fuel supplies and restoration of waste management infrastructure are crucial to mitigating the escalating public health and environmental crisis (UNDP/PAPP 2024).

The widespread failure of wastewater treatment plants and pumping stations in Gaza due to fuel shortages and electricity blockades highlights the critical importance of SDG 7, particularly Target 7.1. The inability to power essential infrastructure has led to severe environmental and health risks, including the uncontrolled discharge of untreated wastewater, contamination of groundwater, and the spread of diseases. Restoring energy access is crucial to re-establishing wastewater management systems and safeguarding public health.

7.4 Adapting to Catastrophe with Alternative Energy Solutions

Electricity supply was disrupted entirely during the recent Israeli war on Gaza. In response, many residents turned to renewable energy, particularly solar panels, which had gained significant popularity even before the war due to frequent outages and the high costs associated with diesel generators and street-sourced electricity. In displacement camps, the primary energy needs shifted to lighting, charging batteries, mobile phones, laptops, and limited commercial uses by vendors. Solar energy emerged as a prevalent alternative, utilizing various types and sizes of solar panels depending on availability. However, the blockade on Gaza prevented the import of such materials and equipment. The price of diesel ranged between $30 and $15 per liter, making the continued use of surviving diesel-powered generators an unsustainable option. This represented a tenfold increase in cost compared to pre-war levels.

In the Al-Buraq displacement camp in Mawasi Khan Younis, researchers observed several families collectively investing in two solar panels to create a shared energy network. This system was used to meet their communal energy needs, including lighting, charging batteries, mobile phones, laptops, and essential electrical devices. An interview with the owner of the system provided several key insights. Approximately 10 families, totaling 70 individuals, depended on the two solar panels for their energy needs. The system enabled them to charge batteries and multiple mobile phones efficiently, offering a cost-effective and time-saving alternative to other charging options. The cost of the system increased significantly during the war, reaching approximately 15,000 shekels (~$4000), compared to just $500 before the war. The solar panels were typically installed on tents or shared facilities, such as restrooms, to supply energy for the participating families.

The owner, Mr. Mohammed Ziad, allowed us to photograph the system from both inside and outside the tent (Fig. 7.8). Entrepreneurs in displacement camps created businesses using one or more solar panels to provide charging services for mobile phones, power banks, and small batteries. These services were offered for a fee—approximately 1 shekel ($0.27) for mobile phones, 2 shekels ($0.54) for power banks, and 3.5 shekels ($1) for small batteries. This initiative generated modest income for two workers and provided a sustainable livelihood.

Expanding these systems with additional panels and charging infrastructure raised project costs to approximately $7000. Entrepreneurs also incorporated solar-powered refrigerators to sell cold water, ice, and ice cream to displaced individuals. Selling ice cream proved to be a highly profitable venture, along with charging services, which operated between 10:00 AM and 4:00 PM, depending on sunlight availability.

Clean energy-based businesses demonstrated significant economic viability during the war, especially with the absence of traditional electricity. For example, the price of a bag of ice increased from 0.5 shekels to 3 shekels, while ice cream rose from 0.5 shekels to 3 shekels, and a bag of carob juice, previously sold for 1 shekel,

Fig. 7.8 A picture from inside the tent showing the charging process in the household use of clean energy and a picture of the clean energy generation system from outside the tent, where two solar panels were placed on top of the shared bathroom in the Al-Buraq camp

reached 5 shekels. These ventures, including water sales with minimal operational costs, became highly profitable. Additional services such as cooling, grinding, and production of essential items further contributed to their success, making these businesses essential for displaced individuals.

Observations and interviews highlighted the profitability of these initiatives and their critical role in supporting displaced communities. This analysis emphasizes the resilience and adaptability of the population in Gaza in utilizing renewable energy to address energy shortages, sustain livelihoods, and meet basic needs during the crisis.

7.5 Environmental Degradation Caused by Energy Insecurity

The recent war on Gaza has severely intensified environmental degradation due to persistent energy shortages, primarily caused by the destruction of critical infrastructure and reliance on diesel generators. A total of 42,500 square meters of solar panels were destroyed, causing significant damage to the renewable energy infrastructure in Gaza and further worsening the energy crisis in the region. The widespread damage to solar installations, which previously supplied power to 20% of households, has resulted in significant environmental contamination (FAO/UN 2024; UNEP 2024). Damaged solar panels release hazardous substances such as lead and cadmium, which can leach into the soil and water, posing long-term risks to agricultural productivity and drinking water quality. Additionally, the burning of damaged panels releases toxic chemicals and heavy metals into the air, further deteriorating air quality and contributing to long-term ecological harm (Keerthivasan et al. 2024). The recent war has turned these once-promising renewable energy solutions into sources of pollution, adding to Gaza's mounting environmental challenges.

Fuel shortages have forced Gaza to rely heavily on diesel generators, which have become the primary source of electricity for homes, hospitals, and businesses. Gaza's only power plant consumes approximately 17,500 L of diesel per hour to meet just 25% of the region's electricity demand (Rettig 2024), while thousands of smaller generators scattered across the Strip contribute further to environmental degradation. The combustion of diesel releases high levels of pollutants, including carbon dioxide, nitrogen oxides, and sulfur dioxide, exacerbating climate change and significantly deteriorating air quality (Perera 2018). In addition to airborne pollutants, oil spills and fuel leaks from generators seep into the soil and groundwater, posing severe risks to Gaza's already fragile water resources.

The energy crisis has also severely disrupted Gaza's waste management system, leading to an accumulation of nearly half a million tons of solid waste in streets and makeshift shelters. The inaccessibility of Gaza's two main landfills—Al-Fukhary and Johr Al-Deek—due to infrastructure damage and fuel shortages has resulted in widespread unregulated dumping. Without proper waste disposal, hazardous medical waste and toxic materials from destroyed infrastructure, including asbestos and heavy metals, are being mixed with municipal waste, further contaminating the environment (UNDP/PAPP 2024; UNEP 2024).

Untreated sewage is being discharged directly into the Mediterranean Sea due to energy shortages affecting wastewater treatment facilities. This pollution not only threatens marine biodiversity but also increases the risk of groundwater contamination, which could have severe long-term consequences for the already depleted water supply in Gaza (Tahmid et al. 2023; Hall et al. 2024).

Another major environmental concern is the destruction of vegetation and agricultural land due to the lack of energy needed for irrigation and farming operations. Farmers, unable to pump groundwater or operate irrigation systems, face declining soil fertility and increased desertification, which further impacts food security and contributes to ecosystem degradation. Without access to sufficient energy, crop production has significantly declined, leading to increased reliance on imported food and further straining Gaza's economy and natural resources. Additionally, military activity and energy shortages have led to the over-extraction of groundwater, further depleting the aquifer and increasing salinity levels, which negatively impacts soil quality and agricultural productivity (Buheji and Al-Muhannadi 2023; UNEP 2024; Holail et al. 2024).

The environmental degradation caused by energy insecurity in Gaza directly undermines SDG 7, particularly Target 7.2, which aims to increase the share of renewable energy in the global energy mix. The destruction of solar panel infrastructure, which previously provided sustainable energy to thousands of households, has forced a return to reliance on polluting diesel generators, exacerbating environmental harm. Additionally, the energy crisis has disrupted essential services such as waste management and wastewater treatment, highlighting the critical need for reliable, clean energy solutions to mitigate environmental damage and ensure sustainable development in the region.

7.6 Critical Health Impacts of War-Induced Energy Shortages in Gaza

The recent war in Gaza has severely damaged energy infrastructure, resulting in widespread shortages with profound implications for public health. The lack of reliable energy supply has disrupted essential healthcare services, limiting access to life-saving treatments and critical medical equipment.

7.6.1 Impact on Healthcare Systems

Energy is fundamental to operating critical healthcare infrastructure. In Gaza, the reliance on inconsistent power supplies and diesel generators has left hospitals struggling to maintain operational standards. For instance, only 13 out of 36 hospitals are operational due to energy deficits, significantly reducing the healthcare capacity in Gaza (Ashour et al. 2024). The hospital faces a critical risk of shutting down due to a severe fuel shortage, jeopardizing its ability to sustain electricity. This crisis has already resulted in numerous tragic incidents, further endangering the lives of patients and straining medical operations (Médecins Sans Frontières 2024; Perugini 2024; Abed Alah 2024).

7.6.2 Increase in Infectious Diseases Due to Sanitation Failures

The link between energy shortages and compromised water sanitation systems has resulted in a significant increase in infectious diseases. Limited access to safe drinking water and critically low water availability in refugee camps have heightened the risk of waterborne illnesses, such as diarrhea, dysentery, and cholera (Farajallah and Farajallah 2024). According to the WHO Public Health Situation Analysis, the inability to operate essential water treatment facilities due to power failures results in unsafe drinking water, contributing to 577,000 cases of diarrheal diseases and a substantial outbreak of acute jaundice syndrome, affecting 107,000 people (OCHA 2024b).

7.6.3 Impact on Vulnerable Populations

Children and the elderly in Gaza are disproportionately affected by the health impacts of energy shortages. The lack of refrigeration facilities due to power cuts threatens vaccine storage, increasing the risk of vaccine-preventable diseases among children (ESCWA 2023; Younis and Tunç 2023). Fuel and aid restrictions in Gaza have put hospitals in a dire situation, endangering children of all ages, including newborns relying on incubators powered by fuel-dependent generators. At Al Shifa hospital, a power outage on November 11 and 12, 2023, forced the relocation of 37 premature babies to an operating room without incubators, and by November 13, six had reportedly died (Arana-Cedeño 2024; Boukari et al. 2024).

Similarly, the elderly and patients with chronic diseases face heightened risks as healthcare providers struggle to offer consistent care and manage chronic conditions effectively, with approximately 350,000 patients with chronic conditions like diabetes and hypertension not receiving necessary medications due to disruptions in the supply chain, which limits routine care for chronic diseases due to the facilities being overwhelmed by war-related injuries (Yang 2024; Ashour et al. 2024).

The disruption of energy infrastructure has severely impacted healthcare facilities, hindering access to life-saving treatments and essential medical equipment. This crisis directly undermines the targets of SDG 7 by highlighting the urgent need for reliable, affordable, and modern energy services to support critical healthcare operations.

7.7 Mitigation Measures and Recommendations

Mitigation measures and recommendations to address energy challenges in Gaza focus on both immediate needs and long-term solutions. Addressing urgent energy demands requires facilitating immediate fuel imports to sustain hospitals, water facilities, and other essential services. International organizations must prioritize funding and logistics to ensure the continued operation of these critical sectors. Additionally, distributing small-scale solar power systems for lighting, phone charging, and basic appliances in displacement camps and households can provide short-term relief and reduce dependence on costly diesel generators. Efforts should also concentrate on the rehabilitation of damaged energy infrastructure, with a particular focus on hospitals and water treatment facilities, to mitigate this humanitarian crisis.

Enhancing energy resilience requires investment in renewable energy projects, such as expanding solar microgrids for schools, hospitals, and community centers to reduce reliance on fossil fuels and provide stable power. Implementing energy efficiency programs, including the distribution of energy-efficient appliances and lighting, can lower consumption and maximize limited resources during crises. Furthermore, developing local energy storage solutions, such as battery systems, will enhance the reliability of renewable energy sources by ensuring continuous power availability during peak demand or periods of limited sunlight.

Policy and international cooperation efforts should focus on advocating for the relaxation of import restrictions on renewable energy equipment, spare parts, and fuel to facilitate recovery efforts in Gaza. International funding and support are crucial for rebuilding energy infrastructure through partnerships between global organizations, NGOs, and local authorities. Providing technical training programs for local engineers and technicians is essential to maintain and expand renewable energy solutions, ensuring long-term sustainability.

Environmental and health considerations must also be addressed by gradually phasing out diesel generators and promoting cleaner alternatives like solar power to reduce air pollution and associated health risks. Waste management programs should be introduced to tackle pollution caused by damaged energy infrastructure and ensure the safe disposal of hazardous materials, such as those from solar panels and generators. Energy recovery efforts should align with public health objectives, as reliable access to electricity is vital for the operation of healthcare facilities and water treatment plants, which in turn helps reduce disease outbreaks.

In the long term, rebuilding renewable energy infrastructure through large-scale projects, such as solar farms and wind turbines, can meet energy needs sustainably while creating job opportunities and reducing environmental degradation. Exploring opportunities for regional energy cooperation, including cross-border energy sharing and joint renewable initiatives, can help stabilize the energy supply in Gaza and foster regional integration. To ensure the effectiveness of these interventions, monitoring and evaluation frameworks should be established to assess progress and guide future strategies.

7.8 Conclusion

The energy crisis in Gaza, exacerbated by war and prolonged blockade, has severely impacted every aspect of life, from healthcare and water access to economic stability and environmental sustainability. The destruction of conventional and renewable energy infrastructure has left millions without reliable electricity, deepening reliance on unsustainable alternatives that are costly and environmentally damaging. Despite these challenges, the resilience and adaptability of Gaza's population have been evident, with people turning to renewable energy solutions such as solar power to address immediate needs and sustain livelihoods. However, these efforts remain insufficient to meet the large-scale energy demands of Gazans.

Achieving sustainable energy solutions in Gaza requires urgent international cooperation, investment in renewable energy infrastructure, and policy reforms to ease import restrictions on essential energy supplies and equipment. Expanding the use of solar energy and other renewable sources, alongside improving energy efficiency and storage capacity, is crucial to reducing dependence on fuel imports and mitigating environmental degradation. Addressing these challenges is essential to achieving SDG 7.

Looking ahead, a comprehensive strategy is essential to rebuild Gaza's energy sector with a focus on long-term sustainability and resilience. This strategy must include enhancing regional energy cooperation, fostering local capacity-building efforts, and developing robust monitoring frameworks to track progress and ensure the effectiveness of energy interventions. The successful recovery of the energy sector also hinges on Israel's commitment to respecting the ceasefire implemented on January 19, 2025, and recognizing the Palestinian people's fundamental rights to energy, water, and a dignified life. A coordinated and sustained effort, grounded in justice and equitable access to resources, is crucial to restoring energy security in Gaza and supporting broader recovery and development across the Strip.

References

Abed Alah M (2024) Echoes of conflict: the enduring mental health struggle of Gaza's healthcare workers. Confl Heal 18:1–6. https://doi.org/10.1186/S13031-024-00577-6/METRICS

Abu Riash A, Humaid M (2023) As people run out of fuel in Gaza, clay ovens are back in demand. Al Jazeera. https://www.aljazeera.com/gallery/2023/11/21/traditional-clay-ovens-in-demand-as-fuel-shortages-continue-in-gaza?utm_source=chatgpt.com. Accessed 18 Jan 2025

Ahmed AK (2023) Israeli authorities' cutting of water leading to public health crisis in Gaza. Hum Rights Watch https://www.hrw.org/news/2023/11/16/israeli-authorities-cutting-water-leading-public-health-crisis-gaza. Accessed 11 Apr 2024

Alagha M, Hussein N (2024) The political economy of Israeli economic warfare on the Gaza strip. In: Elkahlout G (ed) Gaza's cycle of destruction and rebuilding. Palgrave Macmillan, Cham, pp 125–145

Alnabih A, Alnabeh N, Aljeesh Y, Aldabbour B (2024) Food insecurity and weight loss during wartime: a mixed-design study from the Gaza strip. J Health Popul Nutr 43:222

Amnesty International (2024) 'You feel like you are subhuman'—Israel's genocide against Palestinians in Gaza

Arana-Cedeño M (2024) Disease and death: the shock wave of collective punishment against Gaza. Soc Med Pract 17:39–43

Arora NK, Mishra I (2022) Progress of sustainable development goal 7: clean and green energy for all as the biggest challenge to combat climate crisis. Environ Sustain 5:395–399. https://doi.org/10.1007/s42398-022-00257-2

Ashour Y, Jlambo A, Abuzerr S (2024) Patients in Gaza with chronic conditions need urgent interventions. Lancet 403:1847–1848. https://doi.org/10.1016/S0140-6736(24)00705-0

Awosusi AA, Ozdeser H, Seraj M, Adegboye OR (2024) Achieving carbon neutrality in energy transition economies: exploring the environmental efficiency of natural gas efficiency, coal efficiency, and resources efficiency. Clean Techn Environ Policy. https://doi.org/10.1007/s10098-024-02932-w

Bishara A, Elbasri A (2024) Genocide in Gaza: The Ongoing Nakba

Blouin S, Herwix A, Rivers M et al (2024) Assessing the impact of catastrophic electricity loss on the food supply chain. Int J Disaster Risk Sci 15:481–493. https://doi.org/10.1007/s13753-024-00574-6

Boukari Y, Kadir A, Waterston T et al (2024) Gaza, armed conflict and child health. BMJ Paediatr Open 8:e002407. https://doi.org/10.1136/BMJPO-2023-002407

Buheji M, Al-Muhannadi K (2023) Mitigating risks of environmental impacts on Gaza—review of precautions & solutions post (2023 war). Int J Adv Res Eng Technol 14:15–47

Buheji M, Marouf M (2024) Mapping reconstruction of GAZA to overcome donors' fatigue. Gradiva 63:116–130. https://doi.org/10.5281/zenodo.13148160

Dardona Z, Amane M, Dardona A, Boussaa S (2024) Health and environmental impacts of Gaza conflict (2023-2024): a review. One Heal Bull. https://doi.org/10.4103/ohbl.ohbl

Elkahlout G (2024) Sisyphus but the Boulder gets a little heavier every time. In: Elkahlout G (ed) Gaza's cycle of destruction and rebuilding. Palgrave Macmillan, pp 1–16

El-Khozondar HJ, Asfour AA, Nassar YF et al (2024) Photovoltaic solar energy for street lighting: a case study at Kuwaiti roundabout, Gaza strip, Palestine. Power Eng Eng Thermophys 3:77–91. https://doi.org/10.56578/peet030201

Elnaggar M (2023) Useful energy, economic and reduction of greenhouse gas emissions assessment of solar water heater and solar air heater for heating purposes in Gaza. Palestine Heliyon 9:e16803. https://doi.org/10.1016/J.HELIYON.2023.E16803

Ersoy SR, Terrapon-Pfaff J, Brik I (2022) Sustainable transformation of Palestine's energy system: development of a phase model. Friedrich-Ebert-Stiftung-Wuppertal-Institut

ESCWA (2023) War on Gaza: weaponizing access to water, energy and food

Eshtawi TA, Abdellatif M (2024) Spatial vulnerability assessment for water service providers: a case study from the Gaza strip. GeoJournal 89:1–15. https://doi.org/10.1007/S10708-024-11016-7/FIGURES/10

FAO/UN (2024) Agricultural damage assessment in the Gaza strip from October 7th 2023 to September 1st 2024

Farajallah I, Farajallah H (2024) Challenges in access to safe drinking water and its impact on maternal and child health in Gaza. Anatol Clin J Med Sci:52–65. https://doi.org/10.21673/anadoluklin.1575390

Fathi Nassar Y, Yassin Alsadi S (2019) Assessment of solar energy potential in Gaza Strip-Palestine. Sustain Energy Technol Assess 31:318–328. https://doi.org/10.1016/j.seta.2018.12.010

Fischhendler I, Herman L, David L (2022) Light at the end of the panel: the Gaza strip and the interplay between geopolitical conflict and renewable energy transition. New Polit Econ 27:1–18. https://doi.org/10.1080/13563467.2021.1903850

Fischhendler I, Herman L, David L (2023) Gaza's solar power in wartime. Cent Strateg Int Stud 27:1–18. https://doi.org/10.1080/13563467.2021.1903850

Gordon N, Haddad M (2024) The road to famine in Gaza. New York Rev Book 1:1–12

Hall N, Kirschenbaum A, Michel D (2024) The siege of Gaza's water. Cent Strateg Int Stud

Holail S, Saleh T, Xiao X et al (2024) Time-series satellite remote sensing reveals gradually increasing war damage in the Gaza strip. Natl Sci Rev 11:nwae304. https://doi.org/10.1093/NSR/NWAE304

Human Rights Watch (2024) Extermination and acts of genocide Israel deliberately depriving Palestinians in Gaza of water

IEA, IRENA, UNSD, World Bank (2024) Tracking SDG 7 the energy: Progress report 2024. Washington, DC

iMEdD Lab (2024) Gaza: Electricity and water threatened with collapse. https://lab.imedd.org/en/gaza-electricity-and-water-threatened-with-collapse/?utm_source=chatgpt.com. Accessed 19 Dec 2024

IPC (2024) GAZA STRIP: IPC acute food insecurity special snapshot

Juaidi A, Anayah F, Assaf R et al (2022) An overview of renewable energy strategies and policies in Palestine: strengths and challenges. Energy Sustain Dev 68:258–272. https://doi.org/10.1016/j.esd.2022.04.002

Keerthivasan T, Madhesh R, Srinivasan M, Ramasamy P (2024) Photovoltaic recycling: enhancing silicon wafer recovery process from damaged solar panels. J Mater Sci Mater Electron 35:880. https://doi.org/10.1007/s10854-024-12662-w

Khalidi B (2023) The unfolding water catastrophe in Gaza. Oxfam https://views-voices.oxfam.org.uk/2023/11/water-catastrophe-gaza/. Accessed 28 Dec 2024

Kumar P, Upadhyay E, Kannan K, Yadav A (2024) Evaluation of cooking practices in India to achieve SDGs 7.1.2. Phys Chem Earth 136:103774. https://doi.org/10.1016/j.pce.2024.103774

Médecins Sans Frontières (2024) Gaza: Life in a death trip

Middle East Monitor (2024) Gaza estimates $450m losses for electricity sector. https://www.middleeastmonitor.com/20241217-gaza-estimates-450m-losses-for-electricity-sector/. Accessed 18 Jan 2025

Nassar YF, Alsadi SY (2016) Economical and environmental feasibility of the renewable energy as a sustainable solution for the electricity crisis in the Gaza strip. Int J Eng Res Dev 12:35–44

Nassar YF, El-Khozondar HJ, Elnaggar M et al (2024) Renewable energy potential in the State of Palestine: proposals for sustainability. Renew Energy Focus 49:100576. https://doi.org/10.1016/j.ref.2024.100576

OCHA (2024a) Electricity in the Gaza Strip. https://www.ochaopt.org/page/gaza-strip-electricity-supply. Accessed 18 Jan 2025

OCHA (2024b) Gaza humanitarian response update | 22 July-4 August 2024. https://www.unocha.org/publications/report/occupied-palestinian-territory/gaza-humanitarian-response-update-22-july-4-august-2024. Accessed 19 Jan 2025

Palestinian Energy and Natural Resources Authority (2024) Cumulative periodic report issued by the Gaza Electricity Distribution Company (GEDCO), October 7, 2024. https://www.facebook.com/Gedcops. Accessed 18 Jan 2025

PAX (2024) War and Garbage in Gaza, The public health and environmental crisis from widespread solid waste pollution

Perera F (2018) Pollution from fossil-fuel combustion is the leading environmental threat to global pediatric health and equity: solutions exist. Int J Environ Res Public Health 15:2–17. https://doi.org/10.3390/ijerph15010016

Perugini N (2024) Medical Lawfare: the Nakba and Israel's attacks on Palestinian healthcare. J Palest Stud 53:68–91. https://doi.org/10.1080/0377919X.2024.2330366

Procter C (2024) Coerced migration: mobility under siege in Gaza. J Ethn Migr Stud 50:2359–2383. https://doi.org/10.1080/1369183X.2024.2312229

Qutaina B, Shehada A, Yasin A, Alsayed M (2023) Geographical information systems based site selection methodology for renewable energy systems in Palestinian territories. Int J Electr Comput Eng 13:3622–3630. https://doi.org/10.11591/ijece.v13i4.pp3622-3630

Rettig E (2024) Fuel supply in Gaza—needs and uses. BESA Geo Energy Insight:1–5. https://doi.org/10.2139/ssrn.4742438

Rettig E, Spanier B (2024) Striking energy deals in disputed seas: the case of the Gaza marine gas field. J World Energy Law Bus. https://doi.org/10.1093/JWELB/JWAD039

Salah WA, Abuhelwa M (2020) Energy status and practices for efficient energy management to reduce power interruptions: a case study on Tulkarm district in Palestine. Int J Sustain Energy 39:685–699. https://doi.org/10.1080/14786451.2020.1748630

Salah WA, Abuhelwa M, Bashir MJ (2021) The key role of sustainable renewable energy technologies in facing shortage of energy supplies in Palestine: current practice and future potential. J Clean Prod 293:125348. https://doi.org/10.1016/j.jclepro.2020.125348

Salem MZ, Ertz M, Sarigöllü E (2021) Demarketing strategies to rationalize electricity consumption in the Gaza strip-Palestine. Renew Sust Energ Rev 143. https://doi.org/10.1016/j.rser.2021.110956

Shomar B, Rovira J (2023) Human health risks associated with the consumption of groundwater in the Gaza strip. Heliyon 9:e21989. https://doi.org/10.1016/J.HELIYON.2023.E21989

Suzuki H, Sakai K (2024) Gaza Nakba 2023–2024: background, context, consequences. Springer

Tahmid A, Khanam S, Rashid MM, Ibnat F (2023) Reviewing the impact of military activities on marine biodiversity and conservation: a study of the Ukraine-Russia conflict within the framework of international law. Grassroots J Nat Resour 6:15–31. https://doi.org/10.33002/nr2581.6853.060302

Tannira A (2021) The political economy ofthe Gaza strip under Hamas. In: Tartir A, Dana T, Seidel T (eds) Political economy of Palestine. Palgrave Macmillan, pp 135–154

The World Bank (2024a) Gaza strip—interim damage assessment

The World Bank (2024b) Impacts of the conflict in the middle east on the Palestinian economy-December 2024

UNDP (2021) Gaza 2021 infrastructure damage assessment report
UNDP/PAPP (2024) Impact of the war on solid waste management in the Gaza strip
UNEP (2024) Environmental impact of the conflict in Gaza: preliminary assessment of environmental impacts. Nairobi
Wei F (2023) Exclusive: satellite images show Gaza in complete darkness after power supply cutoff, indicating worsening humanitarian crisis. Global Times. https://www.globaltimes.cn/page/202310/1299844.shtml. Accessed 18 Jan 2025
Yang YT (2024) Protecting humanitarian aid workers: an urgent call. Lancet 403:1847. https://doi.org/10.1016/S0140-6736(24)00854-7
Younis SMS, Tunç G (2023) Investigation of electricity distribution, shortages in the Gaza strip, and the role of renewable energy technologies for addressing the electricity crisis. Int J Energy Stud 8:557–579. https://doi.org/10.58559/ijes.1317267

8 Domicide in Gaza: How the Recent War Has Devastated Homes and Lives

Abdo Hassoun, Mariam Fayad, Nida Wasim, Khansa Irfan, Rima Kashash, Khaled Abu Ali, Mones Abu Obayd Allah, and Chantal Youssef

Abstract

This chapter explores the catastrophic impact of domicide—the deliberate destruction of homes—during the recent war (October 2023–January 2025) on Gaza, highlighting the physical, social, and environmental tolls. It examines how systematic urban destruction has rendered vast areas uninhabitable, displaced nearly the entire population, and devastated critical infrastructure, including housing, healthcare, and education facilities. The chapter provides detailed accounts of the psychological trauma, economic collapse, and environmental degradation resulting from this unprecedented crisis, while drawing attention to the implications for sustainable urban development. The discussion aligns with the goals of Sustainable Development Goal 11 (SDG 11), emphasizing the urgent need to rebuild inclusive, resilient, and sustainable communities in Gaza. By leveraging grey literature and peer-reviewed publications, the chapter underscores the necessity of international intervention and coordinated efforts to ensure justice, recovery, and long-term resilience for the affected population.

Keywords

Community resilience · Cultural heritage · Environmental degradation · Gaza reconstruction · Housing crisis · SDG 11 · Sustainable development · Urbanization

A. Hassoun (✉)
Sustainable AgriFoodtech Innovation & Research (SAFIR), Arras, France

Faculty of Agricultural Engineering, University of Aleppo, Aleppo, Syria
e-mail: a.hassoun@saf-ir.com

M. Fayad
Independent Researcher, Gaza, Palestine

N. Wasim
Centre of Excellence for Trauma and Emergencies, Aga Khan University, Karachi, Pakistan
e-mail: Nida.Wasim1@alumni.lshtm.ac.uk

K. Irfan
School of Medicine and Dentistry, University of Central Lancashire, Preston, Lancashire, UK
e-mail: KIrfan@uclan.ac.uk

R. Kashash
Faculty of Health Sciences, American University of Beirut, Beirut, Lebanon

K. A. Ali
Medical sciences department, University College of Science and Technology, Gaza, Palestine

M. A. O. Allah
Gaza Team, Gaza, Palestine

C. Youssef
Saint Joseph University of Beirut (USJ), Beirut, Lebanon

A. Hassoun (ed.), *War on Gaza*, Sustainable Development Goals Series,
https://doi.org/10.1007/978-3-031-88500-6_8

8.1 Introduction

Today, the world faces significant challenges in urban development, housing security, and the protection of vulnerable communities, as rapid urbanization, escalating displacement from conflicts and natural disasters, and growing inequalities have left millions exposed to precarious living conditions. These challenges strain infrastructure, disrupt livelihoods, and erode social cohesion, particularly in regions affected by chronic instability. To tackle these challenges, global frameworks such as the Sustainable Development Goals (SDGs) were established, with SDG 11 specifically targeting the creation of inclusive, resilient, and sustainable communities (da Silva Costa et al. 2023; Dall'Agnolo et al. 2025).

For decades, Gaza has symbolized the intersection of geopolitical conflict and profound human suffering. The repeated destruction of homes during repeated Israeli assaults over the past two decades has deepened the challenges faced by displaced populations. These actions go beyond immediate material loss, inflicting psychological trauma, perpetuating generational displacement, and undermining social cohesion (Agha et al. 2024; Azzouz 2024; Elkahlout 2024; Nijim 2024). The long-term implications, including the immense difficulties of recovery and reconstruction, highlight the far-reaching toll of war on both individual lives and societal systems. Gaza's prolonged blockade and recurring violence have driven a process of de-development, systematically dismantling its economic and social structures (Asi 2020; Alhamawi 2024). Described as the world's largest open-air prison (Jabr and Berger 2022), the region remains trapped in a cycle of destruction and deprivation, with limited access to resources, freedom of movement, and opportunities for sustainable growth (Elkahlout 2024; Klin and Clin 2024).

The recent war on Gaza stands as a stark reminder of the profound human cost of conflict, with the destruction of homes and infrastructure inflicting deep and lasting consequences on individuals, families, communities, and their homes and houses. Housing, beyond being a basic necessity, embodies security, identity, and stability. However, in Gaza, where prolonged blockades and recurrent violence have already strained resources and resilience, the large-scale targeting of residential areas has exacerbated the suffering of a population already living on the edge of survival (Amnesty International 2024; Asi et al. 2024; Human Rights Watch 2024; Milton et al. 2024). The deliberate demolition of homes in Gaza has triggered an unprecedented humanitarian crisis, displacing 1.9 million inhabitants (Kerbaj et al. 2024) and unraveling the social fabric of communities. This act of "domicide" has not only deprived individuals of shelter but also eroded their sense of safety, belonging, and continuity (OHCHR 2023; Ben Saad 2024; Sultany 2024).

This chapter examines the devastating impact of the recent genocidal war on Gaza, focusing on the physical, emotional, and societal consequences of widespread destruction of homes, communities, and critical infrastructure. It aims to provide a comprehensive understanding of the crisis and its broader implications for international humanitarian discourse and sustainable development. By highlighting the urgent need to end the war and rebuild Gaza, the chapter advocates for sustainable urban development, environmental restoration, cultural heritage preservation, and community-driven recovery as essential pillars of a resilient and equitable future.

8.2 Understanding "Domicide"

Having a home is a fundamental human right, often defined and interpreted in various ways. It provides security, psychological safety, peace, and dignity, and serves as a foundation for family gatherings and social connections. It is someone's comfort zone and source of life continuity. The destruction of home highlights its profound significance, as it leads to severe negative impacts on individuals and communities. This understanding has shaped the evolving concept of "domicide," which refers to the deliberate destruction of homes and its far-reaching consequences. Domicide is defined as the devastation of large homes and communities, resulting in human hardship and displacement (Akesson

2014; Akesson et al. 2016; Rajagopal and Pangalangan 2022; Azzouz 2024). The Canadian geographers Porteous and Smith came up with this idea for the first time in 2001 (Porteous and Smith 2001). Originating from the Latin words "domus" (home) and "caedo" (to murder), the term "domicide" refers to both the deliberate demolition of actual homes and the systemic violation of housing rights. The word "domicide," which first appeared in the study of geography, has since been expanded in subsequent research to encompass a variety of processes about housing concerns and life-threatening circumstances.

Domicide was first linked to conflicts and the deliberate devastation of marginalized populations, which resulted in significant casualties. However, the concept has been expanded by Mell Nowicki to include more social impacts of home and degradation along with slow violence (Petković 2023). Throughout history, cities and buildings have consistently been targets of attacks. As wars and conflicts are often fought within urban settings, they result in widespread destruction of economic assets and critical infrastructure, along with the displacement or loss of lives to facilitate occupation. This deliberate damage profoundly impacts individuals, causing significant psychological distress and disrupting their sense of stability and belonging (Azzouz 2023). Housing is a fundamental human right, and forced evictions represent a clear violation of international humanitarian law. This is enshrined in Article 17 of the International Covenant on Civil and Political Rights, which states: *"No one shall be subjected to arbitrary or unlawful interference with their privacy, family, home, or correspondence, nor to unlawful attacks on their honor and reputation"* (UN-HABITAT 2014). Such violations are often intentional, especially during conflicts, where housing is targeted as a weapon of war.

The deliberate destruction of homes, or domicide, has a long history, from the German military's scorched-earth tactics in Eastern Europe to the atomic bombings of Hiroshima and Nagasaki during World War II, and more recently, the systematic destruction of homes in Myanmar and during the Bosnian war, where it was used as an ethnic cleansing tool. Domicide is not only a deliberate wartime strategy but also a systemic issue, evident in peacetime policies such as forced evictions, property confiscation, and laws preventing displaced populations from returning to their homes (Tuathail and Dahlman 2006; UN-HABITAT 2014; Rajagopal and Pangalangan 2022).

Despite its profound consequences, domicide has not been adequately recognized as a standalone crime under international law, leaving significant gaps in addressing housing-related violations. Moreover, selective accountability further undermines justice, as international bodies like the International Criminal Court have prioritized some crises, such as Ukraine, while neglecting others, like Afghanistan and Palestine. This inconsistency weakens the integrity of international law and its ability to protect vulnerable populations (Rajagopal and Pangalangan 2022; Mahomed 2024).

The war on Gaza starkly illustrates how the systematic destruction of residential areas and infrastructure is overshadowed by broader geopolitical debates, effectively sidelining urgent calls for accountability. Despite mounting evidence and repeated international appeals, little has been done to address the catastrophic impact on civilian life and housing rights (Azzouz 2024). Even before the recent devastating war, UN experts have explicitly identified Israel's actions as acts of domicide, underscoring the deliberate targeting of homes and communities as a strategy that exacerbates displacement and humanitarian suffering (OHCHR 2023).

A clear and enforceable framework is urgently needed to hold perpetrators accountable for domicide under international law. Political impasses and selective justice have fostered impunity, undermining trust in global institutions. To address this, treaties and conventions must explicitly classify domicide as a war crime, and the International Criminal Court's mandate should expand to include environmental crimes linked to conflict. Establishing international norms to protect housing and the environment during war is essential to uphold justice and human dignity and ensure accountability for all violations, regardless of context (Akesson et al. 2016; Krever et al. 2024).

8.3 An Overview of SDG 11 and Its Targets

Sustainable Development Goal 11 (SDG 11) aims to make cities and human settlements inclusive, safe, resilient, and sustainable. It provides a comprehensive framework for addressing the challenges of rapid urbanization, inequality, and environmental degradation, offering pathways for creating urban environments that enhance the quality of life for all.

8.3.1 SDG 11 Targets

SDG 11 outlines several interconnected targets to guide sustainable urban development. Target 11.1 focuses on ensuring access to safe, affordable, and adequate housing, addressing the global housing crisis that affects marginalized populations. Target 11.2 promotes sustainable transport systems, emphasizing accessibility, safety, and efficiency, especially for vulnerable groups. Target 11.3 highlights the importance of inclusive and participatory urban planning, ensuring that diverse voices, including those of women and marginalized communities, shape urban development. Target 11.4 seeks to safeguard cultural and natural heritage, recognizing its role in preserving community identity and fostering social cohesion. Additionally, Target 11.5 emphasizes reducing disaster impacts, particularly for vulnerable populations, by enhancing urban resilience and preparedness. Urban environmental management is the focus of Target 11.6, which seeks to minimize pollution and improve waste management systems. Finally, Target 11.7 highlights the need for accessible, safe, and inclusive public spaces that promote recreation, well-being, and social interaction (Thomas et al. 2021; Berisha et al. 2022).

8.3.2 Impact of War and Conflicts on SDG 11

War and conflict profoundly hinder progress toward achieving SDG 11. In regions affected by armed conflicts, such as Ukraine, hostilities exacerbate vulnerabilities, disrupt urban systems, and displace millions of civilians. The destruction of housing, transport infrastructure, and public services severely affects quality of life, cutting off access to essential resources and leaving communities in prolonged crises.

For example, in Ukraine, widespread damage to urban centers and energy infrastructure has left entire cities without power, water, and other essential necessities. The destruction of cultural heritage further erodes community identity and social cohesion, complicating recovery efforts and reducing the sense of stability necessary for rebuilding efforts. Environmental degradation is another critical consequence. In Ukraine, conflicts have polluted rivers and farmland, damaging ecosystems and threatening food and water security. The collapse of waste management systems and industrial accidents caused by shelling exacerbate ecological harm, undermining sustainability efforts and creating long-term health risks. Economic stagnation, weakened governance, and the destruction of critical infrastructure compound the challenges of achieving urban resilience (Pereira et al. 2022; Pasinovych and Myskiv 2023; Shahini et al. 2024; Hryhorczuk et al. 2024).

Addressing these issues requires integrated strategies that balance immediate humanitarian needs with long-term investments in urban infrastructure, environmental restoration, and social cohesion. Without peace and stability, achieving the targets of SDG 11 in conflict-affected areas remains an elusive goal.

8.4 Gaza's Pre-War (Before October 2023) Urban Vulnerabilities

Since 2007, Israel's blockade on Gaza has imposed economic restrictions, movement controls, and material limitations, creating a severe humanitarian crisis. Recurrent escalations (2008, 2012, 2014, 2021, and 2023) have further destroyed homes, public facilities, and infrastructure, worsening socioeconomic conditions and urban vulnerabilities. The following sections examine these impacts in detail.

The Gaza Strip, with a population density of 5138 individuals per km^2 (Sabah 2023), has faced a longstanding housing crisis driven by rapid population growth (Abuelaish 2018), Israeli restrictions on construction materials, recurrent assaults, and limited land and resources (Milton et al. 2024). By 2017, the housing shortage had reached 120,000 units, doubling since 2012 (El-Sawalhi and Lafi 2019). The 2021 escalation caused $35 million in housing sector losses (UNDP 2021). Most residents live in overcrowded multistory buildings, often under unsafe conditions. A 2020 assessment found 28,500 housing units failed to meet minimum standards, with 7000 requiring reconstruction. Additionally, 118,000 people live in flood-prone ground-floor housing due to inadequate drainage systems (OCHA 2020). This persistent housing crisis severely undermines progress toward achieving SDG 11.1 by perpetuating overcrowding, unsafe living conditions, and inadequate urban services.

Palestinian movement in Gaza remains under Israeli control, with Israel granting or denying permissions for infrastructure projects (Weinthal and Sowers 2019; Zaqout et al. 2024). The Oslo Agreement enabled the establishment of a Palestinian airport in 1998, but it was closed and destroyed by Israel after only 2 years (El-Atrash et al. 2014; Daraj 2024). Frequent border closures, the absence of a seaport or airport, and restrictions on importing construction materials have severely hindered infrastructure development. Recurrent escalations, such as in 2021, caused over $22 million in damage to roads and transportation, affecting 240 roads, 1002 vehicles, and public facilities (UNDP 2021). Destruction often involves direct targeting, leaving many areas with unpaved or unfit roads (Weinthal and Sowers 2019; UNDP 2021). This lack of sustainable transport systems and infrastructure presents significant challenges and undermines progress toward achieving SDG 11.2.

Over the past two decades, Gaza has faced severe political and economic challenges that have hindered urban development. Since the election of the Hamas government in 2006, Israel's siege has imposed extensive restrictions on fuel, equipment, and materials, delaying or canceling developmental projects and strategic plans (Abukashif and Riza 2019; Weinthal and Sowers 2019; Zaqout et al. 2024). This has progressively deteriorated the economic situation for residents and weakened local institutions, including municipalities responsible for urban planning. Heavily reliant on external aid, municipalities struggle to deliver essential services and meet the growing demands of Gaza's rapidly increasing population (Abukashif and Riza 2019; UNDP 2021). These political and economic constraints on urban development have profoundly undermined progress toward achieving SDG 11.3.

The cultural heritage of Gaza, including mosques, churches, and ancient sites, remains highly vulnerable to destruction in an unstable region. Since the 2000s, repeated escalations have caused extensive damage to historical landmarks. The prolonged financial blockade has further exacerbated the neglect of the historical sector, limiting funding, expertise, and restoration efforts. Israel's prohibition of geolocation devices has also hindered proper management and preservation of these sites (Ammar and Alramlawi 2020; Center for Engineering and Planning (CEP) 2022; Andreou et al. 2024). The fragility of Gaza's cultural heritage underscores the urgent need to safeguard these landmarks, aligning with the vision of SDG 11.4.

The prolonged blockade and political divisions between Gaza's de facto government (Hamas) and the Palestinian Authority (Fatah) have resulted in fragile, fragmented institutions and a lack of resources, hindering disaster preparedness efforts (Tayeh et al. 2019; UNEP 2024). The absence of a comprehensive disaster preparedness plan significantly increases vulnerability to disasters and reduces resilience, underscoring the need for effective risk reduction measures (SDG 11.5).

The Gaza Strip faces severe environmental degradation, characterized by water contamination, waste mismanagement, and energy shortfalls. The primary water source, the aquifer, has suffered from excessive withdrawal, exposing groundwater to contamination from seawater and sewage. As of 2018, 97% of aquifer water was deemed unfit for human consumption (Seeberg

2024). The 2021 escalation caused extensive damage to 290 water and wastewater infrastructures, while sewage discharge into the sea and the use of cesspits in unconnected areas further pollute soil and groundwater, leading to serious health risks (UNDP 2021; CEOBS 2022). Daily electricity shortages, lasting 12–18 h, hinder the maintenance of essential infrastructure (UNDP 2021). Solid waste management is similarly strained, with 1726 tons of waste generated daily and inadequate systems leading to pollution and overflowing landfills like Johr Edeek. Poor recycling practices and improper medical waste disposal exacerbate contamination and increase health risks (UNEP 2024). These interconnected challenges underscore the critical need for sustainable solutions to mitigate environmental degradation and align with the vision of SDG 11.6.

From the previous analysis, it is clear that even before the recent war, Gaza was already grappling with severe challenges stemming from years of blockade, recurrent escalations, and systemic neglect. These deeply entrenched vulnerabilities have compounded the region's urban, environmental, and institutional fragility, leaving it ill-equipped to withstand or recover from further crises.

8.5 Domicide and the Scale of Urban Destruction in Gaza During the Recent War

Before the recent war, most of Gaza's 2.2 million people lived in its four main cities; Rafah and Khan Younis in the south, Deir al-Balah in the center, and Gaza City, but almost the whole population have now been displaced. Since October 7, 2023, almost 80% of Gaza has become uninhabitable due to widespread damage to homes, businesses, utility infrastructures, and educational and healthcare facilities. According to UNRWA, as of September 30, 2024, 90% of Gaza's population has been displaced due to the hostilities, with nearly 42,000 Palestinian deaths reported during the recent Israeli genocidal war on Gaza (UNRWA 2024). However, other credible studies and reports indicate that the actual number of victims could be significantly higher (Khatib et al. 2024; Jamaluddine et al. 2025), suggesting a more severe humanitarian catastrophe than officially documented. These figures exclude those still buried under rubble, as 70–80% of buildings have been destroyed or severely damaged, suggesting the actual death toll is significantly higher. Nowhere and no one is safe in Gaza, as the recent assaults have left every corner of the territory vulnerable to destruction and its people in constant danger (Asi et al. 2024; Iriqat et al. 2025).

Over the last year and during this prolonged war, the scale of urban destruction in the Gaza strip has been widely documented with geospatial analysis and the live footages from the journalist on the ground (AlJazeera 2024). In just 1 year of war, nearly 60% of Gaza has been damaged or destroyed, according to satellite analysis using Sentinel-1 radar data. The destruction is most severe in Gaza City (73.9%) and North Gaza (69.3%), with significant damage also reported in Deir El-Balah (49.1%), Khan Younis (54.5%), and Rafah (46.3%). The United Nations Institute for Training and Research (UNITAR) and its Operational Satellite Applications Programme (UNOSAT) provide an overview of the extensive destruction in Gaza as of December 2024, with 170,812 buildings damaged or destroyed across an area of 361 km^2 (Fig. 8.1a). Damage analysis highlights the severe impact on municipalities, with Beit Hanoun and Beit Lahiya particularly affected (UNITAR-UNOSAT 2024). The destruction includes civilian infrastructure, severely impairing living conditions and access to essential services. The humanitarian crisis is compounded by the collapse of basic infrastructure, underscoring the urgent need for international intervention and aid. The red zones vividly depict the widespread devastation and the scale of displacement resulting from the war.

UNITAR-UNOSAT's analysis illustrates the extensive destruction of road infrastructure in Gaza as of August 2024 (Fig. 8.1b), with significant damage detected across all governorates. Severely affected areas include North Gaza, Gaza City, and Khan Younis, where roads have been destroyed, heavily disrupting transportation

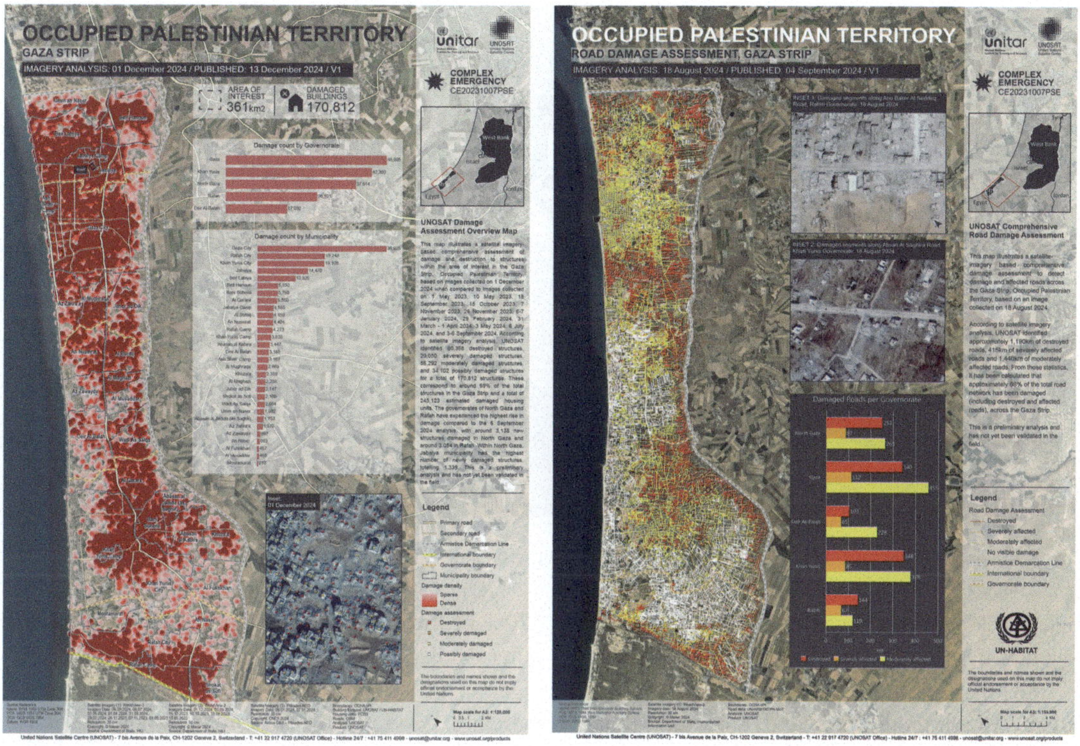

(a) (b)

Fig. 8.1 Assessment of damage and destruction to structures within the area of interest (**a**) and road damage assessment (**b**) in the Gaza Strip. (UNITAR-UNOSAT 2024)

and access to critical services. This widespread damage further exacerbates the humanitarian crisis, limiting the delivery of aid and movement of displaced populations.

In addition to the widespread destruction in Gaza, Israeli forces have further reduced Gaza's boundaries by creating new buffer zones. In May 2024, the Philadelphi Corridor, a nine-mile stretch along Gaza's southern border with Egypt, was occupied. By August 2024, the Netzarim Corridor was established, a six-kilometer route cutting Gaza into northern and southern sections (Sky News 2024). These corridors, created by bulldozing farmland, orchards, and buildings, allow Israeli military movement while denying Palestinians access to their homes, many of which have been destroyed. Analysts from Forensic Architecture report that a third corridor is under construction, threatening further fragmentation of Palestinian territories and restricting movement within Gaza (Fig. 8.2).

Since the beginning of the war, Israel has systematically targeted critical infrastructure, including universities, hospitals, residential areas, and religious sites. In the first month alone, Israeli attacks caused severe damage to 60.8% of health facilities, 68.2% of educational facilities, and 42.1% of water infrastructure. Additionally, 35.1% of health facilities, 40.2% of educational facilities, and 36.8% of water infrastructure were completely destroyed, leaving these essential services largely inoperable (Asi et al. 2024).

The following case studies will highlight the extensive urban destruction in Gaza's cities: Jabaliya, Khan Younis, and Rafah.

Jabaliya, a northern city, has faced relentless missile strikes, intense fighting, and clashes throughout the war, including within the refugee camp. Airstrikes on the 25th day of the war created massive craters in densely populated areas, resulting in a massacre and trapping dozens under rubble. Reports confirmed 50 civilians killed and 150 injured in this attack. Drone footage from

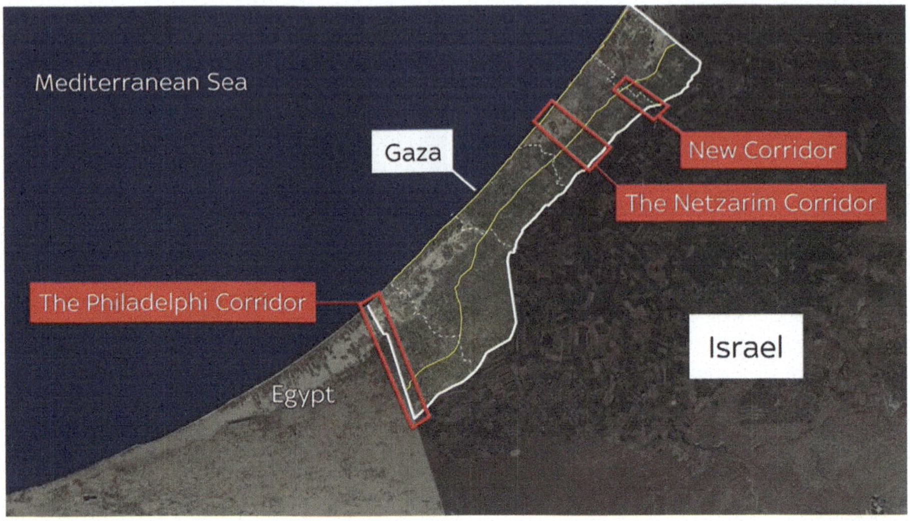

Fig. 8.2 Israeli buffer zones and strategic corridors in Gaza

June 2024 revealed nearly 70% of buildings severely damaged, with residential areas reduced to ruins after a second Israeli assault. In a third attack, Israeli bulldozers and backhoes demolished remaining structures, rendering the area uninhabitable (Fig. 8.3). Schools, hospitals, and universities were wiped out, with Human Rights Watch describing this as intentional destruction aimed at creating buffer zones and corridors for control over northern Gaza (The Guardian 2024a).

The destruction of Rafah, a designated 'safe zone' and refuge for displaced Palestinians, has shocked the world, yet the viral *'All Eyes on Rafah'* campaign failed to prevent the massacres that have been committed there (Fig. 8.4). Rafah served as the primary gateway for aid into Gaza. However, Israeli operations have severed the territory's access to international humanitarian assistance, significantly increasing the risk of famine. On May 26, 2024, Israeli airstrikes resulted in a tent massacre that killed 45 people and injured more than 200 Palestinians in a missile attack that set fire to the camp in Tal al-Sultan, Rafah. This was perpetrated 2 days after the International Court of Justice had demanded Israel stop offense on Rafah. These victims were among the 1.3 million Palestinians sheltering in Rafah, most of whom had been forcibly displaced multiple times in a war that offers no respite for civilians. Over the following months, Rafah endured relentless attacks, even as it served as a refuge for nearly 70% of the displaced population. This led to further displacement toward overcrowded areas like Deir al-Balah and Khan Younis, exacerbating the already overstretched humanitarian response (BBC 2024; The Guardian 2024b).

Khan Younis and Al-Mawasi have experienced similar systematic destruction. After the massacres in Rafah, nearly 900,000 people fled to Al-Mawasi, a narrow coastal stretch along the Mediterranean, as well as Deir al-Balah and Khan Younis. Much of Khan Younis had already been destroyed in an Israeli offensive that ended in April 2024, but many returned only to be displaced again following another evacuation order in May. Al-Mawasi, now sheltering an estimated 1.8 million people within just 50 km^2 (20 miles^2), lacks critical infrastructure and basic services, leaving it overcrowded with tents and makeshift shelters. Overcrowding in UN emergency shelters in central and southern Gaza has far exceeded capacity, while waste piles up around shelters, schools, and dumping sites, posing severe environmental and health risks (Dardona et al. 2024). Acute food shortages have driven some to scavenge for food scraps in garbage heaps, while reliance on makeshift sewage wells threatens groundwater and exacerbates health hazards. Nearly 500,000 people are facing catastrophic hunger, with a high risk of famine throughout Gaza (de Waal 2024; Hassoun et al. 2024; IPC 2024).

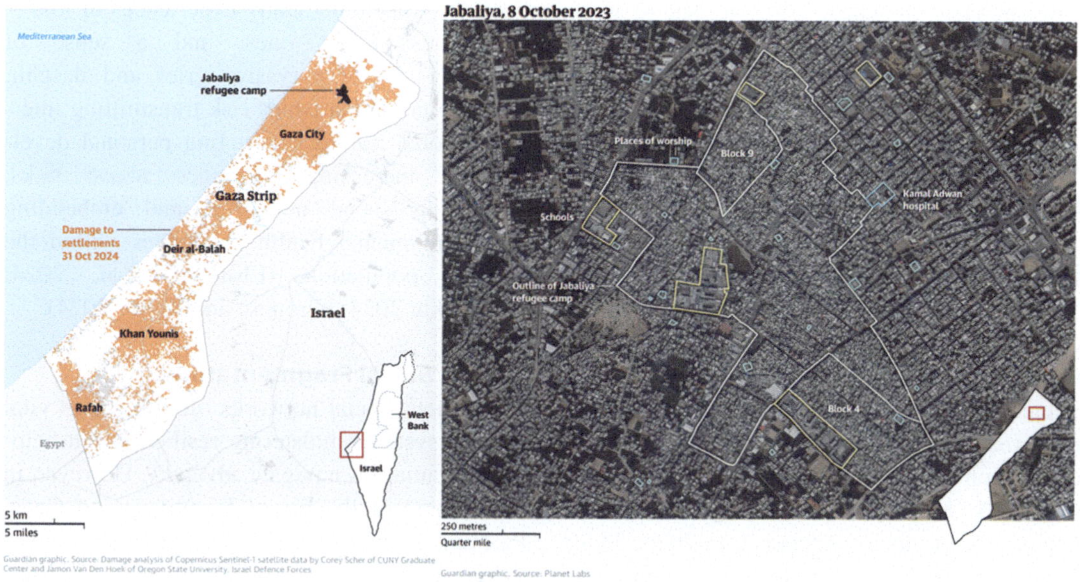

Fig. 8.3 Destruction of Jabaliya

Fig. 8.4 Rafah and the viral 'All Eyes on Rafah' campaign

The war has devastated Gaza's urban landscape, rendering vast areas uninhabitable and forcing the displacement of nearly the entire population. Systematic destruction of homes, critical infrastructure, and entire cities like Rafah, Jabaliya, and Khan Younis has left living conditions unmanageable and essential services inaccessible. Overcrowded displacement camps, severe food shortages, and health risks from poor sanitation have further deepened the humanitarian crisis, compounding the challenges caused by the widespread destruction and collapse of infrastructure.

8.6 Social, Economic, and Environmental Consequences of Domicide

The Israeli war on Gaza has triggered a multidimensional crisis, impacting psychological, social, public health, cultural, economic, and environmental domains. This has led to widespread devastation, including domicide, resulting in mass displacement, loss of safety, and the erasure of community foundations.

8.6.1 Social Consequences

8.6.1.1 Psychological Trauma

The widespread domicide in Gaza has significantly intensified the psychological trauma endured by its population, particularly among vulnerable groups such as children and women, exacerbating a crisis deeply rooted in decades of occupation (Buheji and Buheji 2024; Klin and Clin 2024). Mental health challenges extend well beyond the events of October 7; in 2020, 53.5% of Gazan children were diagnosed with post-traumatic stress disorder (PTSD), highlighting the pervasive psychological toll of war and prolonged conflicts (El-Khodary et al. 2020; Taha et al. 2024). Children's high exposure to traumatic events, such as witnessing violence, losing loved ones, and enduring displacement, has been compounded by the lack of safe spaces, severely disrupted healthcare systems, and critical resource shortages. War-related trauma has led to alarming rates of anxiety, depression, and behavioral disorders, particularly among children and adolescents, underscoring the urgent need for targeted mental health interventions (Abudayya et al. 2023; Veronese et al. 2025).

Women in Gaza face heightened mental health and psychosocial support (MHPSS) needs, exacerbated by gender-based violence, unmet family needs, and lack of access to WASH services. The widespread loss of loved ones, destruction of homes, and caregiving responsibilities have placed immense psychological burdens on women, often leading to feelings of helplessness, guilt, and anxiety. As a result, women in Gaza experience high rates of depression, anxiety, and trauma, underscoring the urgent need for gender-sensitive mental health interventions (Bdier and Mahamid 2024; Klin and Clin 2024). The psychological toll of war extends well beyond immediate trauma, profoundly impacting individuals of all ages, with children and women being particularly vulnerable. Daily experiences of loss—of homes, loved ones, and a sense of normalcy—foster pervasive grief and despair. These ongoing hardships risk transmitting intergenerational trauma, disrupting personal development, increasing substance abuse risks, weakening social networks, and embedding long-term mental health challenges within the broader population (Elsawi et al. 2024; Javanbakht 2024; Veronese and Kagee 2024).

8.6.1.2 Social Fragmentation

In conflicts, social networks often serve as vital support systems, bolstering resilience and helping communities navigate adversity. However, in the context of the recent Israeli war on Gaza, Palestinians are enduring the severe disruption of these traditional support systems—families, friends, and community networks. This disruption is driven by widespread displacement, the destruction of neighborhoods, family separations, and the devastating loss of lives, leaving communities fragmented and vulnerable (Veronese et al. 2022). However, some have reported an increase in the social cohesion through people volunteering in helping each other. This comes in parallel to the increase of violence between people, thefts, etc.; this contradiction is usually a common observation during wars (Hamad et al. 2024; Migdad and Buheji 2024; Migdad et al. 2024).

8.6.1.3 Public Health Crisis

Israel's deliberate attacks on the public health sector in Gaza have intensified the region-wide humanitarian crisis. A WHO report from December 2024 highlights the heightened risk of cholera outbreaks, driven by limited access to clean water and failing sewerage systems. The report also warns about a potential resurgence of polio after Gaza had been polio free since 1999. Contributing factors include widespread poverty, overcrowding, inadequate hygiene, contaminated food, lack of safe drinking water due to the siege, targeted destruction of water pumps, and the collapse of waste management infrastructure. Additionally, outbreaks of measles, acute respiratory infections, hepatitis A, and skin diseases have been reported, reflecting worsening health conditions (WHO 2024). Since the beginning of 2024, two thirds of the children admitted for outpatient treatment were dealing with acute malnutrition in the past 5 months alone (WHO 2024), over 96% of children aged 6–23 months and women are not meeting their nutrient requirements, and more than 60,000 children are estimated to require treatment for acute malnutrition in 2025 (ReliefWeb 2024).

Women and girls face dire shortages of hygiene products and sexual health services, exacerbated by the lack of clean water and other essential necessities. Many are unable to manage basic menstrual hygiene, leading to increased health risks and discomfort. The absence of these essential services further undermines their dignity and well-being in already dire living conditions (Jangda et al. 2024; Sabet et al. 2024). The impact of these public health issues was exacerbated due to the lack of access to treatment and medication, suspension of medical evacuations, attacks on healthcare centers and hospitals, displacement and overcrowding in shelters (Beiraghdar et al. 2023; Shaheen et al. 2024). The situation has worsened significantly due to severe water shortages and a critical WASH crisis. People in Gaza have access to only 4.74 L of water per person per day for all purposes, including drinking, cooking, and washing—a drastic 94% reduction from previous levels. This falls alarmingly short of the internationally accepted minimum standard of 15 L per person per day required for basic survival during emergencies (Abdul Samad et al. 2024).

8.6.1.4 Cultural Heritage Loss

Cultural heritage serves as a vital cornerstone of community identity, collective memory, and social cohesion, playing a crucial role in fostering resilience and preserving the history and traditions of people. Societies sustain their collective identity through social constructs that rely not only on morals, but also on traditions and physical attributes and space (Coakley and McAuliffe 2022). However, in Gaza, the Israeli attacks have systematically targeted cultural heritage, eroding community identity and undermining the social fabric that binds its people together. It is estimated that around 50% of heritage sites in the Gaza Strip have been damaged, with many structures near-demolished (Fradley et al. 2024; Taha 2024). As of November 29, 2024, 75 sites were reported by the UNESCO to be damaged since October 7, 2023, including ten religious sites (mosques and churches), 48 buildings of historical and/or artistic interest, three depositories of movable cultural property, six monuments, one museum and seven archeological sites (UNESCO 2024).

8.6.2 Economic Consequences

The widespread domicide in Gaza has devastated not only homes but also the economic foundations of the region, accelerating an already dire economic collapse. The economy of Gaza, which is already malfunctioning due to the Israeli economic warfare on the Strip (Alagha and Hussein 2024), has completely collapsed during the recent war. According to International Labor Organization (ILO), real GDP in Gaza (constant 2015 USD prices, million) has declined from US$ 2660 million in October 2022–September 2023 to US$ 406 million in October 2023–September 2024, real GDP per capita from US$ 1203 million in October 2022–September 2023 to US$ 181 million in October 2023–September 2024, and unemployment increased to 79.7% compared to a pre-war percentage of 45.3% (ILO 2024).

Movement of goods has been severely restricted by the blockade. Gaza's population mainly relies on agriculture, fishing, and small-scale manufacturing, all affected by the war. Between October 07, 2023 and June 30, 2024, approximately 83% of all plant life in Gaza, 70% of Gaza's agricultural land of fields and orchards, more than 3700 greenhouse structures, more than 47% of groundwater wells, and 65% of water tanks were destroyed (Forensic Architecture 2024). This war has pushed Gaza's economy from a state of crisis to complete obliteration. This widespread destruction has led to a profound reliance on international humanitarian assistance and aid, as food insecurity and poverty have escalated to catastrophic and unprecedented levels (Hassoun et al. 2024; Khalidi and Iwidat 2024; UNCTAD 2024).

8.6.3 Environmental Consequences

Domicide, or the systematic destruction of homes and living environments, has profoundly exacerbated environmental degradation in Gaza, leaving catastrophic consequences for its ecosystem and public health. The war has generated over 50 million tons of debris, much of it contaminated with hazardous materials such as asbestos and heavy metals, which pose long-term risks to soil and water quality (OCHA 2024; Shaheen et al. 2024; UNEP 2024).

The collapse of waste management and sewage systems has led to the discharge of 130,000 cubic meters of untreated sewage into the Mediterranean Sea daily, posing a threat to marine ecosystems and spreading environmental hazards beyond the borders of Gaza. This has caused severe water contamination, with 97% of water in Gaza considered unsafe for human consumption, further intensifying the water crisis in the region (Abd El Hay 2024; Shaheen et al. 2024). Additionally, air pollution from the extensive use of munitions and destroyed infrastructure has significantly deteriorated air quality, contributing to global warming. Within the first 2 months of the war, greenhouse gas emissions from the destruction equaled the burning of 150,000 tons of coal (Buheji and Al-Muhannadi 2023; Dardona et al. 2024).

The destruction of agricultural lands, estimated at 70% of farmland in Gaza, has not only intensified food insecurity but also reduced the region's ability to function as a carbon sink, amplifying atmospheric carbon levels. The loss of biodiversity in nature reserves further compounds the ecological damage, eroding natural resources and reducing resilience in the region (FAO/UN 2024; Shaheen et al. 2024; London et al. 2024). These environmental impacts demand urgent attention, including the development of sustainable reconstruction plans and immediate measures for environmental remediation to mitigate the far-reaching consequences of this war.

8.7 Rebuilding Gaza: Pathways to Achieving SDG 11

Repeated military assaults on Gaza have caused widespread destruction, underscoring the need for people-centered approaches that combine local empowerment, international aid, and sustainable peace efforts. Rebuilding Gaza requires an integrated approach that addresses economic, environmental, and social dimensions. A recent study identified three key factors affecting resource availability for post-disaster housing reconstruction: project stakeholders, project management, and the operational environment. The study highlighted that transparency, accountability, international support, and strong governmental leadership are critical to success, with the operational environment—shaped by factors like the blockade and funding shortages—having the greatest impact (Saleh et al. 2024).

Ensuring safe, affordable housing is crucial for Gaza's recovery. One effective strategy involves using eco-friendly materials, such as compressed earth blocks, and renewable energy to reduce costs and environmental impact. For example, a successful housing project in Gaza, based on an agency-driven approach, constructed thousands of housing units while simultaneously creating thousands of jobs (Elkahlout 2020). Sustainable infrastructure is essential for Gaza's economic recovery and social well-being. Investments in renewable energy, such as solar-powered desalination, can alleviate electricity and water shortages (Nassar et al. 2024). Expanding public transportation improves urban mobility, ensuring accessibility for vulnerable groups (Akuraju et al. 2020).

The nexus approach optimizes resource use by integrating energy and water efficiency into infrastructure projects (Daher et al. 2022). Inclusive and sustainable urbanization demands participatory, integrated planning. Urban resilience frameworks, particularly in post-conflict contexts, emphasize the importance of aligning local initiatives with broader policies to achieve systemic and sustainable change (Bakchan et al. 2021).

One of the aftermaths of war on Gaza is a severely polluted environment; hazardous debris and unexploded ordnance pose significant risks. Pollution control measures are a core component of restoring the environment in Gaza post-conflict, though there is long journey to achieve it. Recycling programs can reduce waste and provide materials for reconstruction (UNEP 2024). Promoting environmental monitoring through advanced techniques like remote sensing could serve as a valuable strategy in the advanced stages of recovery (Meaza et al. 2024).

With over 97% of water in Gaza being unsafe, the need for sustainable solutions such as large-scale desalination becomes critical to ensure access to clean water and address long-term water security challenges in the region. Inclusive approaches, such as Jordan's participatory governance in the energy-water nexus, offer valuable lessons. Collaborative frameworks, integrating Water-Energy-Food (WEF) management, can enhance resource planning and decision-making, fostering resilience and sustainability in Gaza's efforts to combat water scarcity (Daher et al. 2022). Effective waste management systems are essential for sustainable recovery. Community-based recycling initiatives in Gaza, such as organic waste composting for agricultural use, can promote environmental awareness and economic opportunities (Beccarello and Di Foggia 2022).

Cultural heritage sites (including ancient mosques, churches, libraries, and archaeological treasures) are crucial to Gaza's identity and resilience. Their mass destruction in Gaza has led to a "cultural genocide" (Al Jazeera 2024), undermining the importance of rebuilding efforts and restoring these landmarks. Collaborating with organizations like UNESCO and engaging local communities through participatory approaches, such as workshops, enhances preservation efforts. Legal follow-up and plans and programs for recovery and rebuilding must be prioritized to safeguard these sites and restore their cultural and historical significance (Taha 2024).

Addressing the trauma inflicted by this war is of utmost importance. Community-led psychosocial programs such as structured recreational and cultural social activities are associated with improved overall mental health and community resilience (Tol et al. 2011). Education is the cornerstone of sustainable recovery. Rebuilding schools and integrating digital learning tools can ensure continued access to education despite the challenges in Gaza. The "Education Cannot Wait" initiative offers valuable insights into crisis contexts. Community-driven initiatives, vocational training, and microfinance programs foster localized, resilient solutions tailored to Gaza's sociopolitical and environmental context, supporting economic recovery (GCPEA 2022; Faculty of Education University of Cambridge/Centre for Lebanese Studies & UNRWA 2024; Iriqat et al. 2025). Gender-inclusive strategies strengthen economic resilience, enabling greater participation in rebuilding efforts and enhancing long-term sustainability (ILO 2022).

Global partnerships play a critical role in securing the funding and expertise needed to align international aid with SDG 11 targets, supporting sustainable reconstruction efforts. Empowering local authorities and strengthening governance structures are fundamental to achieving the vision of SDG 11. Effective monitoring frameworks are crucial for tracking progress, ensuring accountability, and maintaining transparency. Rebuilding Gaza requires an integrated approach that prioritizes sustainable urban development, environmental restoration, cultural preservation, and community-driven recovery, all underpinned by robust policy frameworks aligned with SDG 11. Collaboration and innovation are indispensable in paving the way for a resilient and sustainable future in Gaza.

8.8 Conclusion

The devastation of urban systems in Gaza through domicide has not only stripped this region of its infrastructure but has also eroded the foundational elements necessary to achieve SDG 11. This deliberate targeting has caused widespread displacement, exacerbated environmental degradation, and led to a cultural loss that undermines community resilience and identity. Sustainable, community-centered rebuilding efforts are essential to reversing these impacts, focusing on housing, cultural heritage, and environmental restoration. Global accountability, supported by international frameworks and collaborative partnerships, remains paramount in ensuring that Gaza's reconstruction aligns with the principles of equity, inclusivity, and long-term sustainability. The international community must prioritize coordinated action to rebuild Gaza, protect human rights, and address the root causes of conflict to prevent further cycles of destruction.

References

Abd El Hay M (2024) The environmental-humanitarian impacts of the Israel Hamas war in Gaza. Arav Inst Environ Stud:1–28

Abdul Samad L, Butcher M, Khalidi B (2024) Water war crimes: how Israel has weaponised water in its military campaign in Gaza. Oxfam Int 6

Abudayya A, Bruaset GTF, Nyhus HB et al (2023) Consequences of war-related traumatic stress among Palestinian young people in the Gaza strip: a scoping review. Ment Heal Prev 32:200305. https://doi.org/10.1016/J.MHP.2023.200305

Abuelaish B (2018) Urban land use change analysis and modeling: a case study of the Gaza strip. In: Camacho Olmedo M, Paegelow M, Mas JF, Escobar F (eds) Geomatic approaches for Modeling land change scenarios. Lecture notes in geoinformation and cartography. Springer, Cham, pp 271–291

Abukashif YJM, Riza M (2019) Urban development of conflict zones: the case of Gaza city. Open House Int 44:45–51. https://doi.org/10.1108/ohi-02-2019-b0006

Agha Z, Esson J, Griffiths M, Joronen M (2024) Gaza: A decolonial geography. Trans Inst Br Geogr 49:e12675. https://doi.org/10.1111/tran.12675

Akesson B (2014) "We may go, but this is my home": experiences of Domicide and resistance for Palestinian children and families. J Intern Displac 4:9–22

Akesson B, Basso AR, Denov M (2016) The right to home: Domicide as a violation of child and family rights in the context of political violence. Child Soc 30:369–383. https://doi.org/10.1111/chso.12174

Akuraju V, Pradhan P, Haase D et al (2020) Relating SDG11 indicators and urban scaling—an exploratory study. Sustain Cities Soc 52:101853. https://doi.org/10.1016/j.scs.2019.101853

Al Jazeera (2024) A 'cultural genocide': which of Gaza's heritage sites have been destroyed? https://www.aljazeera.com/news/2024/1/14/a-cultural-genocide-which-of-gazas-heritage-sites-have-been-destroyed. Accessed 3 Feb 2024

Alagha M, Hussein N (2024) The political economy of Israeli economic warfare on the Gaza strip. In: Elkahlout G (ed) Gaza's cycle of destruction and rebuilding. Palgrave Macmillan, Cham, pp 125–145

Alhamawi M (2024) The philosophy of (De) development and its dichotomy in Palestine. In: Elkahlout G (ed) Gaza's cycle of destruction and rebuilding. Palgrave Macmillan, Cham, pp 189–207

AlJazeera (2024) Israel's war on Gaza: How Israel destroyed the Gaza Strip. https://interactive.aljazeera.com/aje/2024/gaza-before-after-satellite-images/. Accessed 8 Jan 2025

Ammar SMS, Alramlawi NY (2020) Post-war restoration of traditional houses in Gaza. In: Arefian F, Moeini S (eds) Urban heritage along the silk roads. The urban book series. Springer, pp 137–154

Amnesty International (2024) 'You feel like you are subhuman'—Israel's genocide against Palestinians in Gaza

Andreou GM, Elkhoudary Y, Hassouna A (2024) New investigations in Gaza's heritage landscapes: the Gaza Maritime Archaeology Project (GAZAMAP). Antiquity 98:1–9. https://doi.org/10.15184/aqy.2024.68

Asi YM (2020) Achieving food security through localisation, not aid: "de-development" and food sovereignty in the Palestinian territories. J Peacebuild Dev 15:205–218. https://doi.org/10.1177/1542316620918555

Asi Y, Mills D, Greenough PG et al (2024) 'Nowhere and no one is safe': spatial analysis of damage to critical civilian infrastructure in the Gaza strip during the first phase of the Israeli military campaign, 7 October to 22 November 2023. Confl Heal 18:1–13. https://doi.org/10.1186/s13031-024-00580-x

Azzouz A (2023) Domicide: Architecture, war and the destruction of home in Syria: Ammar Azzouz. Bloomsbury Visual Arts, London

Azzouz A (2024) Erased city. City. https://doi.org/10.1080/13604813.2024.2323388

Bakchan A, Hacker ME, Faust KM (2021) Resilient water and wastewater infrastructure systems through integrated humanitarian-development processes: the case of Lebanon's protracted refugee crisis. Environ Sci Technol 55:6407–6420. https://doi.org/10.1021/ACS.EST.0C05630/SUPPL_FILE/ES0C05630_SI_001.PDF

BBC (2024) Gaza war: Battles rage as Israeli troops intensify Jabalia and Rafah operations. https://www.bbc.com/news/world-middle-east-69015371. Accessed 8 Jan 2025

Bdier D, Mahamid F (2024) Womens' mental health and war catastrophes in Palestine. In: Zangeneh M (ed) Essentials in health and mental health, advances in mental health and addiction. Springer, Cham, pp 177–186

Beccarello M, Di Foggia G (2022) Sustainable development goals data-driven local policy: focus on SDG 11 and SDG 12. Adm Sci 12:167. https://doi.org/10.3390/admsci12040167

Beiraghdar F, Momeni J, Hosseini E et al (2023) Health crisis in Gaza: the urgent need for international action. Iran J Public Health 52:2478–2483. https://doi.org/10.18502/ijph.v52i12.14309

Ben Saad H (2024) Urgent humanitarian appeal: protecting the lives of women and children in the Gaza strip (Palestine). J Public Health Res 13:1–2. https://doi.org/10.1177/22799036241229312

Berisha E, Caprioli C, Cotella G (2022) Unpacking SDG target 11.a: what is it about and how to measure its progress? City Environ Interact 14:100080. https://doi.org/10.1016/j.cacint.2022.100080

Buheji M, Al-Muhannadi K (2023) Mitigating risks of environmental impacts on Gaza—review of precautions & solutions post (2023 war). Int J Adv Res Eng Technol 14:15–47

Buheji M, Buheji B (2024) Mitigating risks of slow children development due to war on Gaza 2023. Int J Psychol Behav Sci 14:11–21. https://doi.org/10.5923/j.ijpbs.20241401.02

Center for Engineering and Planning (CEP) (2022) City Center Development Plan for Gaza (CCDP) urban and spatial analysis report

CEOBS (2022) Reverberating civilian and environmental harm from explosive weapons use in Gaza. https://ceobs.org/reverberating-civilian-and-environmental-harm-from-explosive-weapons-use-in-gaza/. Accessed 6 Jan 2025

Coakley S, McAuliffe P (2022) Picking up the pieces: transitional justice responses to destruction of tangible cultural heritage. Netherlands Q Hum Rights 40:311–332. https://doi.org/10.1177/09240519221113121

da Silva Costa TM, Cabral NRAJ, da Silva AC et al (2023) Analysis of the contribution of building rehabilitation to the achievement of SDG 11. J Build Pathol Rehabil 8:1–8. https://doi.org/10.1007/s41024-023-00296-0

Daher B, Hamie S, Pappas K, Roth J (2022) Examining Lebanon's resilience through a water-energy-food nexus lens. Front Sustain Food Syst 6:748343. https://doi.org/10.3389/fsufs.2022.748343

Dall'Agnolo MMY, Jüngling S, Smuts H (2025) Towards a smart city sustainability tracker for achieving SDG 11 in cities. In: Hinkelmann K, Smuts H (eds) Society 5.0. Society 5.0 2024. Communications in computer and information science. Springer Nature, pp 84–97

Daraj (2024) The Gaza Airport: how Israel erases Palestinian sovereignty by controlling travel of all sorts. https://daraj.media/en/the-gaza-airport-how-israel-erases-palestinian-sovereignty-by-controlling-travel-of-all-sorts/?utm_source=chatgpt.com. Accessed 6 Jan 2025

Dardona Z, Amane M, Dardona A, Boussaa S (2024) Health and environmental impacts of Gaza conflict (2023-2024): a review. One Heal Bull. https://doi.org/10.4103/ohbl.ohbl

de Waal A (2024) Famine in Gaza: an example of the global humanitarian crisis. Am J Clin Nutr 119:1383–1385. https://doi.org/10.1016/j.ajcnut.2024.04.015

El-Atrash AA, Salem HS, Isaac JE (2014) Planning towards a sustainable land transportation system in a future Palestinian State. J Nat Sci Sustain Technol 2:305–324

Elkahlout G (2020) Post-conflict housing reconstruction in the Gaza strip: a case study of agency-driven housing in sheikh Hamad city. Int J Hous Mark Anal 13:317–330. https://doi.org/10.1108/IJHMA-03-2019-0034

Elkahlout G (2024) Gaza's cycle of destruction and rebuilding. Understanding the actors, dynamics, and responses. Palgrave Macmillan

El-Khodary B, Samara M, Askew C (2020) Traumatic events and PTSD among Palestinian children and adolescents: the effect of demographic and socioeconomic factors. Front Psych 11:1–11. https://doi.org/10.3389/fpsyt.2020.00004

El-Sawalhi N, Lafi M (2019) Evaluation of the public housing projects role in contributing to mitigation of housing crisis in Gaza strip. Am J Civ Eng 7:88. https://doi.org/10.11648/j.ajce.20190704.12

Elsawi M, Al N, Hospital Q et al (2024) Health in the crossfire-Analyzing and mitigating the multifaceted health risks of the 2023 war on Gaza. Public Heal Res 14:1–11. https://doi.org/10.5923/j.phr.20241401.01

Faculty of Education University of Cambridge/Centre for Lebanese Studies & UNRWA (2024) Palestinian education under attack in Gaza: restoration, recovery, rights and responsibilities in and through education

FAO/UN (2024) Agricultural Damage Assessment in the Gaza Strip from October 7th 2023 to September 1st 2024

Forensic Architecture (2024) A cartography of genocide: Israel's conduct in Gaza since October 2023. https://forensic-architecture.org/investigation/a-cartography-of-genocide. Accessed 7 Jan 2025

Fradley M, Petersen A, Finlayson B (2024) Gaza's cultural heritage has been devastated by the conflict with Israel—this is how we're documenting the damage. https://theconversation.com/gazas-cultural-heritage-has-been-devastated-by-the-conflict-with-israel-this-is-how-were-documenting-the-damage-244247. Accessed 7 Jan 2025

GCPEA (2022) Measuring the impact of attacks on education in Palestine

Hamad BA, Vintges J, Diab R, et al (2024) Shattered lives and dreams: the toll of the war on Gaza on young people

Hassoun A, Al-Muhannadi K, Hassan HF et al (2024) From acute food insecurity to famine: how the 2023/2024 war on Gaza has dramatically set back sustainable development goal 2 to end hunger. Front Sustain Food Syst 8:1402150. https://doi.org/10.3389/fsufs.2024.1402150

Hryhorczuk D, Levy BS, Prodanchuk M et al (2024) The environmental health impacts of Russia's war on

Ukraine. J Occup Med Toxicol 19:1–14. https://doi.org/10.1186/S12995-023-00398-Y/FIGURES/2
Human Rights Watch (2024) Extermination and acts of genocide Israel deliberately depriving Palestinians in Gaza of water
ILO (2022) Gender equality and women's empowerment in the world of work in fragile, conflict and disaster settings
ILO (2024) A year of war in Gaza: impacts on employment and livelihoods in the West Bank and Gaza strip. Bulletin No 5
IPC (2024) IPC famine review committee alert Gaza strip
Iriqat D, Alousi R, Aldahdouh TZ et al (2025) Educide amid conflict: the struggle of the Palestinian education system. Qual Educ All 2:81–99. https://doi.org/10.1108/QEA-10-2024-0120
Jabr S, Berger E (2022) The children of Palestine: struggle and survival under occupation. In: Tiliouine H, Benatuil D, Lau MKW (eds) Handbook of Children's risk, vulnerability and quality of life. International handbooks of quality-of-life. Springer, Cham, pp 279–296
Jamaluddine Z, Abukmail H, Aly S et al (2025) Traumatic injury mortality in the Gaza strip from Oct 7, 2023, to June 30, 2024: a capture—recapture analysis. Lancet 2024:1–9. https://doi.org/10.1016/S0140-6736(24)02678-3
Jangda HN, Farhan K, Akilimali A (2024) Unheard voices: addressing sexual and reproductive health in Gaza's women. Int J Gynecol Obstet 166:139–140. https://doi.org/10.1002/ijgo.15596
Javanbakht A (2024) Gaza's hidden crisis: adults, children, and generations of psychological torment to come. Eur J Psychotraumatol 15:2416824. https://doi.org/10.1080/20008066.2024.2416824
Kerbaj M, Traboulsi J, Maasri Z, Toukan H (2024) Gaza in plain sight: witnessing in solidarity. J Vis Cult 22:1–19. https://doi.org/10.1177/14704129241227847
Khalidi R, Iwidat Q (2024) Assessing the economic and social impacts of Israel's war on Palestine. Arab Cent Res Policy Stud 7:80–90
Khatib R, McKee M, Yusuf S (2024) Counting the dead in Gaza: difficult but essential. Lancet 404:237–238. https://doi.org/10.1016/S0140-6736(24)01169-3
Klin A, Clin A (2024) Behind the rubble: psychological trauma of wars and human rights abuses on women and children in Gaza. Anatol Clin J Med Sci 29:119–136. https://doi.org/10.21673/anadoluklin.1575372
Krever T, Veličković M, Mégret F et al (2024) On international law and Gaza: critical reflections. London Rev Int Law:1–85. https://doi.org/10.1093/lril/lrae012
London L, Watterson A, Mergler D et al (2024) A call from 40 public health scientists for an end to the continuing humanitarian and environmental catastrophe in Gaza. Environ Heal A Glob Access Sci Source 23:1–6. https://doi.org/10.1186/S12940-024-01097-9/PEER-REVIEW
Mahomed S (2024) International humanitarian laws: applicable to all or a privilege for some? SAJBL 17:22–26
Meaza H, Ghebreyohannes T, Nyssen J et al (2024) Managing the environmental impacts of war: what can be learned from conflict-vulnerable communities? Sci Total Environ 927:171974. https://doi.org/10.1016/j.scitotenv.2024.171974
Migdad M, Buheji M (2024) Resilience amidst devastation: qualitative insights from Gaza during the war. Int J Soc Sci Res Dev 6:13–27
Migdad M, Buheji M, Migdad IM (2024) Reviving resilience from the tents—Gaza (2024). Int J Manag 15:93–108
Milton S, Elkahlout G, Attallah S (2024) Shrinking reconstruction space in the Gaza strip: rebuilding after the 2021 and 2022 wars. Conflict Secur Dev 24:49–78. https://doi.org/10.1080/14678802.2024.2314034
Nassar YF, El-Khozondar HJ, Elnaggar M et al (2024) Renewable energy potential in the State of Palestine: proposals for sustainability. Renew Energy Focus 49:100576. https://doi.org/10.1016/j.ref.2024.100576
Nijim M (2024) Gazacide: Palestinians from Refugeehood to ontological obliteration. Crit Sociol 1944:1–23. https://doi.org/10.1177/08969205241286530
OCHA (2020) Humanitarian needs overview
OCHA (2024) Gaza humanitarian response update | 24 November—7 December 2024. https://www.ochaopt.org/content/gaza-humanitarian-response-update-24-november-7-december-2024?_gl=1*1egyngz*_ga*NDE3Nzg4ODEuMTczMzQ2NTkzMQ..*_ga_E60ZNX2F68*MTczNTE0MjExMi4xMC4xLjE3MzUxNDIxMTguNTQuMC4w. Accessed 25 Dec 2024
OHCHR (2023) UN experts say Israel should be held accountable for acts of 'domicide'. https://www.ohchr.org/en/press-releases/2023/02/un-experts-say-israel-should-be-held-accountable-acts-domicide. Accessed 6 Jan 2025
Pasinovych I, Myskiv G (2023) Ukrainian context of sustainable development and the role of business in its achievement. Reg Sci Policy Pract 15:161–180. https://doi.org/10.1111/RSP3.12619
Pereira P, Zhao W, Symochko L et al (2022) The Russian-Ukrainian armed conflict will push back the sustainable development goals. Geogr Sustain 3:277–287. https://doi.org/10.1016/j.geosus.2022.09.003
Petković N (2023) Anthropological approach to the question of domicide. Bastina 392:431–448. https://doi.org/10.5937/bastina33-46775
Porteous JD, Smith SE (2001) Domicide: the global destruction of home. McGill-Queen's University Press
Rajagopal B, Pangalangan RA (2022) The case for the International crime of domicide. https://www.justsecurity.org/83825/the-case-for-the-international-crime-of-domicide/. Accessed 6 Jan 2025
ReliefWeb (2024) Cost of a chicken in Gaza same as Christmas Dinner for four in UK. https://reliefweb.int/report/occupied-palestinian-territory/cost-chicken-gaza-same-christmas-dinner-four-uk. Accessed 7 Jan 2025
Sabah FY (2023) Assessment of residential satisfaction with internationally funded housing projects in

Gaza strip, Palestine. Front Built Environ 9:1289707. https://doi.org/10.3389/fbuil.2023.1289707
Sabet C, Nguyen D, Nada SA et al (2024) Women and girls in Gaza face increasingly dire physical and mental health challenges. BMJ 384. https://doi.org/10.1136/BMJ.Q625
Saleh N, Enshassi A, Sundermeier M (2024) Factors influencing resource availability for post-disaster housing reconstruction: the case of Gaza strip-Palestine. Int J Constr Manag 24:1364–1374. https://doi.org/10.1080/15623599.2023.2174654
Seeberg P (2024) Israeli hydro-hegemony and the Gaza war. Middle East Policy 31:33–46. https://doi.org/10.1111/MEPO.12740
Shaheen A, Dajani R, Zinszer K et al (2024) The war on the Gaza strip and its consequences on global warming. Front Hum Dyn 6:1463902
Shahini E, Shebanina O, Kormyshkin I et al (2024) Environmental consequences for the world of Russia's war against Ukraine. Int J Environ Stud. https://doi.org/10.1080/00207233.2024.2302745
Sky News (2024) Israel establishing infrastructure for long-term military presence in Gaza, satellite images suggest. https://news.sky.com/story/israel-establishing-infrastructure-for-long-term-military-presence-in-gaza-satellite-images-suggest-13206114. Accessed 8 Jan 2025
Sultany N (2024) A threshold crossed: on genocidal intent and the duty to prevent genocide in Palestine. J Genocide Res:1–26. https://doi.org/10.1080/14623528.2024.2351261
Taha H (2024) Destruction of cultural heritage in Gaza. Jerusalem Q 97:45–70
Taha AM, Sabet C, Nada SA et al (2024) Addressing the mental health crisis among children in Gaza. Lancet Psychiatry 11:249–250. https://doi.org/10.1016/S2215-0366(24)00036-1
Tayeh BA, Alaloul WS, Alnairab AF (2019) Challenges facing municipalities to provide wastewater services: a case study of Gaza strip. Int Conf sustain Infrastruct 2019 Lead resilient communities through 21st century—Proc Int Conf Sustain Infrastruct 2019: 428–440. https://doi.org/10.1061/9780784482650.046
The Guardian (2024a) 'Everything is gone': how Israeli forces destroyed Jabaliya refugee camp. https://www.theguardian.com/global-development/2024/dec/18/jabaliya-refugee-camp-gaza-destruction-idf. Accessed 8 Jan 2025
The Guardian (2024b) Israel-Gaza war: Netanyahu says deadly Rafah strike a 'mishap' as UN humanitarian chief warns 'impunity cannot continue'—as it happened. https://www.theguardian.com/world/live/2024/may/27/israel-gaza-war-live-aid-groups-horrified-by-israeli-strike-on-camp-housing-displaced-people-in-rafah?filterKeyEvents=false&page=-with%3Ablock-665484978f08f180c20816c3#block-665484978f08f180c20816c3. Accessed 9 Jan 2025
Thomas R, Hsu A, Weinfurter A (2021) Sustainable and inclusive—evaluating urban sustainability indicators' suitability for measuring progress towards SDG-11. Environ Plan B Urban Anal City Sci 48:2346–2362. https://doi.org/10.1177/2399808320975404
Tol WA, Barbui C, Galappatti A et al (2011) Mental health and psychosocial support in humanitarian settings: linking practice and research. Lancet 378:1581–1591. https://doi.org/10.1016/S0140-6736(11)61094-5
Tuathail G, Dahlman C (2006) Post-domicide Bosnia and Herzegovina: homes, homelands and one million returns. Int Peacekeep 13:242–260. https://doi.org/10.1080/13533310500437647
UNCTAD (2024) Preliminary assessment of the economic impact of the destruction in Gaza and prospects for economic recovery
UNDP (2021) Gaza 2021 infrastructure damage assessment report
UNEP (2024) Environmental impact of the conflict in Gaza: preliminary assessment of environmental impacts. Nairobi
UNESCO (2024) Gaza Strip: damage assessment. https://www.unesco.org/en/gaza/assessment. Accessed 3 Feb 2024
UN-HABITAT (2014) Forced evictions; Fact sheet No. 25/Rev. 1
UNITAR-UNOSAT (2024) UNOSAT Gaza Strip comprehensive damage assessment—December 2024
UNRWA (2024) UNRWA Situation Report #141 on the situation in the Gaza Strip and the West Bank, including East Jerusalem. https://www.unrwa.org/resources/reports/unrwa-situation-report-141-situation-gaza-strip-and-west-bank-including-east-jerusalem. Accessed 8 Jan 2025
Veronese G, Kagee A (2024) Does psychology fail in the face of genocide? Insights about Gaza before and after October 7, 2023. Int J Hum Rights Healthc. https://doi.org/10.1108/IJHRH-07-2024-0053
Veronese G, Pepe A, Diab M et al (2022) Social support, resilience, and mental health in a low-intensity warfare context: the effects of siege on university students in Gaza. J Ment Health 31:383–391. https://doi.org/10.1080/09638237.2021.1979486
Veronese G, Bdier D, Obaid H et al (2025) Agency, life satisfaction, hope, potentially traumatic events, trauma symptoms, and psychological signs. A two waves study with a sample of Palestinian children living in different geographical areas. Child Youth Serv Rev 168:107990. https://doi.org/10.1016/j.childyouth.2024.107990
Weinthal E, Sowers J (2019) Targeting infrastructure and livelihoods in the West Bank and Gaza. Int Aff 95:319–340. https://doi.org/10.1093/IA/IIZ015
WHO (2024) Hostilities in the occupied Palestinian territory (oPt) Public Health Situation Analysis (PHSA)
Zaqout M, Fayad M, Barrington DJ et al (2024) Sanitation is political: understanding stakeholders' incentives in funding sanitation for the Gaza Strip, Palestine. Third World Q 45:1437–1457. https://doi.org/10.1080/01436597.2024.2318476

Seeking Justice and Governance: The Path to Peace in War-Torn Gaza

9

Abdo Hassoun, Samer Abuzerr, Monica Trif, Zuhaib Bhat, Dalal Iriqat, Rana Muhammad Aadil, and Bayan Arouri

Abstract

This chapter explores the complex challenges of attaining peace, justice, and effective governance in Gaza, as outlined in Sustainable Development Goal 16 (SDG 16). It highlights the pivotal role of SDG 16 in promoting stability and addressing the severe humanitarian crisis caused by the recent unprecedented assault on Gaza. The chapter explores the unique struggles faced by Gazans, emphasizing the critical need for accessible justice systems and accountable governance as foundations for peacebuilding. Key themes include Israel's persistent violations of international law, systemic violence, apartheid policies, and actions disproportionately harming children in Gaza. By advocating for the enforcement of international legal frameworks, dismantling apartheid systems, protecting vulnerable populations, and fostering inclusive governance, the work provides recommendations to address humanitarian crises, restore justice, and rebuild resilient institutions essential for sustainable peace in the region.

Keywords

Apartheid · Collective punishment · Conflict · Crime · Forced displacement · Gaza genocide · Humanitarian crisis · Human rights · Palestine · Violence

A. Hassoun
Sustainable AgriFoodtech Innovation & Research (SAFIR), Arras, France

Faculty of Agricultural Engineering, University of Aleppo, Aleppo, Syria
e-mail: a.hassoun@saf-ir.com

S. Abuzerr
Department of Medical Sciences, University College of Science and Technology, Khan Younis, Gaza, Gaza Strip, Palestine

M. Trif
Centre for innovative process engineering (CENTIV), Stuhr, Germany

Z. Bhat
Division of Livestock Products Technology, SKUAST-J, Jammu, India

D. Iriqat
Arab American University of Palestine, Ramallah, Palestine
e-mail: dalal.s.iriqat@aaup.edu

R. M. Aadil
National Institute of Food Science and Technology, University of Agriculture, Faisalabad, Pakistan

B. Arouri (✉)
Doctoral Researcher at Tampere Peace Research Institute, Tampere, Finland
e-mail: bayan.arouri@tuni.fi

A. Hassoun (ed.), *War on Gaza*, Sustainable Development Goals Series,
https://doi.org/10.1007/978-3-031-88500-6_9

9.1 Introduction

The pursuit of peace, justice, and effective governance is fundamental to achieving sustainable development in regions plagued by conflict and instability. As emphasized in Sustainable Development Goal 16 (SDG 16), robust institutions, inclusive governance, and equitable justice systems are critical for fostering long-term stability and social cohesion. However, achieving these ideals in war-torn areas presents a formidable challenge, as violence, forced displacement, and systemic human rights violations erode the foundations needed for sustainable peace. Prolonged conflicts often create environments marked by economic collapse, political fragmentation, and institutional decay, compounding the struggles of affected populations (Bose and Kumar 2021; Küfeoğlu 2022; Milton and Alhamawi 2024).

Violence and genocide remain stark realities in conflict zones, perpetuating cycles of suffering and impunity. Forced displacement disrupts social fabrics, marginalizes vulnerable populations, and exacerbates socioeconomic disparities. In many cases, these conditions are worsened by apartheid-like systems of governance or military occupation that impose restrictions on freedom, movement, and access to essential services. The destruction of infrastructure and legal systems further entrenches inequality, making it exceedingly difficult to establish accountability and rebuild institutions capable of addressing crimes and fostering justice (Dugard and Reynolds 2013; Veronese and Kagee 2024).

The situation in Gaza highlights the dire challenges of achieving SDG 16 in conflict-ridden regions, where safety and stability are increasingly elusive. The ongoing violence, systemic restrictions, and political fragmentation have created a humanitarian crisis of unprecedented scale. Prolonged blockades have crippled access to essential resources, such as food, water, fuel, and medical supplies, exacerbating socioeconomic disparities and eroding governance. Violations of international laws and human rights persist, leaving civilians displaced, impoverished, and vulnerable. Nowhere is safe in Gaza, as relentless bombardments and destruction affect homes, schools, and hospitals alike, depriving people of any semblance of security (Asi et al. 2024; Elkahlout 2024; Human Rights Watch 2024a). Despite numerous United Nations (UN) resolutions and global calls to stop this genocide, these measures remain unfulfilled. This failure exacerbates immense suffering and undermines trust in international governance systems intended to safeguard human rights and uphold justice, further endangering one of the world's most vulnerable populations (Mills et al. 2020; El-Shewy et al. 2024; Sultany 2024).

This chapter provides a comprehensive exploration of the challenges and opportunities in advancing SDG 16 in regions affected by protracted conflicts, with a focus on Gaza as a case study. It begins with an examination of the relevance of SDG 16 to war and conflict, highlighting key targets and their implications for fostering peace, justice, and strong institutions. The discussion then shifts to the impact of violence and armed conflicts on governance and human rights in Gaza, emphasizing the vulnerability of children and marginalized populations to systemic violations. The chapter also addresses the persistent issue of Israeli apartheid, analyzing its detrimental effects on equality, institutional integrity, and societal cohesion. Finally, we suggest mitigation measures and give recommendations for strengthening institutions in conflict-affected regions, offering actionable insights into fostering sustainable peace and justice. By addressing these interconnected themes, this work aims to provide a nuanced understanding of the pathways toward achieving SDG 16 in some of the world's most challenging contexts.

9.2 Sustainable Development Goal 16 (SDG 16): Relevance to War and Conflicts

SDG 16 focuses on promoting peace, justice, and strong institutions, with targets addressing violence prevention, justice, and inclusive governance. Wars and armed conflicts severely undermine these targets by escalating violence, weakening justice systems, and dismantling institutional frameworks essential for stability. Military assaults on Gaza exemplify these challenges, perpetuating cycles of violence, eroding justice mechanisms, and destabilizing the social and institutional fabric in this region.

9.2.1 Targets of SDG 16 Related to War and Conflict

SDG 16 comprises 12 targets that aim to promote peace, justice, and strong institutions. Several of these targets are particularly relevant to addressing the challenges posed by war, armed conflicts, terrorism, and crime. For instance, Target 16.1 seeks to significantly reduce all forms of violence and related death rates globally, addressing not only armed conflicts but also acts of terrorism and organized crime. Similarly, Target 16.3 emphasizes the promotion of the rule of law and equal access to justice for all, which is critical for protecting the rights of individuals and communities affected by violence, terrorism, and criminal activity (Salmon et al. 2018; Hope 2023; O'Reilly 2023). Target 16.6 focuses on developing effective, accountable, and transparent institutions, which are essential for preventing corruption and criminal networks from exploiting post-conflict vulnerabilities. Meanwhile, Target 16.7 calls for inclusive, participatory, and representative decision-making at all levels, fostering resilience in communities and countering extremist narratives. Additionally, Target 16.9 aims to provide legal identity for all, including birth registration, which is crucial for protecting displaced populations and preventing exploitation by criminal groups. Target 16.10 promotes public access to information and safeguards fundamental freedoms, ensuring transparency and combating misinformation often used by terrorist organizations and criminal syndicates to destabilize societies. Notably, Target 16a underscores the importance of international cooperation to equip institutions with the tools and resources needed to address complex global threats effectively, while Target 16.b promotes and enforces non-discriminatory laws and policies for sustainable development (Consuegra 2020; Milton and Alhamawi 2024).

9.2.2 Impact of War and Armed Conflict on SDG 16 Targets

Wars and conflicts severely hinder progress toward achieving SDG 16 by dismantling the institutional frameworks essential for peace, justice, and strong governance. Conflicts devastate judicial systems, rendering them ineffective and unable to uphold fundamental freedoms or human rights. The destruction of governance structures creates environments of impunity, making it challenging to address crimes against civilians or ensure accountability. Prolonged violence undermines efforts to establish inclusive governance, exacerbating cycles of instability and marginalization. These impacts extend beyond the immediate areas of conflict, destabilizing neighboring regions and complicating efforts to rebuild institutions and promote long-term peace and security (Bose and Kumar 2021; Drange 2022).

9.2.3 Relevance to Military Assaults on Gaza

Repeated military assaults on Gaza starkly illustrate the challenges to achieving SDG 16 in this region. Target 16.1 faces extensive harm due to indiscriminate attacks and destruction of infrastructure, including schools and hospitals (Asi et al. 2024; Sultany 2024). Access to justice for displaced and affected populations is nearly nonexistent, violating Target 16.3, while the destruction of governance and institutional capacity (Joronen and Griffiths 2022) directly impacts Target 16.6. Furthermore, Target 16.7 is severely compromised as the destruction of infrastructure, displacement of communities, and violence create an environment where Palestinians, particularly in Gaza, are deprived of opportunities to engage in decision-making about their future. The destruction of basic infrastructure and institutions in Gaza not only affects immediate humanitarian needs but also has long-term consequences for regional stability (Elkahlout 2024; Abed Alah 2024). Addressing these issues will require a concerted effort to apply the principles of SDG 16, focusing on rebuilding institutional trust, ensuring accountability for violations of international law, and enabling the participation of affected populations in decision-making processes.

9.3 Past and Current Violations of International Laws

The prolonged war on Gaza as well as the broader Palestinian territories, has involved persistent and severe violations of international law and UN resolutions, significantly obstructing the achievement of SDG 16. Over decades, the UN General Assembly and the UN Security Council have repeatedly condemned the policies and practices of Israel in occupied Palestinian territories, calling for an end to occupation, settlement expansion, and the use of violence against civilians (Mahomed 2024; Putra and Persadha 2024). These actions, however, have continued unabated, thereby perpetuating a state of instability that directly impedes many of the targets of SDG 16.

Since the establishment of Israel in 1948, multiple UN resolutions have called for respect toward Palestinian rights and a cessation of territorial occupation. One notable example is UN Security Council Resolution 242, passed in 1967, which declares the inadmissibility of acquiring territory by force and explicitly calls for Israel's withdrawal from territories occupied during the Six-Day War (Roy 2016; El-Shewy et al. 2024). The resolution, a cornerstone of international efforts to promote peace in the region, remains unfulfilled, with ongoing occupation and settlement expansion in violation of its terms. Furthermore, the blockade on Gaza, in place since 2007, contravenes international law, specifically Article 33 of the Fourth Geneva Convention, which prohibits collective punishment (Roy 2012; Amnesty International 2024a). This blockade has severely restricted Gaza's access to essential resources, including food, medicine, and clean water, and infringes on the rights of Gazans to freedom of movement (Human Rights Watch 2021, 2023). As a result, SDG 16.2 and SDG 16.10 have been critically hindered, as the blockade fosters conditions that exacerbate violence, poverty, and deprivation.

UN Security Council Resolution 2334 (2016) reaffirms that Israeli settlements in the occupied Palestinian territories, including East Jerusalem, have no legal validity and constitute a flagrant violation of international law (Putra and Persadha 2024). Despite this resolution, Israel has continued to expand these settlements in open defiance of international law and the UN's decision.

UN Security Council has recently attempted to address the military assault in Gaza with resolutions such as Resolution 2728, adopted in March 2024, calling for an immediate ceasefire during Ramadan to facilitate humanitarian relief and reduce hostilities (UN 2024a). This resolution underscores the urgent need for adherence to international law, protection of civilians, and unrestricted access to essential aid for the people of Gaza. In June 2024, the Security Council adopted another resolution (Resolution 2735) proposing a comprehensive three-phase ceasefire agreement to end the war in Gaza, urging both Israel and Hamas to implement it fully, immediately, and without conditions (UN 2024b). Nonetheless, the failure to implement such resolutions, combined with Israel's continued settlement expansion, obstructs efforts to establish inclusive governance and equal access to justice as outlined in SDG 16.3. The enduring denial of basic rights to Palestinians, particularly through policies of segregation and resource restriction, further deepens divisions and fosters resentment, creating a cycle of violence that undermines regional stability and the foundational principles of SDG 16.

The treatment of Palestinian detainees, including minors, presents another critical challenge to achieving peace and justice in Gaza. Thousands of Palestinians, many detained arbitrarily without trial, endure conditions that are widely condemned by human rights organizations as violating international standards of humane treatment and due process. These detainees are often denied legal representation and held in conditions that fall short of the basic standards of human dignity (Addameer 2011; Amnesty International 2023; B'TSELEM 2024; Médecins Sans Frontières 2024). Such practices contravene international norms, including those set forth by the UN and various human rights treaties, and are in direct opposition to, for example, SDG 16.b, which promotes non-discriminatory laws and access to justice, and Target 16.3's aim for fair legal processes for all individuals.

The continued failure to address these violations has not only impeded peace and stability in Gaza but has also strained international relations, highlighting the global ramifications of regional conflict on governance, security, and sustainability.

9.4 Escalating Violence and the Threat to Children's Rights in Gaza

The recent military assault on Gaza has led to a dangerous escalation in violence, with devastating consequences for the rights and well-being of its inhabitants, especially children. Indeed, continuous military actions, blockades, and the lack of a safe environment threaten the very essence of SDG 16.2, which calls for an end to all forms of violence against children. In Gaza, children face physical and psychological harm daily, as well as violations of their rights to safety, education, and health. The increasing intensity of violence places children in vulnerable and traumatic situations, with many experiencing severe injury, displacement, and loss of family members (Taha et al. 2024; Wells et al. 2024; Boukari et al. 2024). Given these catastrophic conditions, numerous experts and UN organizations have emphasized that Gaza has become one of the most perilous places in the world for children, describing it as "a graveyard for children" (Hamad et al. 2024; UNICEF Australia 2024).

9.4.1 Physical and Psychological Violence Against Children in Gaza

The war on Gaza has inflicted severe physical violence on children, with devastating consequences. According to the Palestinian Central Bureau of Statistics, by December 29, 2024, 17,818 children in Gaza had been killed due to the war (PCBS 2024) (Fig. 9.1).

Among those who survive, many suffer from severe injuries, including amputations. Reports reveal that thousands of children have undergone limb amputations, often in dire conditions without adequate medical care, including the absence of anesthesia. These devastating physical injuries not only hinder the ability of children to lead normal lives but also create enduring socioeconomic hardships for their families and communities (Ćerimović 2023; Klin and Clin 2024; Boukari et al. 2024).

Psychological trauma is equally devastating. Children in Gaza are exposed to bombings, the destruction of their homes and schools, and the loss of family members, leading to widespread post-traumatic stress disorder (PTSD), anxiety, and depression. Research has shown that, even before the recent war, approximately 70% of children in Gaza exhibit symptoms of PTSD. This psychological burden is compounded by the lack of access to mental health services, leaving children to cope with their trauma in isolation (El-Khodary et al. 2020; Ferrara et al. 2024; Taha et al. 2024; Schöler et al. 2024; Veronese et al. 2025).

9.4.2 Structural Violence: Attacks on Schools and Healthcare Facilities

Structural violence manifests in Gaza through the deliberate targeting of civilian infrastructure, including schools, healthcare facilities, and other critical infrastructure (Mills et al. 2020; Buheji and Mushimiyimana 2023). Nearly 88% of school buildings in the Gaza Strip (496 out of

Fig. 9.1 Casualties and missing persons in Gaza and West Bank during the recent Israeli war. (PCBS 2024)

564) have been directly hit or damaged (UNRWA 2024). These attacks violate international humanitarian law, which designates schools as protected spaces during wartime.

Many schools that serve as emergency shelters for displaced families are still targeted, further endangering lives of children. The destruction of educational institutions has far-reaching consequences. Children who are unable to attend school face higher risks of exploitation, violence, and long-term poverty (Rabaia and Habash 2024; Hussein et al. 2024; Iriqat et al. 2025). Studies from previous conflicts have shown that children who drop out of school during war often do not return, leading to a loss of educational opportunities that affects entire generations (Justino 2016).

Healthcare facilities in Gaza have also been systematically attacked, exacerbating the already dire health crisis. The blockade and destruction of hospitals have left children without access to essential medical services. Approximately 97% of the water in Gaza is unfit for human consumption, contributing to the spread of diseases like diarrhea and malnutrition (Farajallah and

Farajallah 2024; Alkhaldi and Alrubaie 2024). The prevalence of malnutrition among children has skyrocketed, with acute malnutrition and stunting becoming widespread (Abuzerr et al. 2024; Horino et al. 2024). These conditions not only compromise the physical development of children but also hinder cognitive growth, affecting their future potential (Albelbeisi et al. 2024; Buheji and Buheji 2024).

9.4.3 Denial of Humanitarian Access and Starvation as a Weapon

The blockade on Gaza and the denial of humanitarian access have created conditions that amount to collective punishment. Relief organizations have warned of widespread famine in Gaza, with intentional starvation being used as a method of war. This deliberate deprivation violates international law and highlights the systemic nature of violence against children in Gaza (ESCWA 2023; de Waal 2024; Devereux 2024; Gordon and Haddad 2024; Hassoun et al. 2024; International Crisis Group 2024; Khoodoruth and Khoodoruth 2024). Children are particularly vulnerable to the effects of starvation and malnutrition. The lack of adequate food and medical supplies has led to a surge in preventable diseases, including anemia and severe malnutrition. Research has shown that these conditions have long-term impacts on children's health and well-being, perpetuating cycles of poverty and deprivation (Tanielian 2024; Nasreddine and Jomaa 2024).

9.4.4 Long-Term Implications for Children and Society

The war in Gaza not only affects children in the immediate term but also has long-lasting implications for their development and the future of their society. The loss of educational opportunities, coupled with psychological and physical trauma, decreases the ability of children to contribute to the rebuilding of their communities. This creates a cycle of intergenerational poverty and instability, making it difficult for Gaza to recover even after the cessation of hostilities (El-Khodary et al. 2020; Klin and Clin 2024; Iriqat et al. 2025).

To reduce the impact of the war on children in Gaza, international efforts must focus on key actions. A permanent end to hostilities is crucial to prevent further harm, and Israel must comply with international humanitarian law to ensure the protection of civilian infrastructure. The blockade on Gaza must be lifted to allow the delivery of essential food, medical supplies, and educational resources. Establishing humanitarian corridors can facilitate the safe evacuation of children and families. International donors should support the reconstruction of schools, healthcare facilities, and critical infrastructure, ensuring access to clean water and adequate nutrition. Expanding psychological support programs and training local providers is vital to address mental health needs among children affected by war. Accountability for war crimes, including attacks on children and civilian sites, must be pursued through international legal mechanisms. Sustainable development in Gaza requires long-term investments in education and healthcare to help break the cycle of poverty and violence (El-Khodary et al. 2020; Buheji and Buheji 2024; Elkahlout 2024; Bhutta et al. 2024).

9.5 Apartheid and Its Impact on Governance, Equality, and Human Rights

The apartheid system imposed on Palestinians within the occupied territories gravely undermines their rights and hinders the development of just governance, equality, and human rights, directly opposing SDG 16. Through systematic and institutionalized discrimination, policies and laws are deliberately designed to privilege Israeli citizens while denying Palestinians equal rights, treatment, and protection under the law (Dugard and Reynolds 2013; Amnesty International 2022; Brooks and Griffiths 2024). This structural inequality contradicts, for example, Target 16.3 of SDG 16, which advocates for equal access to justice and the preservation of the rule of law for all people, regardless of nationality or ethnicity, as well as 16b, which calls for the elimination of discriminatory policies to foster sustainable development.

The impacts of apartheid are pervasive across multiple levels of governance and institutional representation. Palestinians are systematically denied political representation in decision-making processes that directly impact their lives, fostering a regime characterized by a lack of transparency, accountability, and inclusivity (B'TSELEM 2021; Human Rights Watch 2021). This discriminatory exclusion from governance structures undermines institutional legitimacy and violates SDG 16.6 and SDG 16.7, which call for inclusive and accountable institutions that genuinely represent all populations under their jurisdiction. This institutionalized inequality leads to a profound lack of trust in governance systems, as Palestinians are alienated from policies that affect their daily lives, thereby perpetuating a cycle of disenfranchisement and marginalization (Iriqat 2024; Jellad 2024; Sultany 2024).

One of the most critical manifestations of this apartheid system is the ongoing blockade imposed on Gaza, which has remained in effect since 2007. This blockade enforces extreme limitations on Palestinians' access to essential resources, services, and economic opportunities, and it has created an unsustainable environment with far-reaching social and economic consequences. Gaza's population endures significant restrictions on imports and exports, severely hampering local industries, healthcare, education, and overall economic growth (Butt and Butt 2016; Human Rights Watch 2021; Huss and Altehe 2024). By blocking access to necessary materials and resources, the system deepens socioeconomic divides, solidifying an environment of poverty and stagnation that starkly contrasts with SDG's targets, especially 16.b.

The forced displacement in Gaza has affected approximately 1.9 million people, accounting for over 90% of the population (Amnesty International 2024b). This mass displacement, occurring under conditions widely recognized as constituting apartheid, has left families without shelter, disrupted livelihoods, and worsened humanitarian conditions, especially during the harsh winter months. Such displacement contradicts many targets of SDG 16, such as Target 16.1, as forced displacement is a direct manifestation of systemic violence, depriving individuals of their homes, security, and dignity while exposing them to severe poverty, escalating health crises, and life-threatening instability. The critical situation in Gaza highlights the urgent need to address the root causes of displacement, terminate apartheid practices, and ensure the protection and well-being of the displaced population, aligning with global commitments to promote peaceful and inclusive societies under the UN's SDG framework.

These inequalities are further reflected in basic human rights violations, including restricted access to housing, healthcare, and employment. Housing shortages and infrastructure limitations are particularly acute in Gaza, where rebuilding efforts have been severely limited due to restrictions on materials such as concrete and steel, creating unsafe living conditions and impeding progress in human development. Health outcomes are similarly impacted, with the region experiencing critical shortages in medical supplies and healthcare infrastructure. Many residents are prevented from traveling to receive medical care unavailable within Gaza, directly contravening the universal right to health as recognized in international human rights agreements, including the International Covenant on Economic, Social and Cultural Rights (Amnesty International 2018; Issa et al. 2021; WHO 2022; Agha et al. 2024; Elkahlout 2024; Human Rights Watch 2024b; Mahomed 2024).

In the legal sphere, apartheid practices extend to the treatment of Palestinian detainees, particularly children, who are frequently detained without formal charges, subjected to inhumane treatment, and denied access to legal representation. These practices violate numerous international legal standards, including those stipulated in the Convention on the Rights of the Child, which guarantees humane treatment and access to fair trial rights (Wells et al. 2024; Boukari et al. 2024). The arbitrary detention of Palestinians, especially minors, undermines SDG Target 16.3's call for equal access to justice and reinforces a culture of systemic impunity, eroding public trust in legal institutions and perpetuating cycles of violence and resentment.

International human rights organizations, including UN, have repeatedly condemned apartheid practices within the occupied Palestinian territories. Reports from Human Rights Watch and Amnesty International have documented extensive evidence of racial discrimination and segregation embedded within Israeli law and policy. These reports align with statements by the UN, which highlight the urgent need for reform to dismantle discriminatory legal structures and ensure that institutions operate inclusively and equitably (Amnesty International 2018, 2022; Human Rights Watch 2021; Gordon 2024). Without significant changes to these systems, achieving SDG 16's vision of a society rooted in peace, justice, and strong institutions remains implausible.

The broader international community holds a critical role in supporting Palestinians' pursuit of justice, human rights, and governance that aligns with SDG 16. International courts, such as the International Court of Justice (ICJ) and the International Criminal Court (ICC), have engaged with cases alleging war crimes and violations of international law in the occupied territories. However, political influences have at times obstructed these cases, hindering accountability and reinforcing cycles of impunity (Gwaya 2024; Hasan and Buheji 2024; Jacob 2024). This lack of accountability not only violates SDG 16.3's call for the rule of law but also perpetuates mistrust among the affected populations, further destabilizing the region.

Apartheid policies, especially through the blockade on Gaza, severely disrupt equality and sustainable development by restricting Palestinians, particularly Gazans, from accessing essential resources, services, and economic opportunities. This system enforces stark inequalities in housing, education, healthcare, and employment, perpetuating poverty and stifling economic growth, directly undermining SDG 16.b's objective of promoting non-discriminatory policies for sustainable development, as well as other SDG 16 Targets. These discriminatory practices violate internationally recognized human rights, restricting freedoms and perpetuating cycles of violence and oppression. Achieving the vision of SDG 16 necessitates dismantling apartheid structures to build a society that upholds human dignity, enables inclusive governance, and promotes sustainable peace and development.

9.6 Mitigation Measures and Recommendations

The humanitarian and governance crises in Gaza necessitate urgent and multifaceted mitigation strategies to address the acute and long-term challenges faced by its population. SDG 16 emphasizes peace, justice, and strong institutions, providing a critical framework for these efforts. Addressing the deep-seated impacts of war, displacement, systemic inequality, and genocide in Gaza requires combined international interventions that prioritize humanitarian support, accountability, and institutional reform.

9.6.1 Prioritizing Humanitarian Aid and Infrastructure Rehabilitation

The immediate cessation of hostilities is critical to prevent further civilian harm and humanitarian devastation in Gaza. In parallel, humanitarian corridors should be established to deliver essential aid, including food, clean water, medical supplies, and shelter for displaced populations. The blockade on Gaza, which restricts access to vital resources and reconstruction materials, must be lifted to allow for rebuilding critical infrastructure such as hospitals, schools, and housing. International donors and organizations must commit to supporting these efforts, ensuring that resources are allocated equitably and transparently (Hassoun et al. 2024; Bhutta et al. 2024).

9.6.2 Strengthening Legal Accountability

Enforcing international law is paramount to mitigating the effects of this unprecedented war and ensuring justice for victims. Key international bodies, including ICC and ICJ, should expedite investigations and prosecute war crimes, such as the targeting of civilians and infrastructure, forced displacement, and violations against children. Transparent and fair legal processes can restore trust in governance and reinforce the rule of law, signaling that violations of international norms will not go unpunished (Brain et al. 2024; Gwaya 2024).

9.6.3 Building Inclusive and Transparent Governance

Reforming governance structures in Gaza is essential for fostering stability and resilience. Anti-corruption measures, inclusive decision-making processes, and participatory governance should be prioritized to empower communities and rebuild trust in local institutions. Strengthening institutional transparency and accountability can mitigate the impacts of systemic inequality and ensure that resources and services are distributed fairly. Research highlights that inclusive governance frameworks, particularly in post-conflict regions, are critical for long-term stability and development (Joronen and Griffiths 2022; Hope 2023; Lahlouh 2024).

9.6.4 Addressing the Needs of Displaced Populations

Forced displacement in Gaza has affected over 1.9 million people, creating dire living conditions, particularly during the harsh winter months. Mitigation efforts must include immediate shelter provision, safe access to health services, and educational opportunities for displaced children. Long-term solutions should focus on reintegration strategies and the reconstruction of destroyed communities. Programs that provide psychosocial support for children and families affected by trauma are vital for breaking the cycle of intergenerational poverty and instability (Human Rights Watch 2024a; Alkhaldi and Alrubaie 2024).

9.6.5 International Cooperation and Enforcement of Resolutions

Global solidarity and collaboration are necessary to address the root causes of crises in Gaza and support sustainable recovery. UN resolutions calling for ceasefires, civilian protections, and humanitarian access must be enforced without delay. International actors should hold accountable those who obstruct these resolutions, ensuring that global governance systems maintain their credibility and effectiveness. Additionally, partnerships between local authorities and international organizations can provide technical expertise and resources for rebuilding resilient institutions and infrastructures (Kondoch 2024; Putra and Persadha 2024).

9.6.6 Promoting Sustainable Development Through Education and Healthcare

Sustainable recovery in Gaza requires investments in education and healthcare systems to address immediate needs and foster long-term resilience. Educational initiatives should prioritize rebuilding schools and providing access to quality learning environments for children, especially those displaced by this war. Healthcare systems must be equipped to manage acute needs and long-term challenges, including mental health services for trauma recovery. These efforts are aligned with SDG 16 targets and contribute to creating a more equitable and inclusive society (Ferrara et al. 2024; Bhutta et al. 2024).

9.7 Remarks on Recent Developments

On January 15, 2025, a ceasefire agreement was brokered between Hamas and Israel, officially ending the war in Gaza on January 19, 2025. However, without addressing the root causes of the conflict, Israel may continue to use the presence of Hamas as justification for future military actions. The durability of this agreement hinges on its scope, the commitment of all parties involved, and the implementation of effective enforcement and monitoring mechanisms. A resolution that fails to adopt a comprehensive approach risks being temporary and unstable.

The long-term outcomes depend on the objectives of the agreement. If the deal includes tangible measures to rebuild Gaza and improve living conditions, including lifting the blockade and ensuring access to essential services like electricity, water, and healthcare, it could bring much-needed relief to the population. Conversely, agreements confined to prisoner exchanges or temporary ceasefires, without addressing fundamental issues, will likely result in a fragile and short-lived peace, leaving the door open to renewed violence and other bloody conflicts in the future. The ongoing pattern of wars followed by temporary truces has shown that addressing this conflict requires tackling key issues, including the blockade, the occupation, and the assurance of political rights for the Palestinian people. Without mechanisms to monitor compliance and hold parties accountable, any agreement risks collapse. Internal Palestinian unity and dialogue are also crucial, as divisions weaken stability and enable policies that undermine Palestinian rights and aspirations for self-determination.

A sustainable and just solution requires a political framework that addresses economic development, ensures political rights, establishes effective governance in Gaza, and includes the withdrawal of the occupation. Without these core elements, any agreement is unlikely to end the cycle of conflict and humanitarian suffering, instead prolonging instability and violence. Equally important is the need for a significant shift in Israeli leadership. The far-right government under Netanyahu has pursued policies and rhetoric that intensify tensions and obstruct peace efforts. Replacing such leadership with moderate, peace-oriented figures is essential to creating an environment conducive to dialogue and reconciliation. A government that prioritizes coexistence and mutual respect would help build trust, encourage meaningful negotiations, and lay the groundwork for a durable resolution that benefits both Palestinians and Israelis.

9.8 Conclusion

Achieving SDG 16 in Gaza presents significant challenges, requiring a combined effort from regional actors and robust international support. The path to peace, justice, and strong institutions is hindered by the recent war, political fragmentation, and economic instability. However, strategies grounded in accountability, transparency, and inclusivity offer a viable way forward. Central to this approach is ensuring violators of international law are held accountable through transparent legal processes, which reinforces human rights protections and rebuilds trust in governance. The enforcement of UN resolutions on ceasefires, humanitarian aid, and civilian protections by the international community is essential to promoting stability and affirming a global commitment to justice and dignity. The systematic targeting of children and civilian infrastructure profoundly undermines international efforts to achieve the SDGs, particularly SDG 16, which aims to end violence and foster peaceful, inclusive societies. Sustainable peace in Gaza requires addressing both immediate hostilities and the structural policies of blockade and deprivation that perpetuate suffering. Ending the violence and apartheid is only the first step; long-term recovery demands dismantling the systemic inequities that drive war and genocide.

The intervention of NGOs, particularly UNRWA, is vital for implementing anti-corruption measures, promoting transparency, and fostering inclusive decision-making processes to ensure the effective delivery of essential services in Gaza and rebuild public trust. Collaboration with global partners can further bolster these efforts by providing the necessary resources and expertise to reconstruct resilient and adaptable infrastructure, addressing both immediate needs and long-term challenges. By integrating principles of justice, human rights, and equitable resource distribution, Gaza can address its immediate humanitarian crises while laying a sustainable foundation for future generations' well-being.

References

Abed Alah M (2024) Shattered hierarchy: how the Gaza conflict demolished Maslow's pyramid of needs. Curr Psychol:1–7. https://doi.org/10.1007/S12144-024-05784-1/FIGURES/1

Abuzerr S, Al-jawaldeh A, Ashour Y et al (2024) The silent crisis: effect of malnutrition and dehydration on children in Gaza during the war. Front Nutr 11:1395903

Addameer (2011) Violations against Palestinian Prisoners and Detainees in Israeli Prisons and Detention Centers

Agha Z, Esson J, Griffiths M, Joronen M (2024) Gaza: A decolonial geography. Trans Inst Br Geogr 49:e12675. https://doi.org/10.1111/tran.12675

Albelbeisi A, Zinszer K, Hamid A, Bilbeisi E (2024) The burden of acute malnutrition among children under five in conflict-afflicted Gaza strip: prevalence and associated factors. Front Nutr 1478485. https://doi.org/10.3389/fnut.2024.1478485

Alkhaldi M, Alrubaie M (2024) Roadmap for rebuilding the health system and scenarios of crisis path in Gaza. Int J Health Plann Manag:1–13. https://doi.org/10.1002/HPM.3861

Amnesty International (2018) Amnesty International Report 2017/18: The state of the world's human rights

Amnesty International (2022) Israel's apartheid against Palestinians: a cruel system of domination and a crime against humanity. https://www.amnesty.org/en/latest/news/2022/02/israels-apartheid-against-palestinians-a-cruel-system-of-domination-and-a-crime-against-humanity/. Accessed 31 Dec 2024

Amnesty International (2023) Human rights in Israel and the Occupied Palestinian Territory. https://www.amnesty.org/en/location/middle-east-and-north-africa/middle-east/israel-and-the-occupied-palestinian-territory/report-israel-and-the-occupied-palestinian-territory/. Accessed 31 Dec 2024

Amnesty International (2024a) 'You feel like you are subhuman'—Israel's genocide against Palestinians in Gaza

Amnesty International (2024b) End Israel's genocide against Palestinians in Gaza. https://www.amnesty.org/en/petition/end-israels-genocide-against-palestinians-in-gaza/. Accessed 31 Dec 2024

Asi Y, Mills D, Greenough PG et al (2024) 'Nowhere and no one is safe': spatial analysis of damage to critical civilian infrastructure in the Gaza strip during the first phase of the Israeli military campaign, 7 October to 22 November 2023. Confl Heal 18:1–13. https://doi.org/10.1186/s13031-024-00580-x

B'TSELEM (2021) A regime of Jewish supremacy from the Jordan River to the Mediterranean Sea: This is apartheid. https://www.btselem.org/publications/fulltext/202101_this_is_apartheid. Accessed 31 Dec 2024

B'TSELEM (2024) Abuse of Palestinians by Israeli soldiers in the center of Hebron

Bhutta ZA, Dominguez GB, Wise PH (2024) When is enough, enough? Humanitarian rights and protection for children in conflict settings must be revisited. BMJ 386:e081515. https://doi.org/10.1136/BMJ-2024-081515

Bose S, Kumar A (2021) Conflict-affected areas and sustainable development goal 16. In: Filho WL, Azul AM, Brandli L et al (eds) Peace, justice and strong institutions, encyclopedia of the UN sustainable development goals. Springer, pp 120–130

Boukari Y, Kadir A, Waterston T et al (2024) Gaza, armed conflict and child health. BMJ Paediatr Open 8:e002407. https://doi.org/10.1136/BMJPO-2023-002407

Brain IY, Journal I, Management L et al (2024) Enforcing the battlefield code: the Geneva conventions and the ICC's role in prosecuting war crimes. Int J Law Manag Humanit 7:1447–1470

Brooks A, Griffiths M (2024) Beyond Apartheid Israel. Polit Geogr 114:103193. https://doi.org/10.1016/j.polgeo.2024.103193

Buheji M, Buheji B (2024) Mitigating risks of slow children development due to war on Gaza 2023. Int J Psychol Behav Sci 14:11–21. https://doi.org/10.5923/j.ijpbs.20241401.02

Buheji M, Mushimiyimana E (2023) Raising Gaza survival capacity as per lesson from best war survival stories in recent history. Int J Adv Res Soc Sci Humanit 07:22–48

Butt KM, Butt AA (2016) Blockade on Gaza strip: a living hell on earth. J Polit Stud 23:157–182

Ćerimović E (2023) At risk and overlooked: children with disabilities and armed conflict. Int Rev Red Cross 105:192–216. https://doi.org/10.1017/S181638312200087X

Consuegra LJ (2020) Democracy and peacebuilding in the framework of SDG 16+. Int IDEA Discuss Pap 1/2020. https://doi.org/10.31752/idea.2020.8

de Waal A (2024) Famine in Gaza: an example of the global humanitarian crisis. Am J Clin Nutr 119:1383–1385. https://doi.org/10.1016/j.ajcnut.2024.04.015

Devereux S (2024) Was there a famine in Gaza in 2024?

Drange B (2022) A tug of war: pursuing justice amid armed conflict. Nord J Hum Rights 40:346–364. https://doi.org/10.1080/18918131.2022.2097787

Dugard J, Reynolds J (2013) Apartheid, international law, and the occupied palestinian territory

Elkahlout G (2024) Gaza's cycle of destruction and rebuilding. Understanding the actors, dynamics, and responses. Palgrave Macmillan

El-Khodary B, Samara M, Askew C (2020) Traumatic events and PTSD among Palestinian children and adolescents: the effect of demographic and socioeconomic factors. Front Psych 11:1–11. https://doi.org/10.3389/fpsyt.2020.00004

El-Shewy M, Griffiths M, Jones C (2024) Israel's war on Gaza in a global frame. Antipode:1–21. https://doi.org/10.1111/ANTI.13094

ESCWA (2023) War on Gaza: weaponizing access to water, energy and food

Farajallah I, Farajallah H (2024) Challenges in access to safe drinking water and its impact on maternal and child health in Gaza. Anatol Clin J Med Sci:52–65. https://doi.org/10.21673/anadoluklin.1575390

Ferrara P, Cammisa I, Zona M et al (2024) The global issue of violence toward children in the context of war. J Pediatr 269:114007. https://doi.org/10.1016/j.jpeds.2024.114007

Gordon N (2024) Between human rights and civil society: the case of Israel's apartheid enablers. Law Soc Inq 49:1426–1452. https://doi.org/10.1017/LSI.2023.41

Gordon N, Haddad M (2024) The road to famine in Gaza. New York Rev Books 1:1–12

Gwaya KR (2024) The authority of the ICC to institute criminal proceedings against Israeli and Hamas leaders: the Gaza conflict. Beijing Law Rev 15:2202–2217. https://doi.org/10.4236/blr.2024.154123

Hamad BA, Vintges J, Diab R, et al (2024) Shattered lives and dreams: the toll of the war on Gaza on young people

Hasan A, Buheji M (2024) A world losing its legitimacy-Gaza from collective punish till ethnic cleansing & genocide. Int J Manag 15:188–209. https://doi.org/10.17605/OSF.IO/24BYN

Hassoun A, Al-Muhannadi K, Hassan HF et al (2024) From acute food insecurity to famine: how the 2023/2024 war on Gaza has dramatically set back sustainable development goal 2 to end hunger. Front Sustain Food Syst 8:1402150. https://doi.org/10.3389/fsufs.2024.1402150

Hope KR (2023) Peace, justice, and inclusive institutions: overcoming challenges to the implementation of sustainable development goal 16 in Africa and beyond. In: Corruption, sustainable development and security challenges in Africa. Advances in African economic, social and political development. Springer, Cham, pp 131–160

Horino M, Zaqqout R, Habash R et al (2024) Food insecurity, dietary inadequacy, and malnutrition in the Gaza strip: a cross-sectional nutritional assessment of refugee children entering the first grade of UNRWA schools and their households before the conflict of 2023–24. Lancet Glob Heal 12:e1871–e1880. https://doi.org/10.1016/S2214-109X(24)00320-6

Human Rights Watch (2021) A threshold crossed: Israeli authorities and the crimes of apartheid and persecution. https://www.hrw.org/report/2021/04/27/threshold-crossed/israeli-authorities-and-crimes-apartheid-and-persecution. Accessed 31 Dec 2024

Human Rights Watch (2023) Israel: starvation used as weapon of war in Gaza

Human Rights Watch (2024a) "Hopeless, starving, and besieged" Israel's forced displacement of Palestinians in Gaza

Human Rights Watch (2024b) Extermination and acts of genocide Israel deliberately depriving Palestinians in Gaza of water

Huss M, Altehe S (2024) On spaciocide and resistance: between Bi'r as-Sab'a and Gaza. Polit Geogr 114:103186. https://doi.org/10.1016/j.polgeo.2024.103186

Hussein A, Wong S, Bright A (2024) History and impact of Israeli siege and attacks on education in Gaza, Palestine. Oxford Res Encycl Educ. https://doi.org/10.1093/ACREFORE/9780190264093.013.1903

International Crisis Group (2024) Stopping Famine in Gaza

Iriqat D (2024) Towards an inclusive political system, Palestinian perspectives on the reconstruction of Gaza. In: Antar U, Asseburg M, Shaban O (eds) Palestinian perspectives on the reconstruction of Gaza. Friedrich-Ebert-Stiftung (FES), Stiftung Wissenschaft und Politik (SWP)

Iriqat D, Alousi R, Aldahdouh TZ et al (2025) Educide amid conflict: the struggle of the Palestinian education system. Qual Educ All 2:81–99. https://doi.org/10.1108/QEA-10-2024-0120

Issa YY, Amro A, Rajabi RK (2021) Environmental and social determinants of health in Palestine. In: Handbook of healthcare in the Arab World, pp 607–649

Jacob C (2024) A new politics of international criminal justice: accountability in Ukraine and the Israel–Gaza war. Int Aff 000. https://doi.org/10.1093/IA/IIAE224

Jellad F (2024) Israel's ethnocratic dynamics. A settler colonialist lens on apartheid-like policies and the marginalization of Palestinian Arabs. J Soc Sci 5:191–213

Joronen M, Griffiths M (2022) Ungovernability and ungovernable life in Palestine. Polit Geogr 98:102734. https://doi.org/10.1016/j.polgeo.2022.102734

Justino P (2016) Supply and demand restrictions to education in conflict-affected countries: new research and future agendas. Int J Educ Dev 47:76–85. https://doi.org/10.1016/j.ijedudev.2016.01.002

Khoodoruth MAS, Khoodoruth WNC kai (2024) From past famines to present crises: the epigenetic impact of maternal malnutrition on offspring health in Gaza. Asian J Psychiatr 95: 103999. https://doi.org/10.1016/j.ajp.2024.103999

Klin A, Clin A (2024) Behind the rubble: psychological trauma of wars and human rights abuses on women and children in Gaza. Anatol Clin J Med Sci 29:119–136. https://doi.org/10.21673/anadoluklin.1575372

Kondoch B (2024) The Gaza conflict: do norms and international institutions matter? J Int Peacekeeping 27:247–248. https://doi.org/10.1163/18754112-27030001

Küfeoğlu S (2022) Peace, justice and strong institutions. In: Emerging technologies, sustainable development goals series. Springer, Cham, pp 58–59

Lahlouh A (2024) Governance and political reform. In: Antar U, Asseburg M, Shaban O (eds) Palestinian perspectives on the reconstruction of Gaza. Friedrich-Ebert-Stiftung, pp 441–460

Mahomed S (2024) International humanitarian laws: applicable to all or a privilege for some? SAJBL 17:22–26

Médecins Sans Frontières (2024) Gaza: life in a death trip

Mills D, Wispelwey B, Muhareb R, Gilbert M (2020) Structural violence in the era of a new pandemic: the case of the Gaza strip. Lancet:19–20. https://doi.org/10.1016/S0140-6736(20)30730-3

Milton S, Alhamawi M (2024) Peace-centred sustainable development: an analysis of SDG 16 in the Arab states. World Dev Perspect 34:100587. https://doi.org/10.1016/J.WDP.2024.100587

Nasreddine L, Jomaa L (2024) Safeguarding the SDG promise to end hunger and leave no one behind: the plight of children in the Gaza strip. Lancet Glob Heal 12:e1748–e1749. https://doi.org/10.1016/S2214-109X(24)00367-X

O'Reilly G (2023) Achieving a sustainable future: the geographical centrality of UN SDG-16, peace, justice and strong institutions. In: De Lázaro Torres ML, De Miguel GR (eds) Sustainable development goals in Europe. Key challenges in geography. Springer, Cham, pp 329–349

PCBS (2024) Martyrs. https://www.pcbs.gov.ps/site/lang__en/1405/Default.aspx. Accessed 30 Dec 2024

Putra ET, Persadha PD (2024) Transforming conflict: analyzing the impact of UN security council resolution 2334 on human security in Israel-Palestine. J Law Sci 6:91–100

Rabaia I, Habash L (2024) The hidden war on higher education: unmasking the 'Educide' in Gaza. In: The war on Gaza and Middle East political science. George Washington University- POMEPS, pp 8–13

Roy S (2012) Reconceptualizing the Israeli-Palestinian conflict: Key paradigm shifts. J Palest Stud 41:71–91. https://doi.org/10.1525/jps.2012.XLI.3.71

Roy S (2016) The Gaza strip: the political economy of de-development, 3rd edn. Institute for Palestine Studies, Washington, DC

Salmon AK, Pérez-Prado A, Morrison K, Iuspa F (2018) SDG 16: peace, justice, and strong institutions, SDG 16 peace, justice, and strong institutions, SDG 4 quality education & SDG 10 reduce inequalities. In: Children's literature aligned with SDGs to promote global competencies. Springer, Cham, pp 174–179

Schöler N, Gal G, Wissow LS, Seita A (2024) Stress and trauma symptoms in young Palestine refugee children following the May 2021 escalation in Gaza. JAACAP Open. https://doi.org/10.1016/J.JAACOP.2024.06.003

Sultany N (2024) A threshold crossed: on genocidal intent and the duty to prevent genocide in Palestine. J Genocide Res:1–26. https://doi.org/10.1080/14623528.2024.2351261

Taha AM, Sabet C, Nada SA et al (2024) Addressing the mental health crisis among children in Gaza. Lancet Psychiatry 11:249–250. https://doi.org/10.1016/S2215-0366(24)00036-1

Tanielian MS (2024) The silent slow killer of famine: humanitarian Management and permanent security. J Genocide Res:1–9. https://doi.org/10.1080/14623528.2024.2310866

UN (2024a) Security Council demands immediate ceasefire in Gaza for month of Ramadan, adopting resolution 2728 (2024) with 14 members voting in favour, United States abstaining. https://press.un.org/en/2024/sc15641.doc.htm. Accessed 31 Dec 2024

UN (2024b) Adopting Resolution 2735 (2024) with 14 Votes in Favour, Russian Federation Abstaining, Security Council Welcomes New Gaza Ceasefire Proposal, Urges Full Implementation. https://press.un.org/en/2024/sc15723.doc.htm. Accessed 31 Dec 2024

UNICEF Australia (2024) The most dangerous place in the world to be a child. https://www.unicef.org.au/stories/gaza-is-the-most-dangerous-place-in-the-world-to-be-a-child?srsltid=AfmBOoruMIigIaSpQRlkAaB39vtZtUKsv6HOajVHEW8bb5Um_kt4jT6M. Accessed 31 Dec 2024

UNRWA (2024) UNRWA situation report #152 on the Humanitarian Crisis in the Gaza Strip and the West Bank, including East Jerusalem. https://www.unrwa.org/resources/reports/unrwa-situation-report-152-situation-gaza-strip-and-west-bank-including-east-jerusalem. Accessed 31 Dec 2024

Veronese G, Kagee A (2024) Does psychology fail in the face of genocide? Insights about Gaza before and after October 7, 2023. Int J Hum Rights Healthc. https://doi.org/10.1108/IJHRH-07-2024-0053

Veronese G, Bdier D, Obaid H et al (2025) Agency, life satisfaction, hope, potentially traumatic events, trauma symptoms, and psychological signs. A two waves study with a sample of Palestinian children living in different geographical areas. Child Youth Serv Rev 168:107990. https://doi.org/10.1016/j.childyouth.2024.107990

Wells K, Cortés-Morales S, Esson J et al (2024) Israel's war on Gaza and the violation of children's rights. Child Geogr:1–4. https://doi.org/10.1080/14733285.2024.2316752

WHO (2022) Right to health: Barriers to health and attacks on health care in the occupied Palestinian territory, 2019 to 2021

Israel's Systematic Weaponization of Water and Sanitation

10

Ghinwa Harik, Zaynab Nemr, and Rami Zurayk

Abstract

This chapter examines Israel's systematic weaponization of water and sanitation in Gaza, illustrating how these essential resources have been transformed into tools of control and deprivation. Gaza's water crisis, exacerbated by repeated conflicts and prolonged blockades, has left millions without access to clean water and adequate sanitation, severely undermining efforts to achieve Sustainable Development Goal 6 (SDG 6): Water and Sanitation for All. The destruction of critical infrastructure, including water treatment plants and desalination facilities, has resulted in widespread contamination, contributing to severe public health crises, environmental degradation, and economic instability. Waterborne diseases such as cholera and dysentery have surged, disproportionately affecting vulnerable populations, particularly children. The chapter explores the dire consequences of water scarcity on health, food security, and livelihoods while emphasizing the need for urgent international intervention to address the crisis. It highlights key factors influencing Gaza's water crisis, including political blockade, military actions, and systemic neglect, which have collectively hindered sustainable development. Recommendations for recovery focus on rebuilding infrastructure, enhancing governance frameworks, and ensuring international accountability. The findings underscore the urgent necessity for a comprehensive, multifaceted approach to water management in Gaza to achieve sustainable resilience and long-term stability.

Keywords

Gaza war · Infrastructure destruction · Public health · Humanitarian aid · Conflict impact · SDG 6 · Environmental degradation

10.1 Introduction

In this chapter, we explore Gaza's Grey Sanitation, Water and Armed Conflicts Nexus highlighting Gaza as a critical example of how water and sanitation has been weaponized by Israel to obstruct progress toward achieving Sustainable Development Goal 6 (SDG 6), Water and Sanitation for All. Water is fundamental to

G. Harik
Faculty of Engineering II, Lebanese University (ULFG2), Beirut, Lebanon
e-mail: gxh00@mail.aub.edu

Z. Nemr (✉) · R. Zurayk
School of Architecture and Design—Maroun Semaan Faculty of Engineering and Architecture, American University of Beirut, Beirut, Lebanon
e-mail: zn40@aub.edu.lb; rzurayk@aub.edu.lb

A. Hassoun (ed.), *War on Gaza*, Sustainable Development Goals Series,
https://doi.org/10.1007/978-3-031-88500-6_10

life and development, an essential resource that sustains ecosystems, economies, and human health. It is enshrined as a basic human right and forms the bedrock of modern public health systems. However, in regions embroiled in conflict, water often becomes a scarce commodity—its access restricted and weaponized by combatants. This chapter explores how the protracted conflict in Gaza has exacerbated an already dire water and sanitation crisis, turning access to clean water into a political weapon. Gaza stands as a case study of how warfare undermines the pursuit of SDG 6, which aims to ensure availability and sustainable management of water and sanitation for all.

SDG 6, part of the United Nations' 2030 Agenda for Sustainable Development, was established in 2015 with a global commitment to ensure universal access to clean water and sanitation. The goal emphasizes the equitable distribution of water resources, the sustainable management of water systems, and the reduction of water scarcity across the globe (United Nations 2017). Achieving SDG 6 would mean every person has access to safe, affordable drinking water and proper sanitation facilities, laying the foundation for healthy communities and thriving economies (United Nations 2017). These objectives are tied to multiple other Sustainable Development Goals, notably SDG 2 (End hunger, achieve food security and improved nutrition, and promote sustainable agriculture), SDG 3 (Good Health and Well-Being), SDG 10 (Reduced Inequalities), and SDG 8 (Decent Work and Economic Growth). Without universal access to clean water and sanitation, progress on education, health, poverty alleviation, and gender equality is severely hindered.

Yet, despite international recognition of water's critical importance, the world is far from achieving SDG 6. According to the World Health Organization (WHO), over 2 billion people globally lack access to safely managed drinking water services, and 4.2 billion people live without safely managed sanitation (WHO 2021). These gaps in service disproportionately affect the world's most vulnerable populations, particularly in conflict zones, where infrastructure is often destroyed, and resources are weaponized. The decades-long siege and the ongoing Israeli genocidal war in Gaza provide a clear and tragic illustration of how conflict renders the achievement of SDG 6 nearly impossible, turning water from a life-sustaining resource into a tool of control and deprivation. In regions affected by conflict, such as Gaza, the destruction of water and sanitation facilities exacerbates the spread of diseases like cholera, dysentery, and typhoid fever, which thrive in environments where clean water is scarce and sanitation systems are compromised. Moreover, communities that lack reliable clean water sources struggle to build resilience against environmental shocks such as droughts and floods, leaving them vulnerable to food insecurity and economic instability (NRC 2024; Refugees International 2024; UNDP 2023).

The situation in Gaza demonstrates how armed conflict compounds these challenges. Gaza's fragile economy, which relies heavily on agriculture, has been devastated by water scarcity and the destruction of essential infrastructure. Agricultural yields have plummeted as seawater intrusion, exacerbated by the destruction of water treatment plants, has contaminated freshwater aquifers. This has severely impacted food security in Gaza, where 80% of the population is already dependent on international food aid (WFP 2024; World Bank 2024a, b, c, d). The intersection of economic deprivation, environmental degradation, and health crises creates a humanitarian emergency, illustrating the deep interconnection between water, development, and human well-being.

The water crisis in Gaza is not merely a consequence of natural resource limitations but a direct result of sustained siege, unbridled military violence, and deliberate control policy. Even before the most recent conflict, Gaza's water and sanitation infrastructure was inadequate. Years of political blockade, insufficient infrastructure investment, and over-extraction of groundwater had left the region's water systems fragile and inefficient. However, the situation has dramatically worsened due to repeated military operations that target civilian infrastructure (Abd El Hay 2024; UNOSAT 2024; Abudayya et al.

2023). As of 2023, 97% of Gaza's freshwater is undrinkable, primarily due to contamination from seawater and untreated sewage, which has seeped into the coastal aquifer that serves as Gaza's primary water source (UNICEF 2023). Israel's military has repeatedly bombed Gaza's water treatment plants, desalination facilities, and sewage treatment systems, crippling the region's ability to provide clean water. This deliberate destruction of infrastructure has turned water into a tool of control, exacerbating the already dire humanitarian situation (Asi et al. 2024; Buheji 2024; Goodarzi et al. 2024; UNOSAT 2024; UNRWA 2024). The children of Gaza bear the brunt of this crisis. Waterborne diseases, such as diarrhea and dysentery, have become rampant, contributing to malnutrition and high child mortality rates. The ongoing war has further intensified the crisis, with daily bombardments destroying what little remained of Gaza's water infrastructure. Public health systems, already stretched to their limits, are now overwhelmed by the surge in water-related illnesses and the growing number of displaced people (Ashour et al. 2024; Boukari et al. 2024; Correia 2024; Elnakib et al. 2024; Gupta et al. 2024; Taha et al. 2024; UNICEF 2024a, b). As a result, Gaza faces a full-scale public health emergency, where the lack of access to clean water has become a matter of life and death.

The overarching objective of this chapter is to critically assess the impact of the current war on Gaza's water and sanitation infrastructure, exploring how the destruction of these systems has led to a public health catastrophe. This analysis will be guided by several key questions:

- How has warfare affected Gaza's water and sanitation infrastructure? The chapter will document the extent of damage to Gaza's critical water infrastructure, analyzing how military actions, environmental degradation, and political blockades have compounded the region's already significant challenges to clean water access.
- What are the public health consequences of Gaza's water crisis? By exploring the effects of water scarcity and inadequate sanitation on public health, particularly among children, the chapter will reveal the immediate and long-term consequences of this crisis. The spread of waterborne diseases, increased malnutrition, and rising mortality rates will be central to this discussion.
- What are the potential pathways for sustainable recovery? Finally, this chapter will consider potential strategies for rebuilding Gaza's water and sanitation systems post-conflict. International cooperation, sustainable water management practices, and investment in infrastructure will be examined as essential components of Gaza's recovery and progress toward achieving SDG 6.

Through this analysis, the chapter aims to provide a nuanced understanding of how conflict transforms essential resources like water into tools of oppression, while highlighting the urgent need for international intervention to protect Gaza's most vulnerable populations from further humanitarian disaster.

In achieving these objectives, we will rely primarily on a systematized review, followed with an in-depth analysis of the findings.

10.2 Systematized Review Methodology

This endeavor involves a comprehensive review of literature, data, and evidence on the weaponization of water infrastructure in conflict zones, with a focus on public health and humanitarian impacts. The review draws from academic journals, research papers, and reports by international organizations and government agencies, aiming to contextualize Gaza's crisis within global water crises in war zones. Using a rigorous methodology aligned with the PRISMA framework (Moher et al. 2009), the review ensures transparency, replicability, and comprehensiveness. A detailed search strategy targeted key databases and sources with specific keywords and Boolean operators to capture relevant studies. Inclusion and exclusion criteria, based on factors like publication date, language, and relevance, were applied to assess study eligibility. Titles and abstracts were screened, and full-text versions of relevant studies were evaluated. A standardized data extraction form captured study characteristics, methodologies, findings, and other variables. Study quality was assessed using appropriate tools, considering factors like design, sample size, and bias. Findings were synthesized to identify themes, patterns, and gaps using qualitative or quantitative methods, providing a nuanced understanding of water weaponization in conflicts and its broader implications.

The comprehensive search was conducted across four renowned multidisciplinary databases: Google Scholar, Agricola, Wiley Online Library, and ScienceDirect, employing a methodology consistent with recent studies on similar topics (Koutsos et al. 2019). To identify pertinent literature, we utilized a combination of search terms such as "Water, Sanitation, Public Health Crisis, Conflict, Water Conflict, Gaza Water Crisis, Infrastructure Destruction, Water Access in Conflict, Water Protection, Water Weaponization, Water Security In Gaza, and Gaza Water Crisis." The search relied on the use of Boolean operators such as AND and OR. A specific set of inclusion and exclusion criteria were developed.

The inclusion criteria focused on several key aspects. Geographically, the review prioritized studies that centered on Gaza or the West Bank. However, it also included research from other conflict zones where water infrastructure had been weaponized or targeted, such as Syria, Yemen, and Iraq, to provide comparative analysis. The time frame was limited to studies published from 1948 to the present days, ensuring the capture of recent and past conflicts and developments in Gaza's water crisis. Older studies, particularly those offering essential historical context, were considered if they provided significant insights into the evolution of water infrastructure in the region. The review focused primarily on studies published in English and Arabic, allowing for access to a broad range of relevant research. Other languages were only included if they provided critical insights and translations were available. The types of studies considered included empirical research—both qualitative and quantitative—as well as peer-reviewed journal articles, conference papers, and reports from governments or international organizations. Reports from NGOs and other forms of grey literature were also included if they provided substantial data on the use of water as a weapon in conflict zones. Additionally, research that discussed international humanitarian law, particularly the Geneva Conventions and their relevance to the protection of water resources during war, was included. Studies linking the water crisis to SDG 6 in conflict regions were also considered.

The exclusion criteria focused on studies that fell outside the relevant geographic scope, particularly those not centered on Gaza or other pertinent conflict zones. Studies that addressed water issues unrelated to armed conflict or siege were excluded, as well as those primarily focusing on topics such as climate change, agriculture, or industrial water use without a connection to conflict. Studies published before 2000 were excluded, unless they provided significant historical context to Gaza's water crisis. Additionally, non-peer-reviewed sources, editorials, and opinion pieces without empirical data were not considered. While newspaper articles or media reports were reviewed, they were only included if they contained verified research or data. Language limitations applied to studies without available translations, and studies with flawed methodologies, such as those with unclear sample sizes or unverified sources, were excluded. Furthermore, studies deemed outdated, or those with a high risk of bias or conflict of interest, particularly when funded by parties with vested interests in the conflict, were excluded from the review.

Throughout the review process, the PRISMA flow diagram (Fig. 10.1) was used to track the number of studies identified, screened, excluded, and ultimately included. This flow diagram enhanced the transparency of the review by visually presenting each stage of the selection process. Finally, the results of the review were written up according to the PRISMA checklist, ensuring that all essential reporting items—such

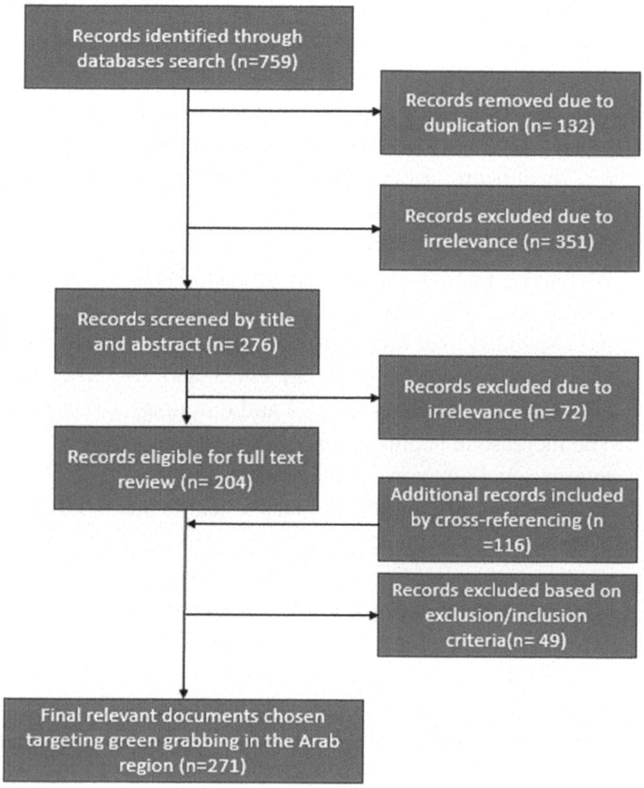

Fig. 10.1 Flowchart of the selection of records based on PRISMA guidelines

as the title, abstract, methods, results, and discussion—were included in the report.

In the initial search, we identified a total of 759 entries. All documents were then screened to meet the inclusion criteria. A total of 132 documents were excluded due to duplication. The screening process then continued with a review of the title and abstract. A total of 351 documents were excluded as they were irrelevant to the topic. The multistage screening process produced 204 documents eligible for a full-text review. In addition, we manually searched the references of the remaining documents to identify any documents not captured during the initial search. This process identified 116 additional documents. The documents were then content-analyzed and selected according to the inclusion/exclusion criteria. This process resulted in the exclusion of 49 documents for the research question. The final screening process yielded 271 documents included in the analysis (Fig. 10.1).

The documents span multiple years, with the following distribution: 80 references are from 2024, 22 are from 2023, 9 from 2022, 10 from 2021, and 12 between 2017 and 2020. Between 2010 and 2016, there are 11 references, 6 from 2000 to 2009, 5 from the 1990s, and 1 from pre-1990. This timeline shows a strong focus on recent publications, with a notable increase in studies from 2023–2024, which likely reflects the ongoing global interest in the Gaza crisis, particularly after recent escalations in conflict.

These documents cover a variety of topics relevant to Gaza, organized into broad categories as follows:

Environmental and Water Crisis: Approximately 50 references focus on issues related to water scarcity, contamination, and the impacts of conflict on water resources. Many studies discuss the intentional use of water as a weapon in the conflict. Notable examples include works by Abd El Hay (2024), Ahmed (2010), Al Jazeera (2022), and Buheji (2024). Over the years, this category has grown significantly, reflecting the worsening environmental conditions and the global awareness of the urgent need for sustainable water management solutions in Gaza.

Health and Humanitarian Impacts: Around 40 references address public health crises, limited healthcare access, mental health challenges, and disease outbreaks resulting from the ongoing conflict. Examples include Abudayya (2023), Al Mezan (2021), WHO (2024), and UN OCHA (2024). This category has expanded over time, paralleling the intensifying humanitarian crisis in Gaza and its effects on health infrastructure and outcomes. The increase in health-focused studies reflects international concern for the severe, multifaceted impacts of conflict on Gaza's healthcare system.

Economic and Social Costs of Conflict: Roughly 20 references analyze the economic sanctions, infrastructure damage, and the social ramifications of the conflict on Palestinian life and economy. Examples include Adnan (2022), Butter (2024), and UNCTAD (2024). Research in this category has become more prevalent as the cumulative impact of economic restrictions and ongoing infrastructure damage have severely strained Gaza's social and economic fabric, further impoverishing the population and affecting the community's resilience.

Historical and Political Context: Around 10 references delve into historical events, political agreements, and the broader geopolitical context affecting Gaza. Examples include works by Barak (2005), Chomsky and Pappe (2011), and Gazit (1998). This category has remained relatively consistent over time, as historical and political analyses continue to provide critical context for understanding the ongoing issues in Gaza.

Food Security and Agriculture: Approximately 15 references examine food insecurity, agricultural damage, and the effects of conflict on local farming practices. Key examples are provided by FAO (2024a, b, c, d), IPC (2023), and Hassoun (2024). This topic has gained more focus in recent years, reflecting the severe food crises in Gaza and the importance of agricultural sustainability amidst conflict.

This analysis underscores a concentration of recent research on environmental and humanitarian topics, with particular emphasis on water crises and public health. The high number of recent publications reflects an acute awareness and response from international agencies, academic institutions, and NGOs to the escalating crisis in Gaza. The research evolution illustrates how early studies provided a foundation on environmental and resource issues, while more recent work emphasizes the compounded effects of conflict on health, economic stability, and humanitarian needs.

10.3 Gaza's Water Crisis: Historical Context

The Gaza Strip, a small and densely populated region along the Mediterranean, has faced a long-standing and acute water crisis exacerbated by political instability, conflicts, environmental degradation, and governance challenges. This crisis is deeply rooted in geographical limitations, unsustainable water management practices, and the legacy of decades of conflict. Gaza's water infrastructure has steadily deteriorated over the years, leaving more than 90% of its population without access to safe drinking water (UNOSAT 2024; UNRWA 2024). The geography of Gaza has always posed significant challenges for water security. Spanning only 365 km^2 and hosting one of the highest population densities in the world, Gaza's natural water resources are inherently limited. As of 2024, the population of Gaza exceeds 2.3 million people, and the region depends almost entirely on the Coastal Aquifer, an underground freshwater source shared with Israel (Al-Quds University, 2023). This aquifer has become severely overexploited and contaminated due to a combination of saline intrusion from the Mediterranean, untreated sewage, and agricultural runoff (World Bank 2018a, b, c). Historically, Gaza relied on rainfall, groundwater, and traditional water management practices. However, annual rainfall—ranging between 200 and 400 mm—has proven insufficient to meet the needs of Gaza's rapidly growing population (Abdul Samad et al. 2024; Asi et al. 2024; Buheji 2024; Goodarzi et al. 2024; UNOSAT 2024).

10.3.1 Pre-1948: Traditional Water Management in Gaza

Before 1948, traditional water management in Gaza maintained a balance between the population and available resources. The main water source was the Coastal Aquifer, a shallow groundwater basin recharged by seasonal rainfall. Despite irregular precipitation, it generally sufficed for the smaller population (UNOSAT 2024; Al-Najjar et al. 2021). Pre-1948 water management had minimal environmental impact, with low-intensity agriculture and limited irrigation or industrial water use.

Communities employed traditional techniques suited to the region's conditions and culture. Rainwater harvesting was vital, with cisterns commonly built to store rainwater for domestic and agricultural use during dry periods. These decentralized, low-maintenance systems were integral to homes and public spaces (Naff 2020). Another significant water management method was the use of shallow wells (approximately 2500 wells) to access the Coastal Aquifer (Trottier 1999). Farmers and households alike relied on wells to meet their water needs, especially for irrigation purposes. Shallow wells were relatively easy to construct and maintain, and they allowed access to the upper levels of the aquifer, which, at the time, had not been heavily exploited (Ekenci 2019). In rural areas, particularly those engaged in agriculture, the use of earthen canals, known as qanats, was prevalent. These underground irrigation systems, which originated in ancient Persia, were effective in transporting water from the aquifer to agricultural lands over long distances with minimal water loss due to evaporation. The qanat system enabled the cultivation of crops, such as grains and citrus fruits, even in arid conditions (Naff 2020).

Before 1948, Gaza's population was about 60,000, with water demand closely aligned to resource availability (Yiftachel 2006). Traditional water management techniques maintained a sustainable balance, as demand rarely exceeded the aquifer's natural recharge rates. Water pollution was minimal due to limited industrial activity and localized wastewater management (Gleick 2019; Kayser et al. 2019; Trottier 1999).

Locally adapted methods like rainwater harvesting, cisterns, and shallow wells reflected a deep understanding of the region's water cycles. Seasonal variations were managed carefully, and over-extraction risks were low due to the small population and limited technology. However, this fragile balance depended on stable environmental and demographic conditions.

Post-1948 geopolitical changes, including a population surge from refugee influx, dramatically increased water demand, straining the Coastal Aquifer and causing long-term environmental consequences (Gleick 2019; Kayser et al. 2019; Brooks et al. 2019).

10.3.2 Post-1948: The Nakba and Refugee Influx

The 1948 Nakba (Arabic for "catastrophe") marked a turning point in the water crisis in Gaza, as the ethnic cleansing policy practiced by the Zionists fundamentally altered the region's demographic and environmental landscape. This period saw the mass displacement of Palestinian populations, with hundreds of thousands fleeing to the Gaza Strip. The sudden influx of refugees drastically increased Gaza's population, straining its already limited natural resources, especially water. The population swelled to over 200,000 as refugees from nearby cities and villages were forcibly displaced and fled to Gaza (Weinthal et al. 2005; Yiftachel 2006).

This exponential population growth created immense pressure on Gaza's water resources. Hastily constructed refugee camps lacked proper infrastructure, exacerbating the demand for basic services such as water, sanitation, and housing. The camps were initially meant to be temporary, but as the conflict persisted, they became permanent fixtures in the region, further overwhelming

Gaza's fragile infrastructure (Zahra 2001; Pappé 2006). The Coastal Aquifer, which had previously supported the population's needs, became the only significant water source, with minimal alternatives available. As a result, the aquifer was subjected to increasing levels of extraction that exceeded its natural recharge rates, leading to a long-term decline in water quality and quantity (Wolf and Ross 1992; Gleick 1993; Neuzil 2007; Baalousha 2009; Messerschmid 2011).

Following the population surge, Gaza became almost entirely reliant on groundwater from the Coastal Aquifer. The over-extraction from this aquifer resulted in severe consequences for both water availability and quality. With so many shallow wells and boreholes tapping into the aquifer, groundwater levels began to drop, leading to the intrusion of seawater into the aquifer, a process known as salinization. Salinization made much of the water undrinkable, worsening the already dire water crisis (Ekenci 2019).

By the 1960s, as the refugee population remained in Gaza, the situation worsened. UNRWA (United Nations Relief and Works Agency for Palestine Refugees in the Near East), which was established to provide assistance to Palestinian refugees, struggled to meet the escalating water needs. Water rationing became commonplace, and the lack of proper sanitation infrastructure in overcrowded refugee camps led to public health concerns, as the risk of waterborne diseases increased (Barhoum 2021; Brooks et al. 2020; Sour et al. 2011; Bashitialshaaer 2014; Efron et al. 2017; Isaac and Sabbah 2017).

The Gaza Strip's dependency on groundwater became increasingly precarious as both Israel and Gaza were extracting water from the shared aquifer without a long-term sustainable plan (Bachmutsky 2017; Bailout 2008; El-Nahhal 2006; Selby 2003). Water became an increasingly strategic resource. Israel's National Water Carrier project, initiated in the 1950s to divert water from the Jordan River and Sea of Galilee to central and southern parts of Israel, reduced water availability for Palestinian territories, including Gaza. This left Gaza at a significant disadvantage, and the refugee influx only heightened the region's vulnerability (Bellisari 1994; Shawwa 1994; Trottier 1999).

The environmental impacts of the post-1948 water crisis were profound. In addition to the depletion and salinization of the Coastal Aquifer, the lack of adequate wastewater management systems in the refugee camps led to contamination of surface and groundwater. Untreated sewage was often mixed with drinking water sources, increasing the risk of disease outbreaks. As the refugee population continued to grow, this issue worsened, creating both environmental and public health crises. The socioeconomic implications were equally dire (ARIJ 2007; Neuzil 2007; Baalousha 2009; Messerschmid 2011). As Gaza's water scarcity intensified, access to clean and reliable water became a daily challenge for many households. Agricultural productivity, which had been a cornerstone of the region's economy, suffered significantly due to water shortages and the deterioration of water quality. Farmers struggled to irrigate their fields, which led to reduced crop yields and increased food insecurity in a region already dependent on humanitarian aid (Shomar and Rovira 2023; Anera 2022; Kemmerling et al. 2022; Akkaya et al. 2011; Al-Juaidi et al. 2011; Bohannon 2006; Zeitoun 2008; Schlütter 2005).

10.3.3 1967–1993: Israeli Occupation and Resource Control

The 1967 Six-Day War marked a turning point for Gaza's water resources. After Israel occupied Gaza and the West Bank, it took control of vital water resources, including the Coastal Aquifer. Policies prioritized Israeli settlers' water needs, leading to severe inequalities in water allocation and infrastructure development (Zreik 2003; Ferragina 2008; Al-Rodhan et al. 2011). Palestinians faced strict restrictions on drilling wells or expanding water infrastructure, with permits rarely granted by Israeli authorities (Efron et al. 2019; Shomar 2011a, b).

Gaza's agricultural sector, heavily reliant on groundwater, suffered due to these restrictions, causing a decline in output and increased food insecurity (Messerschmid 2011; Hass 2014; Romano et al. 2019). In contrast, Israeli settlements utilized advanced irrigation systems to grow water-intensive crops, further depleting the aquifer (Haddad 2011; Gray 2009; Gray and Hilal 2007). By 1992, water allocations for Israeli settlers were nearly five times higher than for Palestinians, deepening water shortages (Roy 2002; Rouyer 1997; World Bank 1992).

Over-extraction led to salinization of the Coastal Aquifer as seawater infiltrated the depleted groundwater, rendering much of it unfit for use. Neglect of Palestinian water infrastructure under Israeli rule worsened access issues, while polluted and saline water sources contributed to public health crises (Efron et al. 2019; Shomar 2011a, b).

Control over water became a tool for political domination, symbolizing broader inequalities under occupation. These dynamics not only exacerbated Gaza's water crisis but also intensified socioeconomic disparities and environmental degradation, with long-term impacts on agriculture, health, and living conditions (Zeitoun 2008; Machairas 2017; Brooks et al. 2020; Kemmerling et al. 2022).

10.3.4 1994–2005: Oslo Accords and the Palestinian Authority's Limited Control

The Oslo Accords (1993, 1995) aimed to improve Israeli–Palestinian water cooperation and recognized Palestinian water rights, including the establishment of the Joint Water Committee (JWC). However, the accords entrenched Israeli control, granting Israel 87% of West Bank groundwater and full control of the Jordan River. Gaza's Coastal Aquifer was excluded from cooperative agreements, leaving Gaza with only 25% of the aquifer's water while Israel took 75%. Restrictions on water transfers and materials for alternative resources worsened Gaza's water shortages (Shlaim 1994; Miller 1998; Weiner 1999).

By 2000, over 80% of Gaza's water was undrinkable due to salinization and sewage contamination, creating a public health crisis. Untreated wastewater frequently mixed with freshwater, causing widespread waterborne diseases (Zreik 2003; Yassin et al. 2006; Qurie 2007; UNEP 2009; Abuzerr et al. 2019). Despite international aid for water infrastructure, efforts were hindered by political instability, Israeli restrictions on materials, and the PA's limited resources. The Second Intifada (2000–2005) further delayed projects as Israeli policies restricted goods and people entering Gaza (Butt and Butt 2016; Bashi 2013).

Israel's control over key infrastructure and JWC veto power meant donor-funded projects were often blocked. While international aid prevented total collapse, the reliance on the Coastal Aquifer and lack of desalination or alternative sources deepened Gaza's crisis (Talhami and Zeitoun 2020; Tutu and Skarphéðinsson 2013; Messerschmid 2011). The unequal water allocation and failing infrastructure fueled resentment, symbolizing the broader power imbalance between Israel and the Palestinians. By the mid-2000s, the Oslo Accords had failed to resolve Gaza's water crisis, leaving it vulnerable to continued environmental and public health challenges (Zeitoun 2008; MacDonald et al. 2016; Shlaim 2016).

10.3.5 2005–2023: Blockade, Recurrent Conflict, and Infrastructure Deterioration

Israel's disengagement from Gaza in 2005, followed by Hamas' election in 2007 and the blockade imposed by Israel and Egypt, devastated Gaza's water infrastructure. Recurrent Israeli attacks in 2008, 2012, 2014, and 2021 caused extensive damage to water treatment plants, sewage systems, and desalination facilities, worsening the region's water crisis (UNICEF 2021; Abdul Samad et al. 2024; Milton et al. 2024). The blockade has restricted essential materials needed for repairs, leaving much of Gaza's water systems in disrepair and limiting access to clean water and functioning sewage systems (Amnesty International 2014a, 2014b; Butt and Butt 2016).

During the 2014 conflict, over 1.2 million residents lost access to clean water, heightening public health risks and increasing waterborne diseases. By 2017, 97% of Gaza's water was unfit for consumption due to contamination by seawater, untreated sewage, and agricultural runoff, creating severe health and humanitarian challenges (Assembly 2018). Repeated delays in projects like the Southern Gaza Desalination Plant, caused by import restrictions and conflict-related damage, have left Gaza's desalination capacity inadequate to meet water needs.

As of 2023, most residents relied on expensive, unregulated private water vendors, whose wells are often unlicensed and provide untreated water, disproportionately impacting poorer households (PWA 2012a, b, c, d; PWA 2013; PEQA 2014; PWA 2014a, b; PWA 2015; Abdul Samad et al. 2024). These systemic issues underscore the severe inequalities in water access and the ongoing humanitarian crisis in Gaza.

10.3.6 2023–To Date

The 2023 Gaza war had a devastating impact on the region's already fragile water infrastructure. Following the Hamas attack on Israel on October 7, 2023, Israel imposed a "total siege" on Gaza, cutting off water, electricity, and fuel supplies. This immediate cutoff drastically worsened the water crisis. By mid-October, over 50% of Gaza's water and wastewater facilities had been damaged due to airstrikes. More than 60 municipal wells were either destroyed or forced to shut down, and Gaza's three main desalination plants were operating at only 25% of their capacity due to fuel shortages (Abdul Samad et al. 2024; Westheime 2024). With only 17% of Gaza's groundwater wells functioning, the region's ability to access clean water was severely diminished. The northern part of Gaza was particularly hard-hit, with no access to clean water. The closure of the Al-Yassin desalination plant, the largest facility in northern Gaza, has severely intensified the region's water crisis. Over 450,000 Palestinians, who previously relied on the plant for clean drinking water, now face acute shortages. The Israeli military's ongoing blockade has prevented the delivery of the fuel required to operate essential facilities like Al-Yassin, forcing its shutdown. This has not only disrupted water supplies but has also placed immense pressure on smaller desalination plants, many of which have also ceased operation due to similar constraints (Younis 2024).

Even before the war, Gazans received just 21 L of safe water per person per day—far below the World Health Organization's recommended 100 L (Abdul Samad et al. 2024; Westheime 2024). After the siege and military operations, this situation worsened as Israel allowed only one of the three main water connections into Gaza to function, and it was only operating at 47% capacity. The destruction of critical infrastructure has left much of the population reliant on tankers to deliver water, placing extreme pressure on service providers, many of whom were killed during airstrikes. The water crisis has been further exacerbated by the total inability to import materials needed to repair and maintain the water systems due to the ongoing blockade (Abdul Samad et al. 2024; Cole 2024; Samad et al. 2024; Westheime 2024). By May 26, 2024, water production in the Gaza Strip had decreased by 84%, primarily due to the destruction of water infrastructure and severe restrictions on fuel, electricity, and the entry of spare parts needed for water and sanitation systems.

Over the following weeks, the situation worsened. By June 3, 2024, five water and sanitation facilities were taken out of service every 3 days as a result of continuous bombardments, leading to a rapid degradation of Gaza's capacity to maintain essential services (Abdul Samad et al. 2024; Westheime 2024; UNOSAT 2024; UNRWA 2024; Añaños Bedriñana et al. 2023; ESCWA 2023). By this same date, Gaza City had lost nearly all of its water production capabilities, with 88% of its wells, 100% of its brackish water desalination plants, and 100% of its seawater desalination facilities damaged or destroyed (Abdul Samad et al. 2024; Cole 2024; Samad et al. 2024; Seeberg 2024).

The Israeli military also targeted key water quality testing laboratories, with the first destroyed in November 2023 and the second on March 21, 2024, further crippling efforts to ensure safe drinking water. Additionally, by June 3, 2024, all water and sanitation warehouses in Gaza City and Khan Younis had been destroyed, along with 70% of sewage pumps. By June 26, 2024, every wastewater treatment plant in Gaza had been completely demolished (Abdul Samad et al. 2024; Cole 2024; Samad et al. 2024; Añaños Bedriñana et al. 2023; ESCWA 2023). Since the Israeli offensive began on October 7, 2023, Gaza's residents have been left with only 4.74 L of water per person per day, covering all needs such as drinking, cooking, and washing. This represents a drastic 94% reduction in the water supply compared to pre-offensive levels. The current amount falls far below the internationally accepted emergency minimum of 15 L per person per day required for basic survival. This severe shortage has pushed Gaza's water crisis to life-threatening levels, severely impacting daily life and health (Abdul Samad et al. 2024; Cole 2024; Samad et al. 2024; Seeberg 2024; Shorrab et al. 2024; Paché 2024; Buheji and Al-Muhannadi, 2023).

10.4 The Impact of War on Gaza's Water Infrastructure

10.4.1 Impact on Water Treatment Facilities

The repeated conflicts in Gaza since 1948, particularly post-2009, have systematically destroyed key water and sanitation infrastructure, causing long-term water scarcity and environmental degradation. Conflicts such as Operation Cast Lead (2008–2009), Operation Protective Edge (2014), and the wars of 2021 and 2023 severely damaged critical facilities, including wastewater treatment plants, desalination plants, and water reservoirs. These losses have forced reliance on contaminated groundwater and unregulated water sources, exacerbating malnutrition and public health risks.

By June 2024, 88% of Gaza's water wells and 100% of desalination plants were rendered inoperative, significantly reducing access to potable water. The destruction of Gaza's main seawater desalination plant, which served 20% of the population, compounded the crisis. With over 96% of groundwater undrinkable, Gaza faces an unsustainable burden on its depleted and contaminated Coastal Aquifer. The broader impacts include soil degradation, loss of irrigation networks, and worsening food insecurity, destabilizing Gaza's fragile economy. Without substantial international intervention, rebuilding this essential infrastructure remains a significant challenge (Beiraghdar et al. 2023; Ramahi 2023; OCHA 2024a; World Bank 2024a, b, c, d) (Table 10.1).

10.4.2 Breakdown of Sanitation Systems

The destruction of Gaza's sanitation systems from 1948 to 2024 has had profound and lasting effects on the region's environment and public health. Repeated military conflicts, including the wars of 1948, 1967, 2008–2009, 2014, 2021, and 2023, severely damaged wastewater treatment facilities and sewage networks, leading to widespread contamination of water resources, agricultural lands, and marine environments. The collapse of sanitation infrastructure has compounded challenges in water and food security, with untreated sewage flooding streets, agricultural fields, and urban areas.

The wars of 2008–2009 and 2014 caused major damage to key facilities, such as the Beit Lahia and Eastern Gaza systems, leaving large areas without access to sanitation services. During the 2021 and 2023 conflicts, Gaza's wastewater treatment capacity was nearly eliminated, with 70% of sewage pumps rendered inoperative and all wastewater plants destroyed. This led to untreated sewage contaminating the Coastal Aquifer and marine environments, degrading biodiversity and threatening livelihoods.

As of 2023, Gaza's WASH (State of Palestine National Water, Sanitation and Hygiene) system has largely collapsed, with water production at less than 5% of its capacity. Damage to desalination plants, water wells, and Mekorot pipelines has reduced water supply to critically low levels. The environmental damage from untreated sew-

Table 10.1 The impact of war on Gaza's water treatment facilities

Facility	Date of destruction and event	Outcome	References
Beit Lahia wastewater treatment plant	Operation Cast Lead (December 2008–January 2009)	Facility collapsed entirely, forcing reliance on contaminated water, damaging sanitation, and causing untreated wastewater to seep into the coastal aquifer. Over half a million people were left without clean water access	To-Hauser (2024); Weinthal and Sowers (2019); Ubeid and Al-Agha (2016); Abd Rabou (2011); Sour et al. (2011); Nembrini (2010); World Bank (2009a, b).
Al-Bureij treatment plant	Operation Protective Edge (2014)	Destruction left over 400,000 people without regular access to clean water, increasing dependence on unregulated sources and contamination risks. By the end of 2014, over half of Gaza's population had limited access to potable water	Amnesty International (2014a, 2015); World Bank (2018a, b, c); Al-Mughrabi (2017); Culverwell (2017); EWASH (2016); Supplement to Readings on Water (2016); Bachmann et al. (2014).
Sheikh Radwan treatment plant	Operation Protective Edge (2014)	Facility destroyed during airstrikes, contributing to the critical state of Gaza's water and sanitation networks	Amnesty International (2014a, 2015); World Bank (2018a, b, c); Al-Mughrabi (2017); Culverwell (2017); EWASH (2016); Supplement to Readings on Water (2016); Bachmann et al. (2014).
Various desalination plants	Operation Guardian of the Walls (2021)	Significant damage to desalination plants and water reservoirs, which were crucial for addressing Gaza's water shortage. Over 96% of water from the coastal aquifer was already unsafe for consumption due to salinity and contamination	Al Mezan Center for Human Rights (2021); AlKhaldi et al. (2021); Mohammed Alashqar et al. (2023).
Main seawater desalination plant	2023 conflict	Severely damaged, resulting in the loss of potable water for 20% of Gaza's population. By June 2024, 88% of water wells and 100% of desalination plants were inoperative, forcing reliance on contaminated groundwater	OCHA (2024a, b, c); Oxfam (2024); World Bank (2024a, b).
Entire water infrastructure	Multiple conflicts since 1948, including the wars of 2021 and 2023	Depletion and contamination of the coastal aquifer due to recurring damage, over-extraction, and seawater intrusion. Agricultural lands irrigated with contaminated water suffered degradation, exacerbating food insecurity and destabilizing Gaza's economy	World Bank (2024a, 2018a, b, c); FAO (2024a, b, c); B'Tselem (2020); Asi et al. (2024); Buheji and Al-Muhannadi (2023); ESCWA (2023); UNOSAT (2024).

age and limited water infrastructure has created a cycle of degradation that threatens Gaza's long-term sustainability (UNEP 2024; World Bank 2023, 2024a, b, c, d; FAO and WFP 2024) (Table 10.2).

The ongoing destruction of Gaza's water and sanitation infrastructure has left residents in a state of dire economic and social hardship. The closure of critical desalination plants, including the Al-Yassin facility, has forced over 450,000 Palestinians to rely on private vendors or untreated well water, which poses severe health risks. The cost of obtaining clean water has sky-rocketed, with a 16-liter container of water now priced at 6 shekels—six times its pre-war cost. For many families, particularly those displaced

Table 10.2 Breakdown of sanitation systems

Facility	Date of Destruction and event	Outcome	References
Wastewater treatment facilities	1948–2024 (multiple conflicts, including 2000, 2008–2009, 2014, 2021, 2023)	Repeated destruction led to untreated sewage flooding streets, agricultural lands, and marine environments. Environmental degradation included groundwater contamination, marine pollution, and soil quality decline, worsening food insecurity, and health risks	Goodarzi et al. (2024); IPC (2024a, b, c); UNDP (2024); Oxfam (2024); Euro-Med (2024); FAO and WFP (2024); UNEP (2009, 2020, 2024)
Beit Lahia wastewater treatment plant	Operation Cast Lead (2008–2009)	Severely damaged, resulting in untreated sewage being discharged into the Mediterranean Sea. Gaza's sanitation systems were overwhelmed, compounding marine pollution and ecosystem degradation	UNEP (2024); World Bank (2024a, b, c, d); FAO (2017); Efron et al. (2017)
Sewage treatment systems	Operation Protective Edge (2014)	Extensive damage to sewage infrastructure in Eastern Gaza, Khan Younis, and Beit Hanoun caused prolonged water shortages and widespread environmental contamination. Municipal wells and water quality were severely affected	UNDP (2024); Efron et al. (2017); FAO (2017); UNEP (2024); Shomar and Rovira (2023); Kemmerling et al. (2022)
Various wastewater treatment plants	2021 conflict (Operation Guardian of the Walls)	Key facilities destroyed, contaminating agricultural lands and the coastal environment. Treatment capacity reduced significantly, exacerbating environmental and health risks	Al Mezan (2021); UNEP (2024); World Bank (2024a, b, c, d); Euro-Med (2024)
All wastewater treatment plants	2023 conflict	Complete destruction of Gaza's six wastewater treatment plants. By June 2024, 70% of sewage pumps were inoperative, leading to untreated sewage flooding streets and further contaminating the Coastal Aquifer	Oxfam (2024); Euro-Med (2024); UNEP (2024); World Bank (2024a, b, c, d)
Gaza coastal aquifer	Ongoing (1948–2024)	Contaminated by untreated sewage and seawater intrusion. Overextraction and lack of treatment led to water being undrinkable, further compounding Gaza's water crisis	UNEP (2009, 2024); Goodarzi et al. (2024); FAO and WFP (2024)
Mekorot water supply connections	October 2023–April 2024	All three connections with Mekorot were disrupted, with partial restorations by April 2024. Northern connection remained offline until mid-2024. Overall water supply remains critically low, at less than a quarter of pre-escalation levels	Euro-Med (2024); World Bank (2024a, b, c, d); UNEP (2024); FAO and WFP (2024)
Agricultural lands	2021–2024	Repeated flooding with untreated sewage degraded soil quality and rendered farmland unsuitable for cultivation, worsening food insecurity and environmental degradation	FAO and WFP (2024); UNEP (2024); Anayah et al. (2021); Gleick (2019); Kayser et al. (2019)
Gaza's sanitation infrastructure	1948–2024 (cumulative conflicts and blockades)	57% of water and sanitation assets destroyed or partially damaged. Infrastructure collapse has left Gaza's population with limited access to safe water and sanitation, deepening the humanitarian crisis	World Bank (2024a, b, c, d); UNEP (2024); FAO and WFP (2024)

by recent military operations, these prices are unaffordable. Fuel shortages further exacerbate the crisis, increasing operational costs for water treatment plants and rendering many facilities nonfunctional (Younis 2024).

10.4.3 Environmental Impact of Sewage and Wastewater Contamination

The long-term environmental damage caused by sewage contamination in Gaza is profound and far-reaching, affecting both terrestrial and marine ecosystems. The collapse of Gaza's sanitation infrastructure, exacerbated by repeated conflicts and a lack of maintenance, has led to severe degradation of these ecosystems, with dire consequences for the region's environment and population.

10.4.3.1 Marine Ecosystem Degradation and Fisheries Collapse

The ongoing conflict and destruction of infrastructure have led to the release of untreated sewage into Gaza's coastal waters, which has severely degraded marine ecosystems. The presence of high concentrations of pollutants such as nitrogen, phosphorus, and pathogens has led to eutrophication and hypoxic conditions (FAO and WFP 2024; UNEP 2024; World Bank 2024a, b, c, d). During the 2010s, the amount of untreated or partially treated sewage/wastewater flowing from the Gaza Strip into the Mediterranean Sea increased steadily from 90,000 cubic meters (m^3) per day in 2012 to 100,000 m^3 per day in 2016 and 110,000 m^3 per day in 2018 (World Bank, European Union and United Nations 2024; World Bank 2023; UNEP 2020). These conditions drastically reduce oxygen levels in the water, creating "dead zones" where marine life cannot survive. Over 120,000 cubic meters of untreated sewage were being discharged daily into the Mediterranean, contributing significantly to the collapse of fish populations along Gaza's coast (Dickie and Withers 2024; Qumsiyeh 2024; UNDP 2024; WHO 2024a, b; UNDP 2021; UNEP 2020; Abualfailat and Qadous 2022; ARIJ 2020).

Fish species that once thrived in these waters have either died off or migrated to cleaner areas. Over 70% of Gaza's coastline is now unsuitable for marine life, primarily due to the high levels of organic and chemical pollutants introduced by untreated sewage and damaged infrastructure (UNEP 2024; Hussein 2023; Hussein et al. 2022, 2024). Gaza's coastal waters, once home to a diverse range of marine species, have experienced a steep decline in fish populations due to the destruction of their habitats and the contamination of the marine environment. Many fish species that were commercially important to Gaza's fishing industry have either disappeared or their numbers have plummeted. Species such as sardines, mackerels, and groupers, which are highly sensitive to changes in water quality, have been particularly affected by the elevated levels of nutrients and pollutants (UNEP 2024; World Bank 2017; Safi 2014; Sarsour 2014; EWASH and Al-Haq 2011; UN OCHA and FAO 2010). Eutrophication caused by excessive nutrient levels has led to algal blooms, which block sunlight and reduce oxygen levels, creating environments where most marine species cannot survive. This disruption to the food web has cascading effects, impacting not only fish but also other marine organisms such as crabs, mollusks, and seabirds that rely on a healthy ecosystem. Fishing has been a traditional livelihood for thousands of families in Gaza, but the degradation of the marine environment has led to a 45% decline in fish stocks since the beginning of the ongoing conflict (Ahmed et al. 2024; Abuhasballah 2023; Hussein et al. 2022; World Bank 2020, 2023a, b, 2024a, b, c, d). Gaza's fishing industry, once a major source of food and income, has been devastated by the loss of marine biodiversity and the inability of fishermen to access cleaner, offshore waters due to the ongoing naval blockade.

The economic consequences of the fisheries collapse have been severe. Fish had been one of the few affordable sources of protein for Gaza's population, playing a crucial role in the region's food security. With fish stocks depleted and those that remain being unsafe for consumption due to contamination, many families have lost their primary source of protein, further exacerbating food insecurity. Gaza's fisheries had been supplying 4000–5000 metric tons of fish per year before the conflicts, but production had fallen to less than 2000 metric tons by 2023. This decline has had devastating consequences for food security, particularly for low-income families who rely on locally sourced fish as a staple in their diets (Al-Riffai 2024; Butter 2024; Khoodoruth et al. 2024; UNCTAD 2024; World Bank 2022, 2023a, b). The contamination of fish stocks is another major concern. Fish caught in Gaza's waters have been found to contain high levels of contaminants, including heavy metals such as lead and mercury, as well as pathogens from untreated wastewater. This has led to public health risks, with the WHO advising against the consumption of locally caught fish due to the high risk of contamination (WHO 2024a, b; UNEP 2018, 2020; UNDP 2021, 2024). The inability to fish safely, coupled with the economic loss from the destruction of boats and fishing gear during conflicts, has left thousands of fishermen without a livelihood (World Bank 2020, 2022, 2023a, b, 2024a, b, c, d).

10.4.3.2 Agricultural Land Degradation in Gaza

The agricultural sector in Gaza has suffered greatly due to ongoing conflicts, with untreated sewage severely contaminating farmlands. The collapse of wastewater treatment systems during and after military operations has exacerbated this issue, rendering large areas of Gaza's farmland infertile. This environmental degradation poses a severe threat to local food security and agricultural productivity, which has historically been a cornerstone of Gaza's economy (Rist et al. 2024; UNCTAD 2022, 2024; UNDP 2023; Qumsiyeh and Abusarhan 2021; UNEP 2020).

Repeated wars in Gaza have led to widespread destruction of infrastructure, including wastewater treatment plants. The release of untreated sewage into agricultural areas has caused significant soil contamination. Over 40% of Gaza's agricultural land has been affected by sewage flooding, making these areas unsuitable for farming due to the degradation of soil quality (FAO 2023a, b, c, d; UNESCWA 2023; UNOSAT 2023). After the 2014 conflict, approximately 35% of Gaza's farmland was damaged or rendered unusable due to contamination from sewage and other pollutants. As the conflicts have continued, this situation has only worsened. By 2024, a significant portion of agricultural land remained unusable, with crop yields in the affected areas dropping by around 60% (FAO 2023a, b, c, d; UNESCWA 2023; UNOSAT 2023; UNRWA 2023).

The contamination of agricultural land has resulted in a dramatic decline in productivity. The Coastal Aquifer, Gaza's main water source for irrigation, has been severely over-extracted and polluted by both seawater intrusion and sewage. This has led to limited access to clean irrigation water, further reducing the capacity of the region's agricultural sector. The environmental impact on Gaza's Coastal Aquifer is also intertwined with agricultural productivity. As farmers face a shortage of clean water for irrigation, they are forced to use brackish water, which further degrades soil quality and reduces crop yields. Salinization has led to a 50% reduction in crop yields in some areas, further straining the food security of a population already suffering from high rates of poverty and unemployment (Goodarzi et al. 2024; Hassoun et al. 2024; Holail et al. 2024; Horino et al. 2024; UNEP 2024; UNDP 2024; World Bank 2013, 2024a, b, c, d; FAO 2023b).

The combined impact of soil degradation and water scarcity has caused Gaza's agricultural productivity to plummet. The reduction in agricultural output has had profound consequences on Gaza's ability to meet its food needs. In fact, food insecurity in Gaza has risen sharply due to the decline in local food production, with many families relying on imported food or international food aid. The combination of contaminated land and insufficient water resources has left farmers unable to sustain the levels of production needed to support the local population (ESCWA/UNDP 2024; FAO 2024b; OCHA 2024; IPC 2023; IPC Famine Review Committee 2023; UNRWA 2023; World Bank 2024a, b, c, d; Ahmed 2010).

The long-term consequences of agricultural land degradation in Gaza are severe; restoring Gaza's agricultural productivity will take decades of sustained effort, with significant investments required to rehabilitate the land and water infrastructure. In addition to the direct environmental impacts, the decline in agriculture has dealt a heavy blow to Gaza's economy. Agriculture, once a significant contributor to the local economy, now plays a diminished role, further weakening Gaza's economic resilience. The damage to Gaza's agricultural land also has broader ecological implications. The degradation of soil quality not only affects crop production but also reduces biodiversity in the region. Agricultural lands in Gaza, which once supported a variety of plant and animal species, have become barren due to contamination and lack of proper management (UNDP 2024; World Bank 2024a, b, c, d; UNDP 2023; Adnan 2022; UNEP 2018, 2020; Ahmed 2010).

10.4.3.3 Coastal Aquifer Contamination in Gaza

The Coastal Aquifer, Gaza's primary source of freshwater, has faced significant degradation due to the combined effects of over-extraction, seawater intrusion, and the infiltration of untreated sewage. This issue, exacerbated by years of conflict and infrastructural damage, has led to a critical water crisis in the region, with far-reaching environmental, economic, and health consequences (UNOSAT 2024; Al Jazeera 2022; Al-Najjar et al. 2021).

As described above, salinization and pollution have contributed to severe pollution of the Coastal Aquifer. Salinity levels often exceed 2000 mg/L, far above the WHO recommended limit of 250 mg/L for drinking water (WHO 2017; UNCTAD 2015). By 2024, over 96% of the water extracted from the Coastal Aquifer was deemed unsafe for human consumption, primarily due to high levels of salinity and contamination by pollutants such as nitrates, pathogens, and heavy metals (UNEP 2024; Boyan 2023). Nitrate levels in the aquifer often exceed 100 mg/L, far surpassing the WHO's recommended limit of 50 mg/L. High nitrate concentrations, along with microbial contamination, are primarily linked to untreated sewage infiltrating the groundwater, as well as agricultural runoff (UNDP 2021, 2024; UNEP 2009, 2018, 2020, 2024; WHO 2024a; Ahmed 2010).

The dependence on such water sources has heightened public health risks, with high rates of waterborne diseases, particularly affecting children (Qumsiyeh 2024; Ahmed 2010). The contamination of the aquifer is also a contributing factor to the increasing prevalence of diseases such as diarrhea, cholera, and other gastrointestinal infections. Waterborne diseases in Gaza account for a significant portion of child mortality, with 25% of all childhood deaths attributed to unsafe water and poor sanitary conditions (UNEP 2024; UNDP 2021, 2024; WHO 2024a; UNICEF 2022).

The cost of accessing potable water has soared in Gaza, creating a cycle of inequality. Families now spend up to one-third of their income on water. Poorer families, who cannot afford desalinated or imported water, often resort to using unsafe water, exposing themselves to severe health risks. This inequality exacerbates existing social vulnerabilities and deepens the economic divide (World Bank 2024a, b, c, d).

10.5 Public Health Crisis: Consequences of Water and Sanitation Deficiencies

The public health crisis in Gaza, driven by chronic water scarcity and contamination, has been exacerbated by ongoing conflict and inadequate infrastructure. These deficiencies have led to severe public health consequences, particularly through the proliferation of waterborne diseases, which disproportionately affect vulnerable populations such as children and the elderly (Ashour et al. 2024; Boukari et al. 2024; Correia 2024; Elnakib et al. 2024).

One of the most critical health risks associated with the high nitrate levels found in Gaza's drinking water is methemoglobinemia, commonly known as "blue baby syndrome." This condition occurs when nitrates interfere with the oxygen-carrying capacity of hemoglobin in the blood, reducing the availability of oxygen in the body.

Infants, whose digestive systems are more sensitive to nitrates, are particularly vulnerable to this condition (Shaban et al. 2023; Qrenawi and Abu Shomar 2020). Several reports have documented cases of methemoglobinemia in Gaza, with infants exhibiting symptoms such as cyanosis (bluish skin), difficulty breathing, and lethargy. They highlighted that the incidence of this condition is notably higher in areas where nitrate contamination is prevalent, and families lack access to alternative water sources. Long-term exposure to high nitrate levels also poses risks to pregnant women, as it can increase the likelihood of birth defects and complications during pregnancy (Shaban et al. 2023; Talhami and Zeitoun 2020; Tutu and Skarphéðinsson 2013; Qrenawi and Abu Shomar 2020; Shomar 2011a, b; Strohmeir and Ybarra 2011; Al Absi 2008).

Moreover, the consumption of saline water over extended periods has been linked to chronic kidney disease (CKD) and other long-term health issues. Excessive salt intake from drinking water is a leading cause of kidney dysfunction in Gaza, contributing to the rising prevalence of CKD among adults. The kidneys, responsible for filtering excess salts and waste from the body, are strained by the continual intake of saline water, which can lead to kidney stones, hypertension, and ultimately kidney failure (Esleem et al. 2015; Qahman and Mokhamer 2013). In addition to kidney dysfunction, high salinity levels have been associated with cardiovascular problems, particularly in adults with preexisting conditions. The persistent intake of saline water exacerbates hypertension, which increases the risk of heart disease and stroke. This creates a vicious cycle where access to clean water becomes a direct determinant of long-term health outcomes in Gaza, affecting not only immediate public health but also the population's overall well-being (Gupta et al. 2024; Taha et al. 2024; UNICEF 2024a, b; Ramon et al. 2023; Tumeh 2022; Davidovitch 2019; Abu Jabal et al. 2017).

The consumption of contaminated water in Gaza has led to widespread outbreaks of waterborne diseases, particularly among children. Diarrheal diseases, including cholera, typhoid, and dysentery, are rampant in areas with poor water quality and inadequate sanitation. The WHO has identified diarrheal diseases as a leading cause of child mortality in Gaza, accounting for a significant proportion of deaths among children under five (Al-Hindi 2021; Efron et al. 2018). Before October 7, 26% of diseases observed in Gaza were water-related, stemming from reliance on the contaminated and brackish water with rates of infection significantly higher in areas where access to clean water is limited. The poor quality of drinking water, combined with the collapse of sanitation infrastructure, has created an environment in which bacteria, viruses, and parasites thrive. Cholera, in particular, poses a serious threat due to the high rate of bacterial contamination in the water supply. The lack of effective water treatment facilities, compounded by frequent power outages, has prevented proper chlorination of the water, increasing the risk of large-scale cholera outbreaks (Al-Hindi et al. 2021; Bou-Karroum et al. 2020; Abuzerr et al. 2019; Hamada et al. 2016; Al Khatib et al. 2015; Wafi 2015).

The closure of desalination facilities after October 7 and the subsequent reliance on untreated well water have led to a significant public health crisis in Gaza. Families are being forced to consume saline, unfiltered water, resulting in a surge of waterborne diseases. Testimonies from affected residents highlight the gravity of the situation. For instance, a father described his inability to provide safe drinking water for his children, who have since suffered from severe intestinal illnesses, including acute diarrhea, due to consuming contaminated water. This crisis is further compounded by the lack of medical supplies and the overcrowding of hospitals, leaving many without access to necessary treatment. Such conditions underscore the profound health risks posed by the water crisis, particularly for vulnerable populations like children, whose weakened immune systems leave them at higher risk of illness and dehydration (Younis 2024).

In summary, the long-term health impacts of water contamination, caused purposely by Israel, in Gaza are substantial. Prolonged exposure to unsafe drinking water can lead to a wide range of chronic illnesses, developmental issues, and reduced life expectancy. Children are particularly vulnerable, as repeated bouts of waterborne diseases during early childhood can cause stunted growth, cognitive impairments, and weakened immune systems. Children exposed to high levels of nitrates and microbial contamination in drinking water are at a greater risk of developmental delays and long-term health complications, including neurological disorders and reduced academic performance (Shaban et al. 2023; Al-Hindi, 2021; Abuzerr et al. 2019; Hamada et al. 2016). In adults, the cumulative effects of consuming saline and nitrate-contaminated water can lead to an increase in non-communicable diseases, such as cardiovascular diseases, cancer, and kidney failure. The public health system in Gaza, already overwhelmed by the demands of emergency care during conflicts, is ill-equipped to manage the growing burden of chronic diseases caused by water contamination (Shaban et al. 2023; Bou-Karroum et al. 2020; Albelbeisi et al. 2018; El-Naeem et al. 2009).

10.6 Rebuilding Gaza's Water Infrastructure: Pathways to Achieving SDG 6

The water crisis in Gaza stands as a pressing humanitarian challenge, exacerbated by decades of conflict, political instability, and restrictions on movement and resources. As a result, Gaza's water infrastructure has suffered extensive damage, leading to severe water scarcity and contamination. These issues have not only jeopardized the health of residents but have also contributed to significant environmental degradation and economic hardships throughout the region (Al-Riffai 2024; Butter 2024; UNCTAD 2022, 2024; World Bank 2023, 2024a, b, c, d; UNDP 2023; Adnan 2022). To address this crisis effectively and align with SDG 6, it is imperative to adopt a comprehensive, multifaceted approach (Table 10.3).

Table 10.3 Rebuilding Gaza's water infrastructure: pathways to achieving SDG 6

Section	Objectives	Strategies and actions
Assessment and immediate stabilization	Identify critical areas for immediate intervention and establish safe interim water sources	Conduct rapid assessments to determine vulnerable communities and contamination levels; deploy emergency water purification units; engage local NGOs to restore minimally damaged water infrastructure for short-term relief
Reconstruction of critical water infrastructure	Restore and modernize Gaza's water infrastructure for safe access and improved wastewater treatment	Upgrade and expand desalination plants with advanced technologies; repair and modernize pipelines and storage facilities; rebuild wastewater treatment plants with nature-based solutions and prioritize reuse systems for agricultural irrigation
Policy and governance reform	Establish transparent, resilient governance frameworks for sustainable water resource management	Create a centralized Water resource Authority; implement legal and regulatory reforms to manage water extraction and contamination; negotiate cross-border agreements for shared water resources with neighboring regions
Community engagement and capacity building	Empower local communities to manage and protect water resources	Launch educational programs on water conservation and sanitation; provide skill development training for local workers; establish community councils for managing local water sources and overseeing small-scale projects
Implementation of sustainable technologies	Integrate green and sustainable technologies for efficient water use and renewable energy	Install solar panels on water facilities; promote rainwater harvesting and greywater recycling; deploy IoT technology for monitoring water quality and consumption patterns

(continued)

Table 10.3 (continued)

Section	Objectives	Strategies and actions
Securing sustainable funding and international support	Establish long-term funding sources and gain global support for water projects in Gaza	Engage with international donors for grants and long-term financing; attract private sector investments through social impact opportunities; leverage crowdfunding and engage the Palestinian diaspora for funding initiatives
Monitoring, evaluation, and adaptive management	Establish mechanisms for continuous improvement, accountability, and transparency in water infrastructure initiatives	Develop and monitor key performance indicators (KPIs); conduct regular audits and impact assessments; implement adaptive management frameworks to adjust projects based on data insights and community feedback

The first step in addressing Gaza's water crisis is to stop Israel's ongoing genocidal war. Subsequent steps involve a thorough assessment to identify critical areas for immediate intervention. This phase aims to establish safe, interim water sources to meet urgent needs. A rapid assessment of water quality and access will help pinpoint the most vulnerable communities and the extent of contamination affecting existing water sources. Collaborating with international organizations, experts can evaluate water quality, focusing on contaminants such as nitrates, heavy metals, and pathogens. In the short term, deploying emergency water purification units, including portable desalination units and mobile water treatment systems, will provide immediate access to clean drinking water. Additionally, engaging local NGOs and community leaders will be essential to identify and restore minimally damaged water infrastructure, such as boreholes and small reservoirs, that can be rehabilitated quickly for short-term relief. By focusing on immediate stabilization efforts, we can mitigate the most pressing water-related challenges faced by Gaza's residents.

The next objective is to restore and modernize Gaza's water infrastructure to ensure safe access to water and improve wastewater treatment systems. This involves upgrading and expanding existing desalination plants through rehabilitation efforts and the integration of advanced technologies, such as reverse osmosis. Partnering with donor organizations will be crucial to secure funding for large-scale desalination facilities that are designed for long-term sustainability. Moreover, the reconstruction of pipelines and storage facilities is vital to reducing water loss and contamination. This includes repairing and modernizing damaged infrastructure while implementing leak detection and monitoring technologies to enhance efficiency. Furthermore, rebuilding wastewater treatment plants using nature-based solutions, such as constructed wetlands, along with advanced technologies like membrane bioreactors, will help prioritize systems for wastewater reuse in agricultural irrigation. This approach not only addresses groundwater depletion but also contributes to improved food security for the region.

To ensure sustainable water resource management, establishing transparent and resilient governance frameworks is paramount. The creation of a centralized Water Resource Authority under Palestinian jurisdiction will oversee water distribution, enforce quality standards, and maintain infrastructure. This authority should include representatives from local municipalities, NGOs, and international experts to promote transparency and inclusivity. In parallel, legal and regulatory reforms must be implemented to regulate water extraction and prevent contamination. Developing guidelines for wastewater reuse in agriculture and establishing restrictions on industrial discharge are critical steps in safeguarding water quality.

Empowering local communities to manage and protect water resources is essential for fostering a culture of sustainable water use. Launching educational programs and awareness campaigns on water conservation, sanitation, and safe drinking water practices will play a key role in this effort. Collaborating with schools, local NGOs, and religious institutions will help reach diverse segments of the community. Skill development and employment opportunities should also be prioritized, offering training programs for local workers in areas such as plumbing, water treatment operations, and environmental monitoring. Partnering with local universities and international NGOs can provide technical training and certifications in water management. Furthermore, encouraging community involvement in managing and monitoring local water sources through the establishment of community councils will empower residents to oversee small-scale projects, such as rainwater harvesting, well maintenance, and communal water storage.

Integrating green and sustainable technologies is crucial to promoting efficient water use and renewable energy in Gaza. One key initiative is the installation of solar panels on desalination plants, pumping stations, and water treatment facilities, ensuring reliable energy sources while reducing costs and carbon emissions. Promoting rainwater harvesting systems in homes and schools, coupled with incentives for implementation, can significantly enhance local water resilience. Implementing greywater recycling systems in public buildings and households for non-potable applications will further bolster water conservation efforts. The deployment of advanced monitoring and data systems, utilizing the Internet of Things (IoT) technology and smart sensors, will enable real-time monitoring of water quality, detection of leaks, and tracking of consumption patterns. Establishing a centralized data system will optimize resource allocation and help identify emerging challenges.

For the continuity and sustainability of water projects in Gaza, establishing long-term funding sources and garnering global support is vital. Engaging with international donors, such as the World Bank, UNICEF, UNDP, and the European Union, will facilitate access to grants and long-term financing. Encouraging multi-year funding agreements will ensure continuous support for water initiatives. Additionally, attracting private sector investment is essential, positioning Gaza's water infrastructure projects as opportunities for social impact. Developing public–private partnerships for desalination, wastewater treatment, and sustainable technology installation can enhance resource mobilization. Leveraging online crowdfunding platforms and engaging the Palestinian diaspora in funding water-related initiatives will further support these efforts. Transparent reporting systems will be critical in building trust and encouraging sustained contributions from both local and international supporters.

Finally, establishing mechanisms for continuous improvement, accountability, and transparency in all water infrastructure initiatives is essential. Developing key performance indicators (KPIs) related to water access, quality, infrastructure uptime, and environmental impact will help align efforts with SDG 6 targets. Regular evaluations of water infrastructure projects, conducted by third-party auditors, will enhance transparency and provide valuable insights for improvement. Implementing a responsive management framework that adapts projects based on data insights, community feedback, and changing conditions will ensure resilience in the face of emerging challenges. Protocols for rapid response to infrastructure damage during periods of conflict or natural disaster will further strengthen Gaza's water management strategies.

10.7 Conclusion

This chapter highlights the systematic weaponization of water and sanitation in Gaza, illustrating how these essential resources have been transformed into tools of oppression amid protracted conflict. Israel's control over Gaza's water infrastructure and recurrent military operations have severely degraded critical facilities, leaving 96% of Gaza's water undrinkable and sanitation systems on the brink of collapse. This has exacerbated public health crises, particularly among vulnerable populations, and contributed to widespread environmental degradation, including the contamination of the Coastal Aquifer, marine ecosystems, and agricultural lands.

The weaponization of water in Gaza not only obstructs progress toward achieving SDG 6 but also perpetuates cycles of poverty, food insecurity, and economic instability. Immediate international intervention is imperative to halt further destruction, rehabilitate critical infrastructure, and ensure equitable access to clean water and sanitation. Addressing these issues through sustainable governance, community engagement, and the integration of green technologies is essential to alleviate the humanitarian crisis and support Gaza's recovery.

References

Abd El Hay M (2024) The environmental-humanitarian impacts of the Israel Hamas war in Gaza. Arava Institute for Environmental Studies, June, 1–28

Abd Rabou AN (2011) Environmental impacts associated with the Beit Lahia wastewater treatment plant, North Gaza strip, Palestine. Middle East J Sci Res 7(5):746–757

Abdul Samad L, Butcher M, Khalidi B (2024) Water war crimes: how Israel has weaponised water in its military campaign in Gaza

Abu Jabal MS, Abustan I, Rozaimy MR, El Najar H (2017) Groundwater beneath the urban area of Khan Younis City, southern Gaza strip (Palestine): assessment for multi-domestic purposes. Arab J Geosci 10:1–15

Abualfailat M, Qadous N (2022) Costea "reuse—wastewater reuse in agriculture" initiative 'Palestine' synthesis report, March 2022

Abudayya A, Bruaset GTF, Nyhus HB, Aburukba R, Tofthagen R (2023) Consequences of war-related traumatic stress among Palestinian young people in the Gaza strip: a scoping review. Mental Heal Prevent 32:200305. https://doi.org/10.1016/J.MHP.2023.200305

Abuhasballah BA (2023) Protecting the marine environment of the Gaza strip under the provisions of international humanitarian law. Int J Glob Comm VI(1):89–100

Abuzerr S, Nasseri S, Yunesian M, Yassin S, Hadi M, Mahvi AH, Nabizadeh R, Al Agha M, Sarsour A, Darwish M (2019) Microbiological quality of drinking water and prevalence of waterborne diseases in the Gaza strip, Palestine: a narrative review. J Geosci Environ Protect 7(4):122–138

Adnan W (2022) From economic integration to near elimination: the economic consequences of isolation. J Dev Stud 58(6):1160–1180. https://doi.org/10.1080/00220388.2022.2029416

Ahmed MR (2010) Water resources and management issues. Palestinian Water Authority, Shiffa street, Gaza strip, Palestine. International conference "transboundary aquifers: challenges and new directions" (ISARM2010)

Ahmed K, Gayle D, Mousa A (2024) 'Ecocide in Gaza': does scale of environmental destruction amount to a war crime? The Guardian. https://www.theguardian.com/environment/2024/mar/29/gaza-israel-palestinian-war-ecocide-environmental-destruction-pollution-rome-statute-war-crimes-aoe

Akkaya S, Fiess N, Kaminski B, Raballand G (2011) Fragility and conflict in Palestine: the costs of the closures regime on West Bank and Gaza. Oxford University Press, Oxford

Al Absi AA (2008) Nitrate contamination of ground water and methemoglobinemia in Gaza strip. Al-Aqsa Univ J 12(1):1–14

Al Jazeera (2022) Impoverished Gaza economy struggles under Israeli blockade. https://www.aljazeera.com/news/2022/6/24/impoverished-gaza-economy-struggles-under-israeli-blockade

Al Khatib G, Hamada M, Sarsour A (2015) Common waterborne diseases in marginalized areas: A case study: Al Shuka area in Gaza strip—Palestine. Int J Sci Res 5:1792–1798

Al Mezan (2021) The Gaza Bantustan—Israeli Apartheid in the Gaza Strip. Al Mezan Center For Human Rights. https://ngo-monitor.org/ngos/al_mezan_center_for_human_rights/:~:text=In%20November%202021%2C%20Al%20Mezan,its%20associated%20violations%2C%E2%80%9D%20as%20well

Albelbeisi A, Shariff ZM, Mun CY, Abdul-Rahman H, Abed Y (2018) Growth patterns of Palestinian children from birth to 24 months. East Medit Heal J 24(3):302–310. https://doi.org/10.26719/2018.24.3.302

Al-Hindi AI (2021) Epidemiology of parasitic diseases in the Gaza strip, Palestine: a continuing health problem. In: Handbook of healthcare in the Arab world. Springer, Cham, pp 2239–2259

Al-Hindi A, Aker A, Al-Delaimy WK (2021) The destruction of Gaza's infrastructure is exacerbating environmental health impacts. Environ Epidemiol 6(1):e186. https://doi.org/10.1097/EE9.0000000000000186

Al-Juaidi AE, Rosenberg DE, Kaluarachchi JJ (2011) Water management with wastewater treatment and reuse, desalination, and conveyance to counteract future water shortages in the Gaza strip. Int J Water Res Environ Eng 3(12):266–282

AlKhaldi M, Abuzerr S, Obaid HA, Alnajjar G, Alkhaldi A, Alkaiyat A (2021) Social determinants of health in fragile and conflict zones before and during the coronavirus pandemic, with a focus on the Gaza strip. In: Handbook of healthcare in the Arab world. Springer, Cham, pp 851–878

Al-Mughrabi N (2017) Gaza's water shortage worsening, no easy solutions seen. Reuters. Retrieved from http://www.reuters.com/article/us-palestinians-gaza-water-idUSKBN15A1FC

Al-Najjar H, Ceribasi G, Ceyhunlu AI (2021) Effect of unconventional water resources interventions on the management of Gaza coastal aquifer in Palestine. Water Supply 21(8):4205–4218. https://doi.org/10.2166/WS.2021.170

Al-Riffai P (2024) The economic and social costs of the war in Gaza (Issue September)

Al-Rodhan NR, Herd GP, Watanabe L, Al-Rodhan NR, Herd GP, Watanabe L (2011) The six-day war and its consequences. In: Critical turning points in the Middle East: 1915–2015, pp 99–115

Amnesty International (2014a) Families under the rubble: Israeli attacks on inhabited homes. Amnesty International

Amnesty International (2014b) Mounting evidence of deliberate attacks on Gaza health workers by Israeli army. Retrieved from https://www.amnesty.org/en/latest/news/2014/08/mounting-evidence-deliberate-attacksgaza-health-workers-israeli-army/

Amnesty International (2015) Unlawful and deadly: Palestinian rocket attacks from Gaza. Amnesty International UK. Retrieved from https://www.amnesty.org.uk/unlawful-and-deadly-palestinian-rocket-attacks-gaza

Añaños Bedriñana KG et al (2023) The human right to water in the Gaza strip from a gender perspective, in the framework of the sustainable development goals. Agua y Territorio/Water Lands 21(21):53–68. https://doi.org/10.17561/AT.21.6503

Anayah F, Al-Khatib IA, Hejaz B (2021) Assessment of water and sanitation systems at Palestinian healthcare facilities: pre- and post-COVID-19. Environ Monit Assess 193(1):1–22. https://doi.org/10.1007/S10661-020-08791-4/TABLES/4

Anera (2022) Before the taps run dry. Responding to Gaza's existential water crisis. Anera reports on the ground in the Middle East. Volume 11. Summer 2022

Applied Research Institute—Jerusalem (ARIJ) (2007) Status of the environment in the occupied Palestinian Territory. Bethlehem, ARIJ

Applied Research Institute—Jerusalem (ARIJ) (2020) Wastewater management infographic, https://www.arij.org/files/arijadmin/weruinfographics.pdf

Ashour Y, Jlambo A, Abuzerr S (2024) Patients in Gaza with chronic conditions need urgent interventions. Lancet 403(10439):1847–1848. https://doi.org/10.1016/S0140-6736(24)00705-0

Asi Y, Mills D, Greenough PG, Kunichoff D, Khan S et al (2024) 'Nowhere and no one is safe': spatial analysis of damage to critical civilian infrastructure in the Gaza strip during the first phase of the Israeli military campaign, 7 October to 22 November 2023. Confl Heal 18(1):1–13. https://doi.org/10.1186/s13031-024-00580-x

Assembly G (2018). GA Res. 73/94, Operations of the United Nations relief and works agency for Palestine refugees in the near East (Dec. 7, 2018)

B'Tselem (2020) Water in Gaza: scarce, polluted and mostly unfit for use. Published 17 Aug 2020. https://reliefweb.int/report/occupied-palestinian-territory/water-gaza-scarce-polluted-and-mostly-unfit-use

Baalousha H (2009) Water scarcity and environmental problems in the Gaza strip, Palestine: origin, impacts and prospects for solutions. In: Water shortages: environmental, economic and social impacts. Nova Science Publishers

Bachmann J et al (2014) Gaza 2014: findings of an independent medical fact finding mission. Physicians for human rights-Israel. Al Mezan Center for Human Rights-Gaza, Gaza Community Mental Health Programme, Palestinian Center for Human Rights-Gaza

Bachmutsky R (2017) Otherwise occupied: the legal status of the Gaza strip 50 years after the six-day war. Va J Int L 57:413

Bailout H (2008) Analysis of nitrate occurrence and distribution in groundwater in the Gaza strip using major ion chemistry. Global NEST J 10(3):337–349

Barak O (2005) The failure of the Israeli–Palestinian peace process, 1993–2000. J Peace Res 42(6):719–736

Barhoum L (2021) Still treading water: reviewing six years of the Gaza reconstruction mechanism and the dire water situation in the Gaza strip

Bashi S (2013) Controlling perimeters, controlling lives: Israel and Gaza. Law Ethic Human Right 7(2):243–282

Bashitialshaaer R (2014) Basic life for the Palestinian: Gaza water and environmental problems. In The "Fourth World Congress for Middle East Studies" (WOCMES), to be held at METU in Ankara, Turkey 18th to 22nd August

Beiraghdar F, Momeni J, Hosseini E, Panahi Y, Negah SS (2023) Health crisis in Gaza: the urgent need for international action. Iran J Public Health 52(12):2478–2483. https://doi.org/10.18502/IJPH.V52I12.14309

Bellisari A (1994) Public health and the water crisis in the occupied Palestinian territories. J Palest Stud 23(2):52–63

Bohannon J (2006) Running out of water—and time

Boukari Y, Kadir A, Waterston T, Jarrett P, Harkensee C, Dexter E, Cinar EN, Blackett K, Nacer H, Stevens A, Devakumar D (2024) Gaza, armed conflict and child health. BMJ Paediat Open 8(1):e002407. https://doi.org/10.1136/BMJPO-2023-002407

Bou-Karroum L, El-Harakeh A, Kassamany I, Ismail H, ElArnaout N, Charide R, Madi F, Jamali S, Martineau T, ElJardali F, Akl EA (2020) Health care workers in conflict and post-conflict settings: systematic mapping of the evidence. PLoS One 15(5):e0233757. https://doi.org/10.1371/JOURNAL.PONE.0233757

Boyan (2023) Gaza water crisis. Bonyan International Organization, Indianapolis, USA. https://bonyan.ngo/palestine/gaza-water-crisis/

Brooks DB, Trottier J, Giordano G (2019) Transboundary water issues in Israel, Palestine, and the Jordan River Basin: an overview

Brooks DB, Trottier J, Giordano G, Brooks DB, Trottier J, Giordano G (2020) The existing Oslo arrangements. Transboundary water issues in Israel, Palestine, and the Jordan River Basin: an overview, 25–37

Buheji M (2024) Addressing human factors in Gaza: challenges and solutions for post-war recovery. Int J Manag 15(3):35–45. https://www.researchgate.net/publication/380457848_Addressing_Human_Factors_in_Gaza_Challenges_and_Solutions_for_Post-War_Recovery

Buheji M, Al-Muhannadi K (2023) Mitigating risks of environmental impacts on Gaza-review of Precautions & Solutions post (2023 war). Int J Adv Res Eng Technol 14(7):15–47

Butt KM, Butt AA (2016) Blockade on Gaza strip: A living hell on earth. J Pol Stud 23:157

Butter D (2024) Economic impact of the Gaza war. In: IEMed: Mediterranean yearbook, pp 273–275

Chomsky N, Pappe I (2011) Gaza in crisis: reflections on Israel's war against the Palestinians. Haymaret Books, p 240

Cole J (2024) Weaponizing a Water crisis: the destruction of water supplies in Gaza. New Lines Institute, pp 1–7

Correia T (2024) Two urgent actions related to international health emergencies amid the escalating conflict in Gaza. Int J Health Plann Manag 39(2):160–163. https://doi.org/10.1002/HPM.3720

Culverwell SM (2017) Israel and Palestine-An analysis of the 2014 Israel-Gaza war from a genocidal perspective. Unpublished manuscript. James Madison University

Davidovitch N (2019) Health risks assessment for the Israeli population following the sanitary crisis in Gaza

Dickie G, Withers A (2024) Gaza conflict has caused major environmental damage, UN says. Reuters. https://www.reuters.com/world/middle-east/gaza-conflict-has-caused-major-environmental-damage-un-says-2024-06-18/

Efron S, Fischbach JR, Giordano G (2017) Gaza's water and sanitation crisis: the implications for public health. In: The crisis of the Gaza strip: A way out. Institute for National Security Studies, Tel Aviv

Efron S, Fischbach JR, Blum I, Karimov RI, Moore M (2018) The public health impacts of Gaza's water crisis. Rand Heal Q 8(10)

Efron S, Fischbach JR, Blum I, Karimov RI, Moore M (2019) The public health impacts of Gaza's water crisis: analysis and policy options. Rand Heal Q 8(3)

Ekenci İ (2019) Water as a tool of oppression: the case of the Israeli-Palestinian conflict. Master's thesis, Namık Kemal Üniversitesi

El-Naeem MFA, Heen ZA, Tubail K (2009) Factors behind groundwater pollution by nitrate in north governorates of Gaza strip (1994–2004). In Thirteenth international water technology conference, IWTC (Vol. 13, p. 2009)

El-Nahhal Y (2006, March) Contamination of groundwater with heavy metals in Gaza strip. In 10th international water technology conference, Alexandria (pp. 1139–1150)

Elnakib S, Fair M, Mayrhofer E, Afifi M, Jamaluddine Z (2024) Pregnant women in Gaza require urgent protection. Lancet 403(10423):244. https://doi.org/10.1016/S0140-6736(23)02835-0

Emergency Water, Sanitation and Hygiene Group (EWASH) & Al-Haq (2011) Israel's violations of the International Covenant on Economic, Social and Cultural Rights with regard to the human rights to water and sanitation in the Occupied Palestinian Territory. http://www2.ohchr.org/english/bodies/cescr/docs/ngos/EWASH-Al-Haq_Israel_CESCR47.pdf

ESCWA (2023) War on ·Gaza: weaponizing access to water, energy and food (Issue November)

ESCWA/UNDP (2024) Gaza war: expected socioeconomic impacts on the State of Palestine. In United Nations (Issue October). https://www.undp.org/arab-states/publications/gaza-war-expected-socio-economic-impacts-state-palestine-october-2024

Esleem F, Abu Heen Z, Aljeesh Y (2015) Evaluation the effect of environmental and social factors on adult patients with chronic renal failure in Gaza governorates: retrospective study. Islamic University-Gaza, 14

Euro-Med (2024) Israel uses water as a weapon of its genocide in Gaza. Euro-Med Human Rights https://euromedmonitor.org/en/article/6393/Israel-uses-water-as-a-weapon-of-its-genocide-in-gaza

EWASH (Emergency Water Sanitation and Hygiene Group) (2016) Water and sanitation in Palestine. Retrieved from http://www.ewash.org/sites/default/files/inoptfiles/160314%20%20WATER%20AND%20SANITATION%20IN%20PALESTINE.pdf

FAO (2017) Evaluation of FAO's programme in West Bank and Gaza Strip, 2011–2015. Food and Agriculture Organization of the United Nations Office of Evaluation. August 2017

FAO (2023a) Damage to agricultural infrastructure due to the conflict in the Gaza strip as of 31 December 2023. Rome. https://www.fao.org/3/cc9402en/cc9402en.pdf

FAO (2023b) Damage to agricultural wells due to the conflict in the Gaza strip as of 31 December 2023. Rome. https://www.fao.org/3/cc9401en/cc9401en.pdf

FAO (2023c) Damage to greenhouses due to the conflict in the Gaza strip as of 31 December 2023. Rome. https://www.fao.org/3/cc9400en/cc9400en.pdf

FAO (2023d) Damage to cropland due to the conflict in the Gaza strip as of 31 December 2023. Rome. https://www.fao.org/3/cc9399en/cc9399en.pdf

FAO (2024a) A rapid geospatial damage assessment of the 2023 conflict in the Gaza strip on agricultural land and infrastructure. Rome https://doi.org/10.4060/cc9556en

FAO (2024b) Damage to agricultural wells due to the conflict in the Gaza strip as of 15 February 2024

FAO (2024c) Agricultural samage assessment in the Gaza Strip from October 7th 2023 to September 1st 2024 (Vol. 10, Issue May)

FAO (2024d) Gaza: geospatial data shows intensifying damage to cropland. New FAO and UNOSAT assessment highlights the impact of conflict on

Gaza's agricultural infrastructure. 03/10/2024. https://www.fao.org/newsroom/detail/gaza-geospatial-data-shows-intensifying-damage-to-cropland/en

FAO and WFP (2024) Monitoring food security in Palestine and the Sudan—A joint FAO/WFP update for the members of the United Nations Security Council, February 2024. Issue no. 13 (Special edition). Rome. https://doi.org/10.4060/cc9698en

Ferragina E (2008) The effect of the Israeli-Palestinian conflict on the water resources of the Jordan River basin. Glob Environ 1(2):152–170

Gazit S (1998) Israel and the Palestinians: fifty years of wars and turning points. Ann Am Acad Pol Soc Sci 555(1):82–96

Gleick PH (1993) Water and conflict: fresh Water resources and international security. Int Secur 18(1):79–112

Gleick PH (2019) Water as a weapon and casualty of armed conflict: A review of recent water-related violence in Iraq, Syria, and Yemen. WIREs Water 6:e1351

Goodarzi H, Badri T, Javadzadeh HR (2024) War, forced displacement, and lack of healthcare in the Gaza strip. Trauma Mon 28(5):945–946. https://doi.org/10.1016/S0140-6736

Gray A (2009) Water development in the Palestinian territories since the Oslo interim agreement in 1995. Water Policy 11(5):525–536

Gray A, Hilal J (2007) Water and security for Palestine. In: Integrated water resources management and security in the Middle East. Springer, Dordrecht, pp 99–117

Gupta N, Grobusch MP, Jokelainen P, Wyllie AL, Barac A, Mora-Rillo M, Gkrania-Klotsas E, Pellejero-Sagastizabal G, Paño-Pardo JR, Duizer E, Lescure F-X (2024) Poliomyelitis in Gaza. Clin Microbiol Infect. https://doi.org/10.1016/J.CMI.2024.10.003

Haddad M (2011) Water scarcity and degradation in Palestine as challenges, vulnerabilities, and risks for environmental security. In: Coping with global environmental change, disasters and security: threats, challenges, vulnerabilities and risks. Berlin/Heidelberg, Springer, pp 409–419

Hamada MS, Sarsour AK, Al Khatib GA (2016) Environmental factors associated with common Water borne diseases among Al Shuka rural Community in the Gaza Strip-Palestine. Int J Sci Res Environ Sci:136–144

Hass A (2014) Israel's control of Water in the West Bank and Gaza. Haaretz

Hassoun A, Al-Muhannadi K, Hassan HF, Hamad A, Khwaldia K, Buheji M, Jawaldeh AA (2024) From acute food insecurity to famine: how the 2023/2024 war on Gaza has dramatically set back sustainable development goal 2 to end hunger. Front Sustain Food Syst 8:1402150. https://doi.org/10.3389/fsufs.2024.1402150

Holail S, Saleh T, Xiao X, Xiao J, Xia G-S, Shao Z, Wang M, Gong J, Li D (2024) Time-series satellite remote sensing reveals gradually increasing war damage in the Gaza strip. Natl Sci Rev 11(9):nwae304. https://doi.org/10.1093/NSR/NWAE304

Horino M, Zaqqout R, Habash R, Albaik S, Abed Y, Al-Jadba G, West KP, Seita A (2024) Food insecurity, dietary inadequacy, and malnutrition in the Gaza strip: a cross-sectional nutritional assessment of refugee children entering the first grade of UNRWA schools and their households before the conflict of 2023–24. Lancet Glob Health 12(11):e1871–e1880. https://doi.org/10.1016/S2214-109X(24)00320-6

Hussein H (2023) Stop violation of international water laws in Gaza. Nature 623(7986):253. https://doi.org/10.1038/D41586-023-03461-0

Hussein DM, Lucchetti MC, Zaqoot HA, Penca J, Hussein MA (2022) Status of fisheries in Gaza strip: past trends and challenges. Int J Euro-Mediterr Stud 15(2)

Hussein S, Ahmed SK, Qurbani K, Fareeq A, Essa RA (2024) Infectious diseases threat amidst the war in Gaza. J Med Surg Publ Heal 2:100067. https://doi.org/10.1016/J.GLMEDI.2024.100067

IPC (2023). Gaza Strip: IPC acute food insecurity | November 2023–February 2024. Rome. https://www.ipcinfo.org/fileadmin/user_upload/ipcinfo/docs/IPC_Gaza_Acute_Food_Insecurity_Nov2023_Feb2024.pdf

IPC (2024a) Gaza Strip: acute food insecurity situation for September–October 2024 and projection for November 2024–April 2025 (Issue October). https://www.ipcinfo.org/fileadmin/user_upload/ipcinfo/docs/IPC_Gaza_Strip_Acute_Food_Insecurity_Feb_July2024_Special_Brief.pdf

IPC (2024b) GAZA STRIP: IPC acute food insecurity special snapshot (Issue September). https://www.ipcinfo.org/fileadmin/user_upload/ipcinfo/docs/IPC_Gaza_Strip_Acute_Food_Insecurity_Feb_July2024_Special_Brief.pdf

IPC (2024c) GAZA STRIP: Risk of Famine persists amidst recent surge in hostilities. https://www.ipcinfo.org/ipcinfo-website/countries-in-focus-archive/issue-112/en/

IPC Famine Review Committee (2023) Gaza strip: famine review of the IPC analysis. Rome. https://www.ipcinfo.org/fileadmin/user_upload/ipcinfo/docs/IPC_Famine_Review_Report_Gaza.pdf

Isaac J, Sabbah W (2017) The intensifying Water crisis in Palestine. Applied research institute–Jerusalem (ARIJ). Available online: http://www.arij.org/files/admin/The_Intensifying_Water_Crisis_in_Palestine.pdf (accessed on 29 September 2019)

Kayser GL, Rao N, Jose R, Raj A (2019) Water, sanitation and hygiene: measuring gender equality and empowerment. Bull World Health Organ 97(6):438–440. https://doi.org/10.2471/BLT.18.223305

Kemmerling B, Schetter C, Wirkus L (2022) The logics of war and food (in)security. Glob Food Sec 33:100634. https://doi.org/10.1016/J.GFS.2022.100634

Khoodoruth MAS, Khoodoruth WNC, kai. (2024) From past famines to present crises: the epigenetic impact of maternal malnutrition on offspring health in Gaza. Asian J Psychiatr 95(March):103999. https://doi.org/10.1016/j.ajp.2024.103999

Koutsos TM, Menexes GC, Dordas CA (2019) An efficient framework for conducting systematic literature reviews in agricultural sciences. Sci Total Environ 682:106–117

MacDonald E, Herrador BG, Hyllestad S, Lund V, Nygard K, Vold L, Lafi M, Ammar W, Iversen B (2016) Water usage in the Gaza strip: recommendations from a literature review and consultations with experts. East Medit Heal J 22(12):910–918

Machairas D (2017) The strategic and political consequences of the June 1967 war. Cogent Soc Sci 3(1):1299555

Messerschmid C (2011) Water in Gaza: problems and prospects

Miller EB (1998) Implementing the Oslo accords. Cardozo J Int Comp L 6:363

Milton S, Elkahlout G, Attallah S (2024) Shrinking reconstruction space in the Gaza strip: rebuilding after the 2021 and 2022 wars. Conf Secur Dev 24(1):49–78

Mohammed Alashqar M, Abdul Rahim A, Abd Aziz AS (2023) War crimes in Gaza strip from year 2008 2021: individual criminal responsibility under the legal framework of Rome statute of the international criminal court. J Int Stud 19(1):61–93

Moher D, Liberati A, Tetzlaff J, Altman DG et al (2009) Preferred reporting items for systematic reviews and meta-analyses: the PRISMA statement. Ann Intern Med 151(4):264–269

Naff T (2020) Water in the Middle East: conflict or cooperation? Routledge

Nembrini PG (2010) The Gaza strip: the state of the water supply after the 2008–2009 war. Cities in war, thirsty cities, 1–28

Neuzil C (2007) The water crisis in the Gaza strip: prospects for resolution. Ground Water 45(6)

NRC (2024) Israel's siege now blocks 83% of food aid reaching Gaza, new data reveals. https://www.nrc.no/news/2024/september/israels-siege-now-blocks-83-of-food-aid-reaching-gaza-new-data-reveals/

OCHA (2024) Hostilities in the Gaza strip and Israel | flash update 10, 16 October 2023. In: OCHA. Geneva, Switzerland. [Cited 22 January 2024]. https://www.ochaopt.org/content/hostilities-gaza-strip-and-israelflash-update-10

OCHA (2024a) Humanitarian situation update 218 | Gaza strip. United Nations Office for the coordination of humanitarian affairs (16 Sep 2024). https://reliefweb.int/report/occupied-palestinian-territory/humanitarian-situation-update-218-gaza-strip-enarhe

OCHA (2024b) Gaza humanitarian response update | 5–18 August 2024. https://www.unocha.org/publications/report/occupied-palestinian-territory/gaza-humanitarian-response-update-5-18-august-2024

OCHA (2024c) Humanitarian situation update 179 | Gaza strip. Broader environmental and water supply impacts (14 Jun 2024)

Oxfam (2024) Oxfam releases report on weaponization of water in Gaza. 3 Sep 2024. The Water Diplomat

Paché G (2024) Israeli-Palestinian conflict: towards a major logistical and environmental crisis? Tech Soc Sci J 53:252

Palestinian Environment Quality Authority (PEQA) (2014) The environmental impacts of the Israeli offensive on the Gaza strip in the period 8 July to 26th of August 2014. PEQA, Gaza

Palestinian Water Authority (PWA) (2012a) Assessment of groundwater supply in Gaza strip for year 2011. Ed. PWA

Palestinian Water Authority (PWA) (2012b) Water supply report 2011. Ed. PWA

Palestinian Water Authority (PWA) (2012c) Status report of water resources in occupied Palestinian territory for 2011

Palestinian Water Authority (PWA) (2012d) Status Report of Wastewater in Occupied Palestinian Territory for 2011

Palestinian Water Authority (PWA) (2013) Status report of water resources in the occupied State of Palestine-2012. Annual water resources status report 2012

Palestinian Water Authority (PWA) (2014a) Water resources directorate, Gaza water resources status report, 2013/2014, December 2014; Palestinian Water Authority

Palestinian Water Authority (PWA) (2014b) Water sector damage assessment report. PWA, Gaza

Pappé I (2006) The ethnic cleansing of Palestine. Oneworld Publications

PW Authority (2015) GAZA STRIP: desalination facility project: necessity, politics and energy. Palestinian Water Authority, Gaza

Qahman K, Mokhamer EAA (2013) Salinity of drinking water and its association with renal failure in Gaza strip, Palestine. J Water Res Ocean Sci 2(5):95–107

Qrenawi L, Abu Shomar R (2020) Health risk assessment of groundwater contamination case study: Gaza Strip

Qumsiyeh MB (2024) Impact of the Israeli military activities on the environment. Int J Environ Stud 81(2):977–992. https://doi.org/10.1080/00207233.2024.2323365

Qumsiyeh MB, Abusarhan MA (2021) Biodiversity and environmental conservation in palestine. In: Biodiversity, conservation and sustainability in Asia: volume 1: prospects and challenges in West Asia and Caucasus, pp 1–22. https://doi.org/10.1007/978-3-030-59928-7_1

Qurie A (2007) The Oslo accords: a Palestinian perspective. Isr J Foreign Aff 1(2):33–39

Ramahi S (2023) The health risks posed by water pollution in the Gaza strip. Middle East Monitor (MEM)

Ramon D, Britzi M, Davidovich N, Tchernov D, Morick D (2023) Accumulation of lead, mercury and cadmium in coastal sediments in the Eastern Mediterranean Sea. In: Lead, mercury and cadmium in the aquatic environment: worldwide occurrence, fate and toxicity. CRC Press, pp 92–125. https://doi.org/10.1201/9781003186441_5

Refugees International (2024) Untangling the reality of famine in Gaza (Issue September)
Rist L, Norström A, Queiroz C (2024) Biodiversity, peace and conflict: understanding the connections. Curr Opin Environ Sustain 68:101431. https://doi.org/10.1016/J.COSUST.2024.101431
Romano D, Stefani G, Rocchi B, Fiorillo C (2019) The impact of assistance on poverty and food security in a fragile and protracted-crisis context: the case of west bank and Gaza strip. Bio-based and applied. Economics 8(1):21–61. https://doi.org/10.13128/bae-8145
Rouyer AR (1997) The water issue in the Palestinian—Israeli peace process. Survival 39(2):57–81
Roy S (2002) Why peace failed: an Oslo autopsy. Curr Hist 101(651):8–16
Safi AS (2014) War on Gaza strip: participatory environmental impact assessment. Palestinian environmental NGOs network—FoE Palestine. Maan Development Center
Samad LA, Butcher M, Khalidi B (2024) Water war crimes: how Israel has weaponised water in its military campaign in Gaza
Sarsour A (2014) Impact of the last Israeli military operation on the vital environmental sectors in Gaza governorates. http://www.tlaxcala-int.org/article.asp?reference=13881
Schlütter B (2005) Water rights in the West Bank and in Gaza. Leiden J Int Law 18(3):621–644
Seeberg P (2024) Israeli hydro-hegemony and the Gaza war. Middle East Policy 31(2):33–46
Selby J (2003) Water, power, and politics in the Middle East: the other Israeli-Palestinian conflict. I.B. Tauris
Shaban J, Al-Najar H, Kocadal K, Almghari K, Saygi S (2023) The effect of nitrate-contaminated drinking water and vegetables on the prevalence of acquired methemoglobinemia in Beit Lahia City in Palestine. Water 15(11):1989
Shawwa IR (1994) Water situation in the Gaza strip. In: Studies in environmental science, vol 58. Elsevier, pp 251–259
Shlaim A (1994) The Oslo accord. J Palest Stud 23(3):24–40
Shlaim A (2016) 13 the rise and fall of the Oslo peace process. Int Relat Middle East 285
Shomar B (2011a) Retracted article: groundwater contaminations and health perspectives in developing world case study: Gaza strip. Environ Geochem Health 33:189–202
Shomar B (2011b) Water scenarios in the Gaza strip, Palestine: thirst, hunger and disease. Int J Environ Stud 68(4):477–493
Shomar B, Rovira J (2023) Human health risks associated with the consumption of groundwater in the Gaza strip. Heliyon 9(11):e21989. https://doi.org/10.1016/J.HELIYON.2023.E21989
Shorrab A, Nassef M, Subhi A, Giwa B, Buheji M (2024) Health in the crossfire-analysing and mitigating the multifaceted health risks of the 2023 war on Gaza. Publ Heal Res 14(1):1–11
Sour S, Khalil A, Omran AN (2011) Water crisis in Gaza strip, Palestine. Acta Tech Corvin Bull Eng
Strohmeir EM, Ybarra M (2011) Gaza water problem contributing factors preventing solution. Walden University. PSYC 8700 Social Change
Supplement to Readings on Water (2016) Water in the Israeli-Palestinian Conflict-European Parliament—Europa. Europarl. Retrieved from http://www.burlington.org/departments/library/circulation/docs/Supplement%20%20to%20Readings%20on%20Water.pdf
Taha AM, Mahmoud H, Nada SA, Abuzerr S (2024) Controlling the alarming rise in infectious diseases among children younger than 5 years in Gaza during the war. Lancet Infect Dis 24(4):e211. https://doi.org/10.1016/S1473-3099(24)00067-7
Talhami M, Zeitoun M (2020) The impact of attacks on urban services II: reverberating effects of damage to water and wastewater systems on infectious disease. Int Rev Red Cross 102(915):1293–1325
To-Hauser K (2024) Compounded nature of civilian harm: infrastructure damage in the 2008–2009 and 2014 wars in Gaza. Master's thesis
Trottier J (1999) Hydropolitics in the West Bank and Gaza strip. Palestinian Academic Society for the Study of International Affairs (PASSIA)
Tumeh DMM (2022) Upper gastrointestinal complications and treatment among patients with chronic diseases: a cross sectional study from Palestine. Doctoral dissertation, Faculty of Graduate Studies, An-Najah National University
Tutu D, Skarphéðinsson Ö (2013) The Oslo accords 1993–2013: a critical assessment. IB Tauris
Ubeid KF, Al-Agha MR (2016) Assessment of impact sewage effluents on the coastal water quality around the mouth of Wadi Gaza (Gaza and Middle Governorates, Gaza Strip, Palestine). Assessment 6(11)
UN OCHA, FAO (2010) Farming without land, fishing without water: Gaza agriculture sector struggles to survive. Fact Sheet https://unispal.un.org/DPA/DPR/unispal.nsf/0/9A265F2A909E9A1D8525772E004FC34B
UNCTAD (2015). Report on UNCTAD assistance to the Palestinian people: Developments in the economy of the Occupied Palestinian Territory, Geneva, September 2015
UNCTAD (2022) Report on UNCTAD assistance to the Palestinian people: Developments in the economy of the Occupied Palestinian Territory (Vol. 1397, Issue August). https://data.worldbank.org/indicator/NY.GDP.PCAP.PP.CD
UNCTAD (2024) Preliminary assessment of the economic impact of the destruction in Gaza and prospects for economic recovery (Issue January)
UNDP (2021) Gaza 2021 infrastructure damage assessment report (Issue May)
UNDP (2023) Gaza War: Preliminary Findings on the Socio-Economic and Environmental Impact on Lebanon (Issue December). http://en.wikipedia.org/w/index.php?title=Gaza_War&oldid=571863943

UNEP (2009). Environmental assessment of the Gaza Strip following the escalation of hostilities in December 2008 – January 2009. Palestinian Water Authority, PWA (2014) Water Resources Directorate, Gaza Water Resources Status Report, 2013/2014, December 2014
UNEP (2020) State of environment and outlook report for the occupied Palestinian territory
UNESCWA (2023) War on Gaza: weaponizing access to water, energy and food, 23 November 2023. Beirut, Lebanon. https://www.unescwa.org/sites/default/files/pubs/pdf/war-gaza-weaponizing-access-waterenergy-food-land-english.pdf
UNICEF (2021) State of Palestine 'Escalation in the Gaza Strip, the West Bank and Israel' Report No 4, 3 June 2021
UNICEF (2022) Humanitarian action for children. State of Palestine
UNICEF (2023) Drinking water. Universal access to safe drinking water is a fundamental need and human right. https://data.unicef.org/topic/water-and-sanitation/drinkingwater/
UNICEF (2024a) Children face severe water shortages and a dire hygiene situation in the Gaza strip. UNICEF is on the ground to provide water, sanitation, and hygiene for every child in the Gaza Strip. https://www.unicef.org/sop/stories/children-face-severe-water-shortages-and-dire-hygiene-situation-gaza-strip
UNICEF (2024b) Children in Gaza need life-saving support. https://www.unicef.org/emergencies/children-gaza-need-lifesaving-support
United Nations (2017) Gaza ten years later. https://www.un.org
United Nations Development Programme (2021) Gaza infrastructure damage assessment report. UNDP https://www.undp.org/papp/publications/gaza-infrastructure-damage-assessment-report
United Nations Development Programme (2024) A silent threat: Gaza's struggle with solid waste management, 28 March. https://stories.undp.org/a-silent-threat-gazas-struggle-with-solid-waste-management
United Nations Environment Programme (2009) Environmental assessment of the Gaza Strip- following the escalation of hostilities in December 2008 to January 2009. https://www.unep.org/resources/assessment/environmental-assessment-gaza-strip
United Nations Environment Programme (2018) Technical report: mosul debris management assessment https://wedocs.unep.org/20.500.11822/42306
United Nations Environment Programme (2020) State of environment and outlook report for the occupied Palestinian territory. https://www.unep.org/resources/report/state-environment-and-outlook-report-occupied-palestinian-territory-2020
United Nations Environment Programme (2024) Environmental impact of the conflict in Gaza: preliminary assessment of environmental impacts. Nairobi. http://wedocs.unep.org/20.500.11822/45739unep.org/resources/report/environmental-impact-conflict-gaza-preliminary-assessment-environmental-impacts
UNOSAT (2023) Gaza Strip agricultural damage assessment—December 2023. https://unosat.org/products/3772
UNOSAT (2024) UNOSAT Gaza strip comprehensive damage assessment—September 2024 (Issue September)
UNRWA (2023) Situation report 20, 1 November 2023. Gaza. https://www.unrwa.org/resources/reports/unrwa-situation-report-no-20-situation-gaza-strip-and-west-bankincluding-east-Jerusalem
UNRWA (2024) Situation report 112 on the situation in the Gaza Strip and the West Bank, including East Jerusalem (Issue June)
Wafi HN (2015) Assessment of heavy metals contamination in the Mediterranean Sea along Gaza Coast–a case study of Gaza Fishing Harbor. In Al-Azhar University-Gaza. https://www.alazhar.edu.ps/arabic/He/files/20112125.pdf
Weiner J (1999) Israel, Palestine, and the Oslo accords. Fordham Int LJ 23:230
Weinthal E, Sowers J (2019) Targeting infrastructure and livelihoods in the West Bank and Gaza. Int Aff 95(2):319–340
Weinthal E, Vengosh A, Marei A, Gutierrez A, Kloppmann W (2005) The water crisis in the Gaza strip: prospects for resolution. Groundwater 43(5):653–660
Westheime N (2024) The devastating impact of Gaza's acute and prolonged water crisis. March 14, 2024. Middle East Institute. https://www.mei.edu/publications/devastating-impact-gazas-acute-and-prolonged-water-crisis
WFP (2024) Gaza updates: Hunger deepens as aid plummets. https://www.wfp.org/stories/gaza-updates-hunger-deepens-aid-plummets
Wolf A, Ross J (1992) The impact of scarce water resources on the Arab-Israeli conflict. Nat Res J 32:919
World Bank (1992) The West Bank and Gaza strip: economic developments in the 1980s and prospects for the 1990s
World Bank (2009a) West Bank and Gaza assessment of restrictions on Palestinian water sector development. Sector note April 2009
World Bank (2009b). West Bank and Gaza assessment of restrictions on Palestinian water sector development
World Bank (2013) West Bank & Gaza—Gaza emergency Water project (English). World Bank Group, Washington, DC. http://documents.worldbank.org/curated/en/723291474939249677/West-Bank-Gaza-Gaza-Emergency-Water-Project
World Bank (2017) The performance of Palestinian local governments: an assessment of service delivery outcomes and performance drivers in West Bank and Gaza. World Bank
World Bank (2018a) Securing water for development in West Bank and Gaza
World Bank (2018b) Securing Water for development in West Bank and Gaza. World Bank, Washington, DC
World Bank (2018c) Toward Water security for Palestinians: West Bank and Gaza Water supply, sani-

tation, and hygiene poverty diagnostic. WASH poverty diagnostic. World Bank, Washington, DC

World Bank (2020) Rapid fishery and aquaculture sector diagnosis using fishery performance indicators in the Gaza Strip, May 2020. https://openknowledge.worldbank.org/entities/publication/64cc2a35-d26f-5e20- a4e9-10ee640459c7

World Bank (2022) Agriculture, forestry, and fishing, value added (% of GDP)—West Bank and Gaza, Euro area. In: World Bank data. Washington, DC. [Cited 22 January 2024]. https://data.worldbank.org/indicator/NV.AGR.TOTL.ZS?view=chart&locations=PS-XC

World Bank (2023) Impacts of the conflict in the Middle East on the Palestinian economy. p. 2. Washington, DC. https://thedocs.worldbank.org/en/doc/7600aee6c75eff4cf9b71ea1fe4016db0280012023/original/NoteWBG-dec11-CLEAN.pdf

World Bank (2023a) West Bank and Gaza country climate and development report. World Bank Group, Washington DC

World Bank (2023b) Water security development—Gaza central desalination program—associated works phase I project (P168739)—implementation status & results report, 12 September. https://documents.worldbank.org/en/publication/documents-reports/documentdetail/099091223183039258/p1687390b32bf40c4086d10cbb140f2770b

World Bank (2024a) Gaza damage assessment bi-weekly (Gaza strip bi-weekly damage summary: 28 December 2023 to 10 January 2024)

World Bank (2024b) Note on the impacts of the conflict in the Middle East on the Palestinian economy—September 2024 (issue September)

World Bank (2024c). Gaza Strip—interim damage assessment summary note—March 29, 2024

World Bank (2024d) Joint World Bank, UN report assesses damage to Gaza's infrastructure. PRESS RELEASE April 2, 2024. https://www.worldbank.org/en/news/press-release/2024/04/02/joint-world-bank-un-report-assesses-damage-to-gaza-s-infrastructure:~:text=The%20water%20and%20sanitation%20system,of%20children%20out%20of%20school

World Bank, European Union and United Nations (2024) Gaza strip interim damage assessment, 3 April. https://palestine.un.org/en/265025-gaza-strip-interim-damage-assessmentMarch

World Health Organization (WHO) (2017) 2.1 billion people lack safe drinking water at home, more than twice as many lack safe sanitation. 12 July 2017 News release GENEVA. https://www.who.int/news/item/12-07-2017-2-1-billion-people-lack-safe-drinking-water-at-home-more-than-twice-as-many-lack-safe-sanitation

World Health Organization (WHO) (2021) Billions of people will lack access to safe water, sanitation and hygiene in 2030 unless progress quadruples – warn WHO, UNICEF. 1 July 2021 Joint News Release Geneva/New York. https://www.who.int/news/item/01-07-2021-billions-of-people-will-lack-access-to-safe-water-sanitation-and-hygiene-in-2030-unless-progressquadruples-warn-who-unicef

World Health Organization (2024) Preventing famine and deadly disease outbreak in Gaza requires faster, safer aid access and more supply routes. https://www.who.int/news/item/15-01-2024-preventing-famine-anddeadly-disease-outbreak-in-gaza-requires-faster%2D%2Dsafer-aid-accessand-more-supply-routes

World Health Organization (WHO) (2024a) WHO's response in the occupied Palestinian territory. WHO Office for the occupied Palestinian territory Othman Ibn Affan Street, 24 Sheikh Jarrah

World Health Organization (WHO) (2024b) oPt Emergency Situation Update Issue 46 7 Oct 2023–25 Sep 2024

Yassin MM, Amr SSA, Al-Najar HM (2006) Assessment of microbiological water quality and its relation to human health in Gaza Governorate, Gaza Strip. Publ Heal 120(12):1177–1187

Yiftachel O (2006) Ethnocracy: land and identity politics in Israel/Palestine. University of Pennsylvania Press

Younis A (2024, December 18):لا بديل من «المياه السامة»: الاحتلال يقتل الغزيين... عطشاً. *Al-Akhbar*. https://shorturl.at/JMyCh

Zahra BAAA (2001) Water crisis in Palestine. Desalination 136(1–3):93–99

Zeitoun M (2008) Power and water in the Middle East: the hidden politics of the Palestinian-Israeli water conflict. I.B. Tauris

Zreik R (2003) The Palestinian question: themes of justice and power: part I: the Palestinians of the occupied territories. J Palest Stud 32(4):39–49

11 Environmental Consequences of Conflict: Lessons from the Recent War on Gaza

Abdo Hassoun, Khawla Al-Muhannadi, Amira Haddarah, Farah Slim, Dalal Iriqat, Ahmed Hamad, Alix Faddoul, and Mohamed Buheji

Abstract

The recent war on Gaza (October 2023–January 2025) has caused severe damage to the natural environment and has had far-reaching global consequences for the future of life in this region and its surrounding areas. The study intends to analyze a wide range of sources, including peer-reviewed and grey literature, to assess the environmental sustainability challenges posed by this war, focusing specifically on SDGs 6 (Water and Sanitation), 13 (Climate Action), 14 (Marine Ecosystems), and 15 (Terrestrial Ecosystems), with an emphasis on climate action. Our findings indicate that, in addition to its significant human death toll, this war substantially contributes to severe environmental damage due to the extensive use of chemical warfare, including white phosphorus and other toxic chemicals and explosives. There is increasing evidence that air, soil, and water pollution, as well as biodiversity loss and other potential ecological damages, are irreversible and irreparable consequences of the recent war, exacerbating Gaza's vulnerability amid climate change. This war is leaving a toxic legacy for generations to come, causing severe, long-lasting damage to the environment and anticipated long-term consequences for human health and ecosystems. These findings highlight the critical importance of learning from this devastating war to better understand and address the environmental impacts of large-scale violence and to develop strategies for environmental recovery and resilience in affected regions.

A. Hassoun
Sustainable AgriFoodtech Innovation & Research (SAFIR), Arras, France

Faculty of Agricultural Engineering, University of Aleppo, Aleppo, Syria
e-mail: a.hassoun@saf-ir.com

K. Al-Muhannadi (✉)
Environment Friends Society & Reem Program for Young Environmental Leaders, Manama, Bahrain

A. Haddarah · F. Slim
Doctoral School of Sciences and Technology, Lebanese University, Hadath, Lebanon
e-mail: amira.haddarah@ul.edu.lb

D. Iriqat
Arab American University of Palestine, Ramallah, Palestine
e-mail: dalal.s.iriqat@aaup.edu

A. Hamad
Department of Food Hygiene and Control, Faculty of Veterinary Medicine, Benha University, Benha, Egypt
e-mail: ahmed.alhussaini@fvtm.bu.edu.eg

A. Faddoul
World Bank, Washington, DC, USA

M. Buheji
Socioeconomic Institute for Advanced Studies (SIAS), Kigali, Rwanda

Inspiration Economy Institute, Manama, Bahrain

A. Hassoun (ed.), *War on Gaza*, Sustainable Development Goals Series,
https://doi.org/10.1007/978-3-031-88500-6_11

Keywords

Biodiversity loss · Carbon emission · Climate crisis · Ecocide · Ecosystem · Environment · Natural resources · Pollution · SDG 13 · Sustainability · Sustainable development

11.1 Introduction

The United Nations (UN) has adopted 17 Sustainable Development Goals (SDGs) to end poverty and hunger, fight inequalities, discrimination, and injustices, protect the planet, and achieve peace and prosperity by 2030. These SDGs were proposed to address the three dimensions of sustainable development: economic, social, and environmental (Fonseca et al. 2020; Franc-Dąbrowska and Drejerskai 2022; Stefanini and Vignali 2024; Yang et al. 2024). However, recent crises, especially the coronavirus pandemic in 2019 (COVID-19) and Russia's war on Ukraine, have significantly perturbed the UN 2030 Agenda for sustainable development and led to various socioeconomic and environmental degradations (Rawtani et al. 2022; da Costa et al. 2023; Leal Filho et al. 2023; Saharti et al. 2024).

More recently, the 2023–2025 war on Gaza has resulted in even more severe consequences for both human life and the environment, given the enormous death toll and the extensive scale of destruction (Asi et al. 2024; Hassoun et al. 2024; London et al. 2024). This assault is rooted in a broader historical context of policies implemented by Israel since its establishment in 1948, often described as expansionist. These policies include settlement activities, appropriation of land and resources, displacement of populations, and actions criticized as violations of international resolutions and Palestinian rights (Hughes et al. 2023; Khalidi and Seikaly 2023; Tanous 2023; Buheji and Al-Muhannadi 2023). The imposed siege on Gaza by Israel since 2007 has devastated the strip's economy and led to what was described as de-development and unviability (Roy 2016). The situation had already been described as "slow-motion genocide" due to the nature of Israeli state control over the Palestinian population (Nijim 2023). In more recent publications, the terms "Gazacide" (Nijim 2024) and second Nakba (Choonara 2024) have been introduced to describe Israel's policies and actions in Gaza over the years.

Since the conflict escalation following October 7, 2023, Israel has launched unprecedented attacks from the air, sea, and land. Over the course of the first year of this unprecedented war, these assaults have resulted in the deaths of more than 42,000 Palestinians, predominantly women and children (Hamad et al. 2024; Khatib et al. 2024; McMahan 2024). According to the Lancet's calculations (Khatib et al. 2024), the actual number could be at least 186,000 by July 2024. Many scholars, activists, and nongovernmental organizations and agencies have called for a ceasefire in order to prevent acts of genocide against the Palestinian people in Gaza (Browne 2024; Buheji and Ahmed 2024; Buheji and Al-Muhannadi 2023; Faddoul et al. 2023; Hassoun et al. 2024; UN 2023; Wells et al. 2024).

Since the beginning of the war in Ukraine, some scholars have investigated its impact on the environment (Pereira et al. 2022a; UNEP and GRID-Arendal 2022; Appiah-Otoo and Chen 2023) and questioned whether that could be considered as an 'ecocide' (Gardashuk 2022; Sousa et al. 2022; Shahini et al. 2024; Hryhorczuk et al. 2024). However, little has been reported about the consequences of the recent war on Gaza on the environment. A preliminary assessment released by the United Nations Environment Program (UNEP) provides a summary of the impacts of the conflict in Gaza on the environment (UNEP 2024). Beyond describing the 'known and visible' impacts, this chapter underscores the challenges of reporting on critical conflict-related issues due to the severe access restrictions currently imposed on the Gaza Strip. It emphasizes the pressing need for comprehensive field-based assessments to gain a deeper understanding of the situation. Despite sustained efforts from scholars and legal experts, 'ecocide' is not yet recognized as an international crime. In June 2021, a panel of experts offered the following definition: "unlawful or wanton acts committed with knowledge that there is a substantial likelihood of severe and either widespread or long-term damage to the environment being caused by those acts" (Mehta et al. 2021).

Given that the data on the recent war on Gaza has not yet been widely published or disseminated through peer-reviewed sources, we opted for a grey literature search plan. Following the best practices of conducting grey literature recommended by Paez (2017), our search plan was developed incorporating various searching strategies, including customized Google search, targeted websites, news reports from reputable organizations, and technical and working papers produced by regional and international organizations, such as Food and Agriculture Organization of the United Nations (FAO), World Health Organization (WHO), United Nations Relief and Works Agency for Palestine Refugees in the Near East (UNRWA), United Nations Economic and Social Commission for Western Asia (ESCWA), United Nations Conference on Trade and Development (UNCTAD), United Nations Educational, Scientific and Cultural Organization (UNESCO), United Nations International Children's Emergency Fund (UNICEF), World Food Programme (WFP), United Nations Development Programme (UNDP), Relief Web (ReliefWeb) a humanitarian information service provided by the United Nations Office for the Coordination of Human Affairs (OCHA), and United Nation Environment Program (UNEP).

The grey literature search was supplemented with relevant peer-reviewed articles extracted from the Scopus database (and similar resources, e.g., PubMed and Web of Science) and Google Scholar database using relevant search terms, such as "war on Gaza," "environment or environmental," "ecology or ecological," "sustainability," and "Sustainable Development Goals." These terms were chosen to capture the broadest range of literature relevant to the topics investigated in this study. Using multiple databases and a combination of search terms ensured that a comprehensive and diverse range of literature on the discussed perspectives was identified and studied.

11.2 Exploring the Framework of Ecological SDGs

Sustainability is built on three interconnected pillars—environmental, social, and economic—and the environmental pillar serves as the foundation for achieving the other two. This is particularly evident in the ecological SDGs, such as SDGs 6, 13, 14, and 15, which directly address environmental sustainability to support broader development goals (Fig. 11.1).

The impact of war on sustainability is a critical area of study, particularly in understanding how armed conflicts can hinder progress toward ecological goals. According to the "wedding cake" model, the most highlighted ecological SDGs are SDGs 6, 13, 14, and 15 (Fassio and Tecco 2019; Trappey et al. 2023). Growing literature shows that these 4 SDGs are highly impacted by wars and hostilities. For example, military operations and warfare have been reported to cause significant loss and damage to soil resources (Kucher and Kucher 2024), detrimental impacts on water resources and infrastructure (Shumilova et al. 2023), and contribute to climate change (Saxena 2023). Such environmental damages and human-made ecological disasters have been qualified as "ecocide" due to their far-reaching consequences on ecosystems and natural resources (Wenning and Tomasi 2023; Esonye et al. 2023).

SDG 6 aims to ensure clean water and sanitation for all by 2030 through six key targets: safe and affordable drinking water (6.1), equitable sanitation and hygiene (6.2), improved water quality (6.3), water-use efficiency (6.4), sustainable water management (6.5), and protection of water-related ecosystems (6.6). Clean water is essential not only for health and hygiene but also for food production and sustainable agriculture (Huan and Liu 2023; Mehta et al. 2023). It plays a pivotal role in the water–food–energy nexus, highlighting its interconnectedness with food security and energy generation (Simpson and Jewitt 2019; David et al. 2022; Huan and Liu 2023).

Fig. 11.1 Overview of the ecological SDGs

SDG 14 aims to conserve and sustainably use oceans, seas, and marine resources through targets like reducing pollution, protecting ecosystems, addressing ocean acidification, achieving sustainable fishing, and conserving marine areas. Fisheries and aquaculture significantly contribute to multiple SDGs, including poverty reduction, food security, economic growth, and reducing inequalities. However, achieving SDG 14 faces challenges such as ocean acidification, warming, overfishing, plastic pollution, and greenhouse gas (GHG) emissions, all of which hinder progress toward environmental sustainability (FAO 2018; Laso et al. 2022; Troell et al. 2023).

SDG 15 is a crucial component of the SDGs that aims to achieve sustainable terrestrial ecosystems, focusing on combating desertification, land degradation, and biodiversity loss. This goal has 9 targets, mainly including conservation of terrestrial ecosystems (Target 15.1), ending deforestation (Target 15.2) and land degradation (Target 15.3), and ensuring the conservation of biodiversity (Target 15.4). SDG 15 is fundamental and interconnected with other SDGs and sustainability topics, such as poverty alleviation, food security, gender equality, water and energy access, and economic growth (Måren 2019; Adhikari et al. 2023).

SDG 13, which is the main focus of this chapter, aims to take actions to tackle climate change urgently through the reduction in GHG emissions and the promotion of climate-resilient development. This goal aims to enhance resilience to climate-related events (Target 13.1), integrate climate change measures into national policies (Target 13.2), and promote education and awareness-raising on climate change mitigation and adaptation (Target 13.3). In recent years, the importance of achieving net zero emissions and carbon-neutral future has been highlighted in many studies to minimize GHG emissions and the carbon footprint (Wang et al. 2021; Rolnick et al. 2022).

The following sections will place particular emphasis on SDG 13, which focuses on taking urgent action to combat climate change and its impacts. The environmental destruction caused by the recent war has significantly contributed to climate instability, including the release of greenhouse gases, the destruction of vegetation and carbon sinks, and the disruption of ecosystems critical to climate regulation. By centering on SDG 13, this chapter aims to shed light on how such large-scale environmental damage exacerbates global climate challenges and hinders progress toward achieving climate action goals.

11.3 Environmental Damage Caused by the Recent War on Gaza

The war on Gaza has caused significant environmental damage and has pushed back progress toward the four ecological SDGs, especially SDG 13.

11.3.1 Implications of the War on Gaza for SDG 13

Armed conflicts worldwide pose significant challenges to achieving SDG 13, as they intensify climate change through escalating air pollution, GHG emissions, and the destruction of ecosystems. Military operations, such as the use of explosives, heavy vehicles, and the widespread destruction of infrastructure, have generated significant emissions of pollutants like nitrogen dioxide (NO_2) and sulfur dioxide (SO_2). These emissions arise from bomb detonations, fires at petroleum facilities, and the decomposition of war-related waste (Malarvizhi et al. 2023; Bun et al. 2024; Han et al. 2024).

The recent war in Gaza has had a profound and detrimental impact on SDG 13, severely impeding progress toward climate action and sustainability goals (Fig. 11.2). Already vulnerable to climate change before the recent war (Mason et al. 2011), the region now faces an environmental crisis of extraordinary proportions. Within just 60 days, the war emitted 281,315 metric tons of CO_2, far surpassing the 77 metric tons released in the first 18 months of the Ukraine war. This figure equates to burning 150,000 tons of coal and exceeds the annual emissions of over 135 countries combined (Neimark et al. 2023; Bun et al. 2024; Lakhani 2024). Rebuilding the infrastructure of Gaza using conventional methods could generate an additional 30 million metric tons of CO_2, further highlighting the long-term climate implications (Shaheen et al. 2024). These emissions severely hinder Indicator 13.2, which calls for integrating climate change measures into national policies, strategies, and planning, as all governance structures and resources in Gaza are currently overwhelmed by the humanitarian crisis.

The use of chemical weapons, such as white phosphorus bombs, has added another layer of environmental and health hazards. These toxic substances contaminate soil, water, and air, leaving a legacy of long-term harm to ecosystems and human health, significantly complicating recovery and rebuilding efforts in the region (Buheji and Al-Muhannadi 2023; Evereux 2024; Khafaei et al. 2024). This environmental degradation demonstrates how this war impedes the global goal of building institutional capacity and public awareness for climate change mitigation and adaptation, as outlined in Target 13.3.

The destruction of agricultural lands, coastal habitats, and waste management facilities has further degraded the environment of Gaza. Damaged sewage and water treatment systems have resulted in untreated waste flooding into water sources, exacerbating water pollution and health risks. Climate change-induced rising temperatures and reduced rainfall, combined with the destruction of agricultural infrastructure, have worsened food insecurity in the region (Elgendy 2023; Ahmed et al. 2024; Dardona et al. 2024). These impacts illustrate how the war obstructs efforts to enhance adaptive capacity to climate-related hazards, central to Target 13.1, while also undermining future efforts to secure financial and technological resources for climate action as called for in Targets 13.a and 13.b.

The recent war on Gaza, coupled with years of blockade and repeated cycles of aggression, has not only limited access to natural resources and altered ecosystems but also hindered efforts to enhance resilience against climate-related events such as floods and storms. These events can dra-

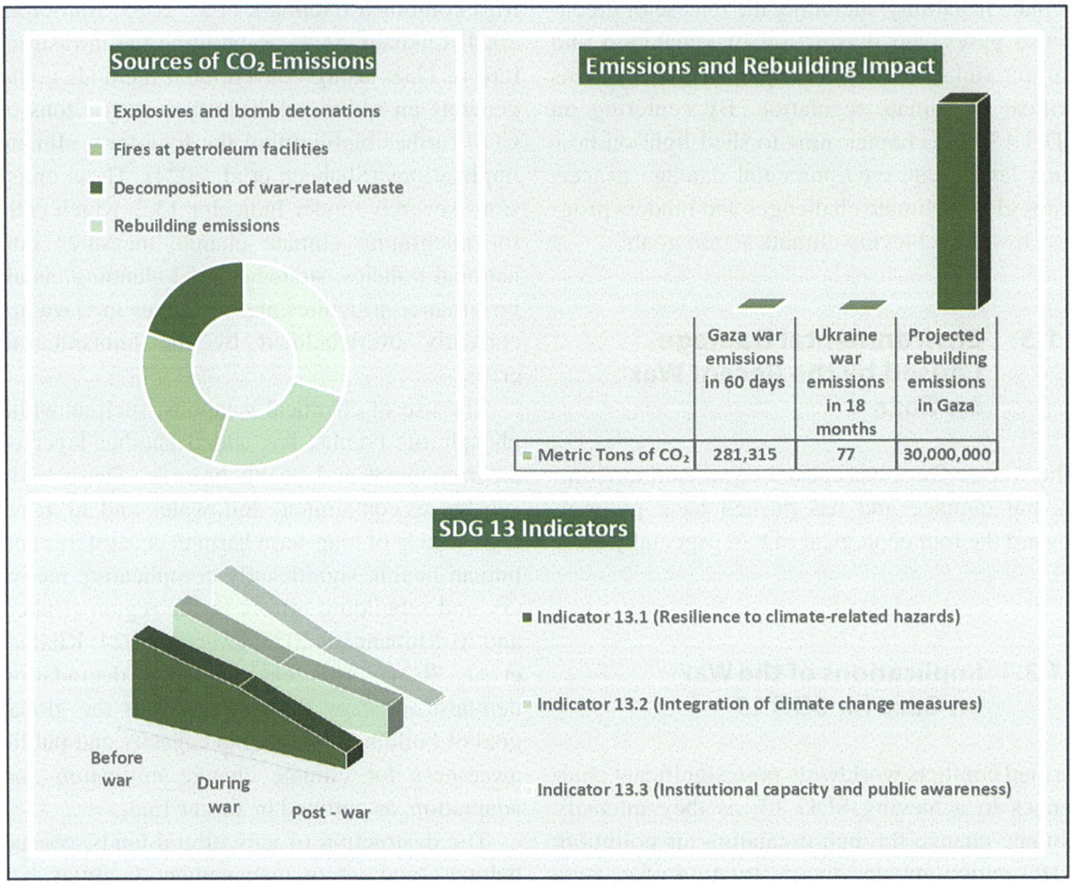

Fig. 11.2 The environmental crisis in Gaza and its implications for SDG 13

matically affect thousands of Palestinians who have lost their homes due to bombardment and are now living in makeshift tents. Adding to the environmental concerns are decaying bodies and contaminated water, which could lead to the spread of pandemic diseases (Saber 2023; Kholoud 2024).

Before the war, residents of Gaza demonstrated remarkable adaptability to climate challenges, utilizing renewable energy sources like solar panels (Hamed and Peric 2020; Elnaggar 2023). However, the destruction of this infrastructure has undermined efforts to mitigate and adapt to climate change. This loss of clean energy options directly affects Target 13.1, which focuses on strengthening resilience to climate-related hazards, leaving the region increasingly vulnerable to environmental challenges. Despite these devastating consequences, Israel has engaged in greenwashing, promoting itself as environmentally conscious while disregarding the environmental destruction caused in Gaza (Hughes et al. 2023; Williams 2024). This narrative further complicates efforts to address the region's climate crisis.

The war on Gaza has not only halted progress toward SDG 13 but has also undermined its long-term sustainability efforts. The compounded effects of military aggression, environmental degradation, and resource shortages present critical challenges for the future climate resilience and sustainability of Gaza. Permanent cessation of hostilities and substantial international intervention are essential to rebuild infrastructure in Gaza and reintegrate climate action into its recovery process.

11.3.2 Implications of the Recent War on Gaza for the Other Ecological SDGs

11.3.2.1 Implications for Water and Sanitation (SDG 6)

Water resources are essential for sustainable development and are crucial for achieving various sustainable goals, particularly SDG 6, which emphasizes the sustainable management of water and sanitation. Armed conflicts severely impact water resources, jeopardizing access to safe water and undoing progress toward sustainability.

Israel's blockade and control over Palestinian water resources have weaponized access to this critical resource. Barriers, restrictions, and the deliberate targeting of water infrastructure have prevented Palestinians from accessing their water sources. Even before October 2023, Gaza faced a severe water crisis, with 97% of its water unfit for human consumption (UNICEF 2018). Due to poor water, sanitation, and hygiene (WASH) conditions in households, waterborne diseases were widespread, exacerbating public health challenges and disproportionately affecting vulnerable populations (Abuzerr et al. 2020; Arana-Cedeño 2024; Seeberg 2024). The war has compounded this crisis, reducing Gaza's water supply by 95% and leaving 75% of the population reliant on salinized and contaminated water. This has drastically increased the risk of waterborne diseases and other health crises (ESCWA 2023; Hassoun et al. 2024).

Israel weaponizes water in Gaza by systematically destroying critical water and sanitation infrastructure while enforcing a blockade that restricts the entry of essential supplies needed for repairs. Actions include the cutting off of the Mekorot water supply, which once provided 12% of Gaza's water, and targeting desalination plants, wells, and reservoirs, leaving many water wells and all desalination plants out of service. Additionally, restrictions on fuel and materials prevent repairs and operations of water desalination and treatment plants, exacerbating the humanitarian crisis (Abdul Samad et al. 2024; Cole 2024).

11.3.2.2 Implications for Marine Environment (SDG 14)

Fisheries and aquaculture contribute considerably to various SDGs, especially zero poverty (SDG 1) and food security (SDG 2) (Lynch et al. 2020; Hussein et al. 2022). However, the ongoing occupation and blockade of Gaza have severely impeded the development and modernization of its harbor, port, and fishing zones, leaving the enclave unable to fully utilize its maritime resources (Qafisheh 2024). Moreover, during the recent war, Israel has restricted Gazans' access to ports and fishing resources by destroying much of Gaza's fishing fleet. Recent intensified attacks on fishers while fishing coincide with accelerated famine in Gaza and the unavailability of food, compelling people to resort to fishing despite the associated risks. Gazan fishers' activities are limited to 6 miles by Israel occupation army (Hussein et al. 2022), although fishing range should be 20 miles according to Oslo Agreement (B'tselem 2013). As a result, the fishing community in Gaza is being driven into poverty due to reduced catches and the diminished quality of fish, which receive very little demand even at lower prices (UNRWA 2013). According to a recent statistics released by Gaza's Ministry of Agriculture, since October 2023, 200 fishers have been killed by Israel's forces, with loss of 4600 tons of fish production, forcing about 4000 fishers supporting 50,000 dependents out of employment, and the fishing sector suffering $20 million loss (Middle East Monitor 2024).

The recent war on Gaza has significantly damaged the marine environment and life below water. Displaced Gazans along the coast are compelled to resort to the sea for washing and other sanitation activities, a practice that is likely to contribute to the pollution of seawater. Water treatment systems have been destroyed because of the indiscriminate bombardment leading to sewage being dumped directly into the sea, aggravating the sea pollution (Abd El Hay 2024; The World Bank 2024; UNEP 2024). Pollution and acidification of the sea can have detrimental effects on marine resources and biodiversity along the coast of the Gaza Strip, as well as in other marine areas within the Mediterranean region.

11.3.2.3 Implications for Life on Land (SDG 15)

The war on Gaza has destroyed its terrestrial ecosystems, which are already scarce and fragile due to the high population density and urbanization. Damaged agricultural areas and croplands as well as increased physical and chemical land and soil degradation are among the first consequences of the intensive bombing, carpet bombing, and heavy military vehicles traveling over farming soil in Gaza (Buheji and Al-Muhannadi 2023; Dardona et al. 2024). In the last months, deliberate direct destruction of old olive trees and other trees was filmed and shared over social media, and declaration by more than 1 million olive trees have been uprooted since October 2023 (OCHA 2024), together with destruction of forest patches, and crops plantations. This not only worsens food insecurity for Gazans, but also threatens the survival of fauna and flora, destroying their most important long-lived ecosystems and species habitats.

The implications of war for deforestation, loss of biodiversity, and destruction of habitats for wild plant and animal species have also been reported in several studies (Pereira et al. 2022a, b; Shahini et al. 2024). Although Gaza has not been extensively studied due to the challenging security situation, the brutality of the war, and the persistent assaults and blockade, the damage is expected to be disproportionately severe (Gunawan 2024; Paché 2024). This is attributed to Gaza's limited geographical size, the scale and intensity of military aggression, and the destructive weaponry employed in the conflict.

The war in Gaza is estimated to have generated approximately 50 million tons of debris, according to the latest update from ReliefWeb (ReliefWeb 2024a). Furthermore, the WASH system is now defunct with the shutdown of wastewater treatment plants, resulting in sewage contamination of water resources, wetlands, and the soil with pathogens, microplastics, and hazardous waste, threatening marine and terrestrial ecosystems, habitats, and biodiversity in addition to Gaza's life (UNEP 2024). The impacts are both immediate and long term. Wildlife such as amphibians of Gaza are reported to be sensitive to pollution and salinity of the underground water that supports its survival in the plantation areas in Gaza, that can lead to severe decrease or extinction of amphibian species and populations (Abd Rabou 2019).

Gaza has multiple clean energy projects and buildings using solar energy, including the UN-funded ones. Solar panels were targeted and destroyed by Israeli occupation forces, as part of their announced plan for depriving Gaza of energy. The deliberate energy cuts have forced hundreds of thousands of displaced people to burn wood, waste, and plastic for cooking and heating during cold winters while sheltering under inadequate conditions. This raises serious concerns about human health, air quality, and the climate. Additionally, the bombing of solar panels risks lead and heavy metal leakage, leading to water and soil contamination (ReliefWeb 2024b).

11.3.3 Impact of the Recent War on Gaza on the Other SDGs

Besides its impact on the environmental SDGs (i.e., SDGs 6, 13, 14, and 15), the war on Gaza has been catastrophic for social and economic sustainability and has pushed back all the other SDGs (Fig. 11.3). These dire circumstances pose a threat to future development and regional security, exacerbating violence, eroding governance structures, and impeding peace-building initiatives. This significantly obstructs the realization of SDG 16 (peace, justice, and strong institutions), which is crucial for advancing progress across all other SDGs (Hope 2020).

From social perspectives, the war had led to extreme poverty (SDG 1) and acute food insecurity (SDG 2), especially by preventing small-scale food producers (SSFP) from safe access, utilization, and development of the sea and farmlands. SSFP were reported to be targeted and

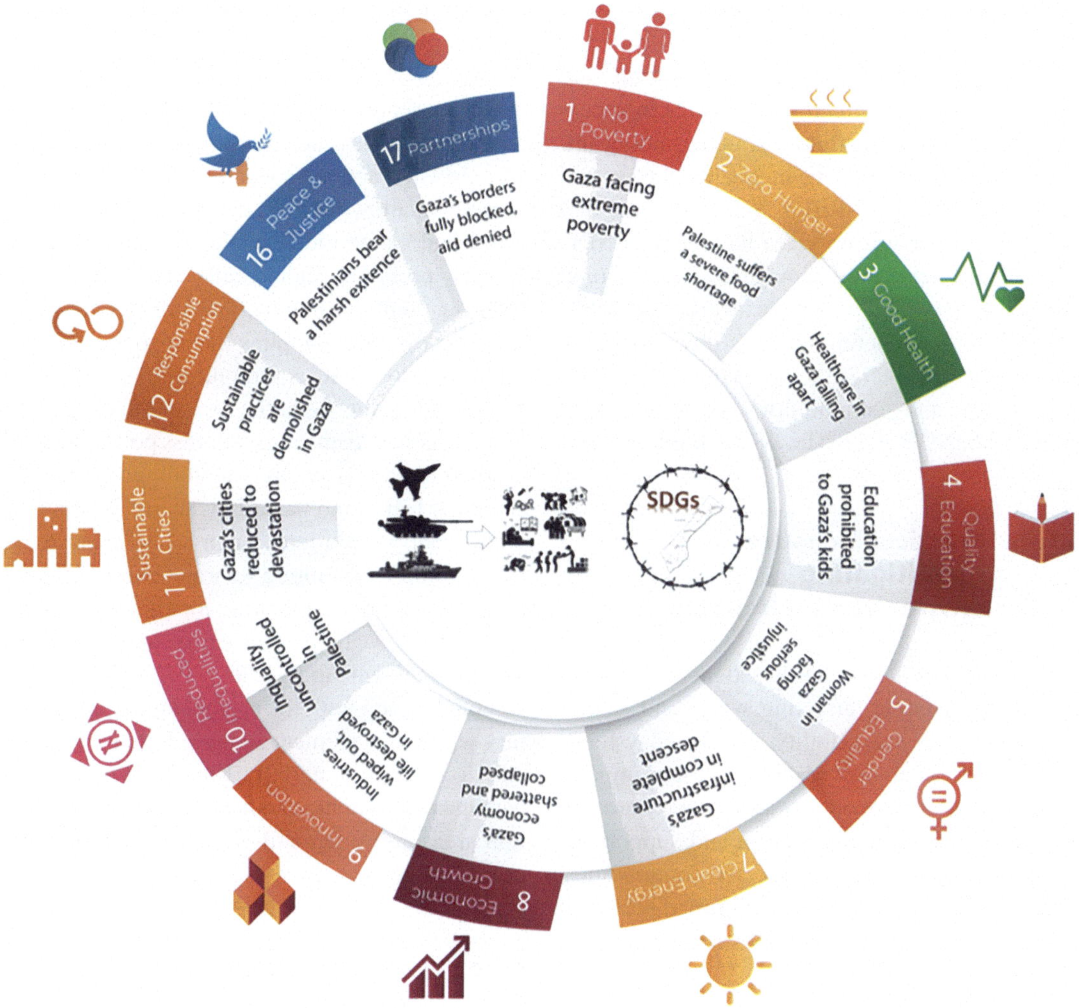

Fig. 11.3 Impact of the recent war on Gaza on social and economic SDGs

even killed. Healthcare in Gaza (SDG 3) is crumbling under the weight of war and dire humanitarian conditions. The war on Gaza has resulted in "educide" (SDG 4) marked by the widespread destruction of schools, universities, and other educational institutions (Rabaia and Habash 2024; Iriqat et al. 2025). While the war on Gaza has hindered progress toward SDG 5 by exacerbating gender inequality through the destruction of infrastructure, the deepening economic crisis, and increased barriers to women's empowerment, women have also emerged as pivotal figures within their families and communities, demonstrating remarkable resilience in the face of profound hardships (Al-Muhannadi and Buheji 2024a, b; Buheji et al. 2024).

Regarding the economic pillar of sustainability, this war on Gaza has hindered efforts to advance SDG 7 by disrupting infrastructure and energy access, exacerbating power shortages, and impeding sustainable development in the region (Kholoud 2024; Rantissi et al. 2024). The war on Gaza has impeded progress toward both SDG 8 and SDG 9 by disrupting infrastructure development and economic growth, exacerbating unemployment and poverty, and undermining efforts to foster sustainable industrialization and innovation in the region (Butter 2024; UNCTAD 2024).

The war on Gaza has hindered efforts to advance SDG 10 by deepening social and economic inequalities, exacerbating marginalization and exclusion, and perpetuating disparities in access to resources and opportunities, undermining initiatives aimed at reducing inequality within and among countries (Al-Riffai 2024; Al Attar and Brik 2024). The war on Gaza has set back SDG 11 by causing extensive damage to infrastructure, displacing populations, and exacerbating urban poverty and vulnerability, thereby impeding efforts to create inclusive, resilient, and sustainable cities and communities in this region (ESCWA/UNDP 2024; The World Bank 2024).

11.4 Practical Measures for Mitigating Environmental Impacts

The war in Gaza has caused extensive environmental degradation, affecting water and sanitation systems, climate stability, marine ecosystems, and terrestrial habitats. Mitigating these impacts requires targeted, evidence-based measures to address the scale of destruction and support the recovery and resilience of affected communities. Efforts should be divided into immediate relief actions and long-term recovery strategies.

11.4.1 Immediate Relief Measures

Urgent actions must prioritize the restoration of water supply systems and sanitation facilities to ensure the population of Gaza has access to sufficient clean water, a fundamental human right. Providing adequate water is critical not only to prevent outbreaks of waterborne diseases but also to uphold public health and dignity amid the crisis. The repair and reconstruction of damaged water-related infrastructure must involve specialized engineers and humanitarian organizations, with a focus on ensuring the swift and efficient delivery of aid. Portable water purification systems should be deployed to meet the immediate needs of displaced populations, while wastewater recycling technologies must be implemented to reduce reliance on increasingly scarce freshwater sources. Additionally, it is imperative to prevent the weaponization of water resources, a practice that exacerbates suffering and undermines recovery efforts. The international community must take decisive action to ensure Israel ceases such practices, allowing for unrestricted access to water and sanitation services. A comprehensive WASH program should also be established, focusing on promoting hygiene practices and distributing essential supplies like soap and sanitizers, ensuring that communities can maintain health and resilience in the face of adversity.

Emergency teams must address pollution from destroyed infrastructure, including oil spills, chemical leaks, and hazardous waste. Temporary green energy solutions, such as solar-powered water pumps and mobile solar generators, should be introduced while planning for climate-resilient reconstruction. To protect ecosystems, immediate containment and cleanup of pollutants are essential, along with the establishment of temporary waste management systems to handle debris and hazardous materials. Rapid environmental damage assessments can help prioritize intervention areas and safeguard biodiversity during this critical period.

11.4.2 Long-Term Recovery and Sustainability

For a sustainable recovery, upgrading water and sanitation infrastructure is imperative to ensure consistent access to clean water and sanitation for all Gazans. Renewable energy sources, such as solar panels and wind turbines, should power homes and public buildings, reducing reliance on fossil fuels. Sustainable building practices and materials must be adopted during reconstruction to enhance resilience and durability, paving the way for a greener future.

Comprehensive waste management systems should be established to reduce pollution, including recycling programs and safe disposal of hazardous materials. Sustainable agricultural practices, such as crop rotation, organic farming, and natural pest control, can minimize environmental impact while enhancing food security. Large-scale tree-planting initiatives can combat desertification, restore land, and support biodiversity, helping to protect endangered species and habitats disrupted by the war. To maintain marine ecosystems, sustainable fishing practices should be enforced, preventing overfishing and ensuring the regeneration of marine life.

11.4.3 Community Engagement and Monitoring

Engaging local communities in planning and implementing environmental measures is essential for long-term sustainability and capacity-building. Systems to monitor environmental indicators should be established to evaluate the effectiveness of implemented measures. International bodies, such as the United Nations, should prioritize efforts to enforce Gaza's water rights and environmental protection laws. Adopting renewable energy alternatives, such as wind-powered desalination plants, can provide sustainable solutions to water scarcity while reducing carbon emissions. These technologies should be integrated with existing water infrastructure, and local technicians should receive training to ensure their long-term viability.

11.4.4 Innovations and Global Collaboration

At advanced stages of recovery, technological innovations, including Industry 4.0 technologies, can enhance sustainability and resilience (Hassoun et al. 2022; Bigliardi et al. 2024). Strengthening international collaboration with global environmental organizations, donor agencies, and neighboring countries is crucial for securing technical expertise, funding, and support for long-term recovery. Advocacy for strong environmental policies and governance structures is essential to promote environmental justice, accountability, and transparency, ensuring that environmental considerations are integrated into all levels of decision-making.

11.5 Implications and Limitations

This study provides a comprehensive review that serves as a valuable reference for documenting the connections between the war on Gaza and its devastating impact on environmental sustainability within the framework of the SDGs. It highlights the ecocide unfolding in Gaza, exacerbated by the failure of the international community to address the atrocities committed in this region. This war is leaving a toxic legacy for future generations, with severe and long-lasting damage to the environment, human health, and ecosystems. The findings underscore that while the immediate victims of the war are the local and regional environments, the long-term global consequences for the planet are expected to be significant.

In addition to its environmental focus, this study emphasizes that the war has also profoundly affected economic and social dimensions. For example, land destruction and soil degradation not only harm human health but also reduce land productivity, leading to cascading effects on societal well-being and economic stability. These interconnected impacts demonstrate the urgent need for integrated approaches to address the consequences of this war.

However, this study faces limitations due to the early stage of research on the recent war's impact on the environmental SDGs in Gaza. The findings are based on collective observations, fragmented reports, and emerging news, which may not fully capture the scope of the damage. Restrictions and severe security threats have prevented researchers from conducting detailed investigations on the ground, further limiting the availability of comprehensive data.

Additionally, as this analysis was conducted during the early stages of the war, it is likely that environmental damage and ecocide will extend far beyond what is described in this work. The study also acknowledges uncertainty about the full extent of the long-term consequences, as these will depend on the progression of the war and the response of the international community. As a result, the estimates provided in this study are likely conservative and approximate in nature.

Despite these limitations, the findings underline the urgent need to address the humanitarian crisis in Gaza and to prioritize the development of robust post-war recovery plans. These plans must focus on effective environmental restoration, sustainable land management, and the rebuilding of ecosystems in Gaza and surrounding regions. The study calls for immediate international action to address the environmental, economic, and social impacts of this war and to prevent further long-term harm to the planet.

11.6 Lessons from Gaza and Conclusion

The devastation of Gaza's ecosystems—its water, soil, air, and sea—serves as a stark reminder of the catastrophic environmental consequences of modern warfare. The indiscriminate bombardment and use of chemical and other prohibited weapons during the recent war have inflicted irreparable harm on the besieged enclave. While many reports describe the crisis in Gaza as one of the most severe humanitarian disasters of our time, the lessons from Gaza reveal that the environmental and human toll will extend far beyond the immediate war. The toxic chemicals and explosive remnants of this recent military assault are expected to leave a legacy, with repercussions that will persist for generations.

Environmental destruction in Gaza is not confined to its borders; it poses significant regional and global concerns. Long-term consequences of contaminated ecosystems, including diminished agricultural productivity and polluted water sources, will likely manifest in the coming years, affecting the broader region and amplifying global environmental challenges. These lessons underscore the urgent need for international cooperation to address the ecocide and prevent similar crises in the future.

Moreover, the experience of Gaza highlights critical areas of focus for global environmental policy. First, it emphasizes the importance of integrating environmental safeguards into international humanitarian law to prevent large-scale ecological destruction during armed conflicts. International frameworks must evolve to address the deliberate targeting of ecosystems, water resources, and agricultural land, which exacerbate human suffering and hinder recovery efforts. Gaza exemplifies the dire need for robust mechanisms to monitor, document, and address environmental crimes, ensuring accountability for violations that devastate both human and natural systems.

Second, the Gaza case reinforces the necessity of sustainable reconstruction strategies in post-conflict settings. The war on Gaza has caused unprecedented destruction, surpassing previous war and affecting nearly all essential infrastructure. Earlier rebuilding efforts focused on restoring what was lost using conventional methods, leaving Gaza vulnerable to future crises. This time, sustainable reconstruction is critical to address compounded challenges like resource scarcity, energy shortages, and environmental degradation. By prioritizing renewable energy, waste management, and water restoration, Gaza can build resilience, adapt to climate impacts, and break the destructive cycle of rebuilding after each war.

Third, the lessons from Gaza demonstrate the crucial role of community-based approaches in addressing environmental and humanitarian crises. The resilience of Gaza's people—illustrated by their use of renewable energy and innovative water management solutions before the war—underscores the value of empowering local populations in recovery efforts. Engaging affected communities in decision-making processes fosters inclusive solutions that align with cultural and social priorities, ensuring that recovery initiatives are both effective and sustainable.

Finally, Gaza serves as a sobering case study on the intersection of environmental and social justice. The destruction of ecosystems in Gaza reflects broader patterns of environmental inequity, where vulnerable populations bear the brunt of ecological damage. Addressing these injustices requires a global commitment to environmental justice that ensures equitable access to resources, restoration of damaged ecosystems, and accountability for environmental crimes.

In conclusion, the lessons from Gaza highlight the profound and interconnected impacts of war on the environment, human health, and sustainable development. They underscore the urgent need for international solidarity, innovative solutions, and unwavering commitment to addressing the environmental crises caused by conflicts. Learning from this tragedy in Gaza offers a pathway toward more resilient, equitable, and sustainable futures for all.

References

Abd El Hay M (2024) The environmental-humanitarian impacts of the Israel Hamas war in Gaza. Arav Inst Environ Stud:1–28

Abd Rabou AFN (2019) The mammalian, reptilian and amphibian fauna of Al-Mawasi ecosystem, south-western Gaza Strip-Palestine. Res Technol Open Access J. https://doi.org/10.19080/ARTOAJ.2019.23.556233

Abdul Samad L, Butcher M, Khalidi B (2024) Water war crimes: how Israel has weaponised water in its military campaign in Gaza. Oxfam Int 6

Abuzerr S, Nasseri S, Yunesian M et al (2020) Water, sanitation, and hygiene risk factors of acute diarrhea among children under five years in the Gaza strip. J Water Sanit Hyg Dev 10:111–123. https://doi.org/10.2166/WASHDEV.2019.072

Adhikari B, Urbach D, Chettri N et al (2023) A multi-methods approach for assessing how conserving biodiversity interacts with other sustainable development goals in Nepal. Sustain Dev 31:3239–3253. https://doi.org/10.1002/SD.2582

Ahmed F, Alhajeeli N, Badheeb M et al (2024) A call for global intervention: the pressing requirement to address the health emergency in Gaza health crisis. Avicenna 2024:1–6. https://doi.org/10.5339/avi.2024.5

Al Attar HMJ, Brik AB (2024) The historical tragedy: evaluating social protection programs in Gaza. In: Brik AB (ed) Public policy in the Arab World, pp 213–243

Al-Muhannadi K, Buheji M (2024a) Value-based resilience- stories from Gaza (during war 2023). Int J Manag 15:15–37. https://doi.org/10.17605/OSF.IO/KP2DX

Al-Muhannadi K, Buheji M (2024b) Surviving despite intentional food & water scarcity—cases of resilience from Gaza (2023–2024). Int J Manag 15:65–76

Al-Riffai P (2024) The economic and social costs of the war in Gaza

Appiah-Otoo I, Chen X (2023) Russian-Ukrainian war degrades the total environment. Lett Spat Resour Sci 16:1–17. https://doi.org/10.1007/S12076-023-00354-8/FIGURES/7

Arana-Cedeño M (2024) Disease and death: the shock wave of collective punishment against Gaza. Soc Med Pract 17:39–43

Asi Y, Mills D, Greenough PG et al (2024) 'Nowhere and no one is safe': spatial analysis of damage to critical civilian infrastructure in the Gaza strip during the first phase of the Israeli military campaign, 7 October to 22 November 2023. Confl Heal 18:1–13. https://doi.org/10.1186/s13031-024-00580-x

B'tselem (2013) Lift the restrictions on the Gaza fishing range. https://www.btselem.org/gaza_strip/20130324_restrictions_on_fishing_should_be_lifted. Accessed 1 Mar 2024

Bigliardi B, Filippelli S, Pini B et al (2024) Industry 4.0 and food sustainability: role of automation, digitalization, and green technologies. In: Hassoun A (ed) Food industry 4.0: emerging trends and technologies in sustainable food production and consumption. Academic Press, pp 15–33

Browne BC (2024) The Coloniality of enforced starvation: Reading famine in Gaza through an Gorta Mór. J Palest Stud. https://doi.org/10.1080/0377919X.2024.2379709

Buheji M, Ahmed D (2024) Keeping the boycott momentum - from 'War on Gaza' till 'Free - Palestine'. Int J Manag 14:205–229. https://doi.org/10.17605/OSF.IO/NC75D

Buheji M, Al-Muhannadi K (2023) Mitigating risks of environmental impacts on Gaza - review of precautions & solutions post (2023 war). Int J Adv Res Eng Technol 14:15–47

Buheji M, Migdad M, Muqdad SI, Muqdad AI (2024) Palestinian women in the shadow of genocide in Gaza (2024). Int J Soc Sci Res Dev 6:157–172

Bun R, Marland G, Oda T et al (2024) Tracking unaccounted greenhouse gas emissions due to the war in Ukraine since 2022. Sci Total Environ 914:169879. https://doi.org/10.1016/j.scitotenv.2024.169879

Butter D (2024) Economic impact of the Gaza war. IEMed Mediterr Yearb:273–275

Choonara J (2024) A second Nakba? Int Social

Cole J (2024) Weaponizing a water crisis: the destruction of water supplies in Gaza. New Lines Inst:1–7

da Costa JP, Silva AL, Barcelò D et al (2023) Threats to sustainability in face of post-pandemic scenarios and the war in Ukraine. Sci Total Environ 892. https://doi.org/10.1016/j.scitotenv.2023.164509

Dardona Z, Amane M, Dardona A, Boussaa S (2024) Health and environmental impacts of Gaza conflict (2023-2024): a review. One Heal Bull. https://doi.org/10.4103/ohbl.ohbl

David LO, Nwulu NI, Aigbavboa CO, Adepoju OO (2022) Integrating fourth industrial revolution (4IR) technologies into the water, energy & food nexus for sustainable security: a bibliometric analysis. J Clean Prod 363:132522. https://doi.org/10.1016/j.jclepro.2022.132522

Elgendy K (2023) Climate casualties: how the Gaza war threatens Mideast climate action. In: Middle East Inst. https://www.mei.edu/publications/climate-casualties-

how-gaza-war-threatens-mideast-climate-action. Accessed 26 Feb 2024
Elnaggar M (2023) Useful energy, economic and reduction of greenhouse gas emissions assessment of solar water heater and solar air heater for heating purposes in Gaza, Palestine. Heliyon 9:e16803. https://doi.org/10.1016/J.HELIYON.2023.E16803
ESCWA (2023) War on Gaza: weaponizing access to water, energy and food
ESCWA/UNDP (2024) Gaza war: Expected socioeconomic impacts on the State of Palestine—October 2024
Esonye C, Esonye CO, Agha EO et al (2023) The 2022 Russia invasion on Ukraine: the biofuel energy security challenge on Ukraine and some related countries. Heliyon 9:2405–8440. https://doi.org/10.1016/j.heliyon.2023.e21483
Evereux VD (2024) Some insights on application of the international law to the damaged environment during the war in Gaza. ACTA Univ CAROLINAE—Iurid 4:177–194. https://doi.org/10.14712/23366478.2024.167
Faddoul A, Shannon G, Asghar K et al (2023) The health dimensions of violence in Palestine: a call to prevent genocide. Lancet. https://doi.org/10.1016/S0140-6736(23)02751-4
FAO (2018) Impacts of climate change on fisheries and aquaculture. Rome, Italy
Fassio F, Tecco N (2019) Circular economy for food: a systemic interpretation of 40 case histories in the food system in their relationships with SDGs. Systems 7:43. https://doi.org/10.3390/SYSTEMS7030043
Fonseca LM, Domingues JP, Dima AM (2020) Mapping the sustainable development goals relationships. Sustain For 12:3359. https://doi.org/10.3390/SU12083359
Franc-Dąbrowska J, Drejerskai N (2022) Resilience in the food sector—environmental, social and economic perspectives in crisis situations. Int Food Agribus Manag Rev 25:757–770. https://doi.org/10.22434/IFAMR2022.0010
Gardashuk T (2022) Is Russian aggression in Ukraine ecocide? Envigogika 17:1–6. https://doi.org/10.14712/18023061.642
Gunawan Y (2024) Environmental devastation in Gaza: analysing Israeli attacks through the view of international environmental law. E3S Web Conf 594:04005
Hamad BA, Vintges J, Diab R, et al (2024) Shattered lives and dreams: the toll of the war on Gaza on young people
Hamed TA, Peric K (2020) The role of renewable energy resources in alleviating energy poverty in Palestine. Renew Energy Focus 35:97–107. https://doi.org/10.1016/j.ref.2020.09.006
Han L, Zhou Z, Shi B, Wang Y (2024) Challenges to environmental governance arising from the Russo–Ukrainian conflict: Evidence from carbon emissions. J Environ Manag 349:119481. https://doi.org/10.1016/j.jenvman.2023.119481
Hassoun A, Prieto MA, Carpena M et al (2022) Exploring the role of green and industry 4.0 technologies in achieving sustainable development goals in food sectors. Food Res Int 162:112068. https://doi.org/10.1016/j.foodres.2022.112068
Hassoun A, Al-Muhannadi K, Hassan HF et al (2024) From acute food insecurity to famine: how the 2023/2024 war on Gaza has dramatically set back sustainable development goal 2 to end hunger. Front Sustain Food Syst 8:1402150. https://doi.org/10.3389/fsufs.2024.1402150
Hope KR (2020) Peace, justice and inclusive institutions: overcoming challenges to the implementation of sustainable development goal 16. Glob Chang Peace Secur 32:57–77. https://doi.org/10.1080/14781158.2019.1667320
Hryhorczuk D, Levy BS, Prodanchuk M et al (2024) The environmental health impacts of Russia's war on Ukraine. J Occup Med Toxicol 19:1–14. https://doi.org/10.1186/S12995-023-00398-Y/FIGURES/2
Huan S, Liu X (2023) Network modeling and stability improvement of the water-energy-fertilizer-food nexus flows based on global agricultural trade. Sustain Prod Consum 39:480–494. https://doi.org/10.1016/j.spc.2023.05.034
Hughes SS, Velednitsky S, Green AA (2023) Greenwashing in Palestine/Israel: settler colonialism and environmental injustice in the age of climate catastrophe. Environ Plan E Nat Sp 6:495–513. https://doi.org/10.1177/25148486211069898
Hussein DMA, Lucchetti MC, Zaqoot HA et al (2022) Status of fisheries in Gaza strip: past trends and challenges. Int J Euro-Mediterranean Stud 15:179–216
Iriqat D, Alousi R, Aldahdouh TZ et al (2025) Educide amid conflict: the struggle of the Palestinian education system. Qual Educ All 2:81–99. https://doi.org/10.1108/QEA-10-2024-0120
Khafaei M, Panahi Y, Abolghasemi H et al (2024) Injuries caused by recurrent white phosphorus bombing as a crisis in Gaza strip. Trauma Mon 29:1052–1057. https://doi.org/10.30491/tm.2024.452290.1719
Khalidi RI, Seikaly S (2023) From the editors. J Palest Stud 52:1–8. https://doi.org/10.1080/0377919X.2023.2304504
Khatib R, McKee M, Yusuf S (2024) Counting the dead in Gaza: difficult but essential. Lancet 404:237–238. https://doi.org/10.1016/S0140-6736(24)01169-3
Kholoud M (2024) The impact of Gaza war on environmental security. Serbian J Eng Manag 9:1–5. https://doi.org/10.5937/sjem2401001m
Kucher A, Kucher L (2024) War loss and damage to soil resources: towards sustainable land management. Int J Environ Stud. https://doi.org/10.1080/00207233.2023.2296769
Lakhani N (2024) Emissions from Israel's war in Gaza have 'immense' effect on climate catastrophe. In: Guard. https://www.theguardian.com/world/2024/jan/09/emissions-gaza-israel-hamas-war-climate-change. Accessed 26 Feb 2024

Laso J, Ruiz-Salmón I, Margallo M et al (2022) Achieving sustainability of the seafood sector in the European Atlantic area by addressing eco-social challenges: the NEPTUNUS project. Sustain For 14:3054. https://doi.org/10.3390/SU14053054

Leal Filho W, Viera Trevisan L, Simon Rampasso I et al (2023) When the alarm bells ring: why the UN sustainable development goals may not be achieved by 2030. J Clean Prod 407. https://doi.org/10.1016/j.jclepro.2023.137108

London L, Watterson A, Mergler D et al (2024) A call from 40 public health scientists for an end to the continuing humanitarian and environmental catastrophe in Gaza. Environ Heal A Glob Access Sci Source 23:1–6. https://doi.org/10.1186/S12940-024-01097-9/PEER-REVIEW

Lynch AJ, Elliott V, Phang SC et al (2020) Inland fish and fisheries integral to achieving the sustainable development goals. Nat Sustain 3:579–587. https://doi.org/10.1038/s41893-020-0517-6

Malarvizhi AS, Liu Q, Trefonides TS et al (2023) The spatial dynamics of Ukraine air quality impacted by the war and pandemic. Int J Digit Earth 16:3680–3705. https://doi.org/10.1080/17538947.2023.2239762

Måren IE (2019) Food systems for sustainable terrestrial ecosystems (SDG 15). Food Ethics 2:155–159. https://doi.org/10.1007/S41055-018-00032-2/METRICS

Mason M, Zeitoun M, El Sheikh R (2011) Conflict and social vulnerability to climate change: lessons from Gaza. Clim Dev 3:285–297. https://doi.org/10.1080/17565529.2011.618386

McMahan J (2024) Proportionality and necessity in Israel's invasion of Gaza, 2023–2024. Anal und Krit 46:387–407. https://doi.org/10.1515/auk-2024-2024

Mehta J, Sands KC P, Sow DF, et al (2021) Stop ecocide foundation: independent expert panel for the legal definition of ecocide—commentary and core text

Mehta L, Ringler C, Varghese S (2023) Water for life and food: synergies between SDGs 2 and 6 and human rights. In: The environment in global sustainability governance, pp 127–148

Middle East Monitor (2024) 200 fishermen killed in Gaza, fishing sector suffers $20m loss. https://www.middleeastmonitor.com/20241212-200-fishermen-killed-in-gaza-fishing-sector-suffers-20m-loss/. Accessed 18 Dec 2024

Neimark AB, Bigger P, Otu-larbi F, Larbi R (2023) A multitemporal snapshot of greenhouse gas emissions from the Israel-Gaza conflict the carbon costs of conflict. SSRN Electron J

Nijim M (2023) Genocide in Palestine: Gaza as a case study. Int J Hum Rights 27:165–200. https://doi.org/10.1080/13642987.2022.2065261

Nijim M (2024) Gazacide: Palestinians from Refugeehood to ontological obliteration. Crit Sociol 1944:1–23. https://doi.org/10.1177/08969205241286530

OCHA (2024) Humanitarian Situation Update #224 | Gaza Strip. https://www.unocha.org/publications/report/occupied-palestinian-territory/humanitarian-situation-update-224-gaza-strip-enarhe. Accessed 18 Dec 2024

Paché G (2024) War-induced environmental crises: a reality too critical to ignore. Cris Response J 19:46–49

Paez A (2017) Gray literature: an important resource in systematic reviews. J Evid Based Med 10:233–240. https://doi.org/10.1111/JEBM.12266

Pereira P, Bašić F, Bogunovic I, Barcelo D (2022a) Russian-Ukrainian war impacts the total environment. Sci Total Environ 837. https://doi.org/10.1016/j.scitotenv.2022.155865

Pereira P, Zhao W, Symochko L et al (2022b) The Russian-Ukrainian armed conflict will push back the sustainable development goals. Geogr Sustain 3:277–287. https://doi.org/10.1016/j.geosus.2022.09.003

Qafisheh MM (2024) From Constantinople to Oslo to Gaza: developments of Palestine's maritime legislation under colonization. Mar Policy 160:105954. https://doi.org/10.1016/j.marpol.2023.105954

Rabaia I, Habash L (2024) The hidden war on higher education: unmasking the 'Educide' in Gaza. In: The war on Gaza and Middle East political science. George Washington University- POMEPS, pp 8–13

Rantissi T, Gitis V, Zong Z, Hankins N (2024) Transforming the water-energy nexus in Gaza: a systems approach. Glob Chall 8:2300304. https://doi.org/10.1002/GCH2.202300304

Rawtani D, Gupta G, Khatri N et al (2022) Environmental damages due to war in Ukraine: a perspective. Sci Total Environ 850:157932. https://doi.org/10.1016/j.scitotenv.2022.157932

ReliefWeb (2024a) Gaza humanitarian response update | 24 November - 7 December 2024. https://reliefweb.int/report/occupied-palestinian-territory/gaza-humanitarian-response-update-24-november-7-december-2024. Accessed 18 Dec 2024

ReliefWeb (2024b) Environmental impact of the conflict in Gaza: Preliminary assessment of environmental impacts—occupied Palestinian territory. https://reliefweb.int/report/occupied-palestinian-territory/environmental-impact-conflict-gaza-preliminary-assessment-environmental-impacts. Accessed 18 Dec 2024

Rolnick D, Kaack LH, School H et al (2022) Tackling climate change with machine learning. ACM Comput Surv 55:1–96. https://doi.org/10.1145/3485128

Roy S (2016) The Gaza strip: the political economy of De-development, third. Institute for Palestine Studies, Washington, DC, Washington, DC

Saber IF (2023) Is Israel's Gaza bombing also a war on the climate? In: Aljazeera. https://www.aljazeera.com/news/2023/12/5/is-israels-war-on-gaza-also-hurting-the-climate. Accessed 26 Feb 2024

Saharti M, Chaudhry SM, Pekar V, Bajoori E (2024) Environmental, social and governance (ESG) performance of firms in the era of geopolitical conflicts. J Environ Manag 351:119744. https://doi.org/10.1016/j.jenvman.2023.119744

Saxena A (2023) Deteriorating environmental quality with special reference to war and its impact on climate change. Natl Acad Sci Lett 1:1–4. https://doi.org/10.1007/S40009-023-01279-Y/METRICS
Seeberg P (2024) Israeli hydro-hegemony and the Gaza war. Middle East Policy 31:33–46. https://doi.org/10.1111/MEPO.12740
Shaheen A, Dajani R, Zinszer K et al (2024) The war on the Gaza strip and its consequences on global warming. Front Hum Dyn 6:1463902
Shahini E, Shebanina O, Kormyshkin I et al (2024) Environmental consequences for the world of Russia's war against Ukraine. Int J Environ Stud. https://doi.org/10.1080/00207233.2024.2302745
Shumilova O, Tockner K, Sukhodolov A et al (2023) Impact of the Russia–Ukraine armed conflict on water resources and water infrastructure. Nat Sustain 6:578–586. https://doi.org/10.1038/s41893-023-01068-x
Simpson GB, Jewitt GPW (2019) The development of the water-energy-food nexus as a framework for achieving resource security: a review. Front Environ Sci 7:422789. https://doi.org/10.3389/FENVS.2019.00008/BIBTEX
Sousa R, da Silva JP, Douda K, Mammola S (2022) The cost of war for biodiversity: a potential ecocide in Ukraine. Front Ecol Environ 20:394–396. https://doi.org/10.1002/fee.2548
Stefanini R, Vignali G (2024) The influence of industry 4.0 enabling technologies on social, economic and environmental sustainability of the food sector. Int J Prod Res 62:3800–3817. https://doi.org/10.1080/00207543.2023.2248523
Tanous O (2023) "You, as of now, are someone Else!": Minoritization, settler colonialism, and indigenous health. J Palest Stud 52:68–86. https://doi.org/10.1080/0377919X.2023.2171806
The World Bank (2024) Gaza Strip—interim damage assessment
Trappey CV, Trappey AJC, Lin HJ, Chang AC (2023) Comparative analysis of food related sustainable development goals in the North Asia Pacific region. Food Ethics 8:1–24. https://doi.org/10.1007/S41055-023-00132-8/TABLES/10
Troell M, Costa-Pierce B, Stead S et al (2023) Perspectives on aquaculture's contribution to the sustainable development goals for improved human and planetary health. J World Aqua Soc 54:251–342. https://doi.org/10.1111/JWAS.12946
UN (2023) Gaza Strip: States are obliged to prevent crimes against humanity and genocide, UN Committee stresses. https://www.ohchr.org/en/press-releases/2023/12/gaza-strip-states-are-obliged-prevent-crimes-against-humanity-and-genocide. Accessed 10 Feb 2024
UNCTAD (2024) Preliminary assessment of the economic impact of the destruction in Gaza and prospects for economic recovery
UNEP (2024) Environmental impact of the conflict in Gaza: preliminary assessment of environmental impacts. Nairobi
UNEP, GRID-Arendal (2022) The environmental impact of the conflict in Ukraine: a preliminary review
UNICEF (2018) GVC and UNICEF help improve access to water in the Gaza Strip. https://www.unicef.org/sop/press-releases/gvc-and-unicef-help-improve-access-water-gaza-strip. Accessed 18 Dec 2024
UNRWA (2013) Gaza fishermen struggle to eke out a livelihood. https://www.unrwa.org/newsroom/features/gaza-fishermen-struggle-eke-out-livelihood. Accessed 1 Mar 2024
Wang F, Harindintwali JD, Yuan Z et al (2021) Technologies and perspectives for achieving carbon neutrality. Innovations 2. https://doi.org/10.1016/j.xinn.2021.100180
Wells K, Cortés-Morales S, Esson J et al (2024) Israel's war on Gaza and the violation of children's rights. Child Geogr 1–4. https://doi.org/10.1080/14733285.2024.2316752
Wenning RJ, Tomasi TD (2023) Using US natural resource damage assessment to understand the environmental consequences of the war in Ukraine. Integr Environ Assess Manag 19:366–375. https://doi.org/10.1002/IEAM.4716
Williams J (2024) Greenwashing: appearance, illusion and the future of 'green' capitalism. Geogr Compass 18:e12736. https://doi.org/10.1111/GEC3.12736
Yang R, Li L, Chen J et al (2024) Assessing provincial environment governance efficiency in China: a multi-agents participation perspective. Environ Sci Pol 160:103838. https://doi.org/10.1016/j.envsci.2024.103838

Marine Ecosystems in Peril: The Impact of the Recent War on Gaza's Coastal and Marine Life

12

Nour Alhaj Abdullah, Ammar Hdaifeh, Khaled Babaadoune, Ali Jebreen, Nour Awad, Abderahman Rejeb, Yehia El-Temsah, and Abdo Hassoun

Abstract

This chapter explores the severe impacts of the recent war (October 2023–January 2025) on Gaza's coastal and marine ecosystems, with a focus on threats to marine biodiversity, environmental health, and the livelihoods that rely on these ecosystems, while addressing the challenges to achieving Sustainable Development Goal 14 (SDG 14). It examines the environmental effects on coastal communities, including habitat destruction, biodiversity loss, pollution from debris, oil spills, chemical contaminants, and the degradation of water quality. The socio-economic repercussions are equally significant, encompassing increased poverty, loss of income, soaring unemployment, and worsening food insecurity. The discussion extends to the implications for SDG 14, emphasizing the urgent need for comprehensive environmental monitoring, targeted restoration initiatives, and strengthened international support to rehabilitate Gaza's marine ecosystems. The chapter underscores the importance of peacebuilding, sustainable management practices, and global cooperation in safeguarding these critical ecosystems and ensuring their resilience in the face of war. This analysis is grounded in both grey literature and peer-reviewed publications, offering a robust foundation for understanding the multifaceted challenges and proposing actionable solutions to support long-term recovery and sustainability.

N. A. Abdullah (✉) · A. Hassoun
Sustainable AgriFoodtech Innovation & Research (SAFIR), Arras, France

Faculty of Agricultural Engineering, University of Aleppo, Aleppo, Syria
e-mail: n.alhaj@saf-ir.com; a.hassoun@saf-ir.com

A. Hdaifeh
Sustainable AgriFoodtech Innovation & Research (SAFIR), Arras, France

Université Clermont Auvergne, Clermont-Ferrand, France

K. Babaadoune
Sustainable AgriFoodtech Innovation & Research (SAFIR), Arras, France

A. Jebreen
Department of Therapeutic Medical Nutrition, Faculty of Allied Medical Sciences, Ahliya University, Bethlehem, Palestine

N. Awad
Faculty of Engineering, Food Engineering Department, Ondokuz Mayis University, Samsun, Turkey

A. Rejeb
Faculty of Business and Economics, Széchenyi István University, Győr, Hungary
e-mail: rejeb.abderahman@sze.hu

Y. El-Temsah
Centre for Biosafety, Tromsø, Norway

A. Hassoun (ed.), *War on Gaza*, Sustainable Development Goals Series,
https://doi.org/10.1007/978-3-031-88500-6_12

Keywords

Marine ecosystems · Ecological balance · SDG 14 · Gaza fisheries · Biodiversity loss · Pollution · Socioeconomic challenges · Overfishing · Restoration strategies · Resilience

12.1 Introduction

Marine ecosystems are fundamental to planetary health since they regulate the climate through carbon sequestration, produce oxygen essential for life, and support a vast array of biodiversity. These ecosystems also provide food, livelihoods, and economic opportunities through fishing, tourism, and maritime trade. Acting as natural buffers against climate change, they absorb nearly a third of the carbon dioxide emissions generated by human activity. Despite their critical importance, marine ecosystems face escalating threats from pollution, habitat destruction, and overexploitation, which undermine their ecological and economic functions (Pawar 2016; Botterell et al. 2023; Cheung et al. 2024).

The coastal and marine environments of the Gaza Strip face significant challenges due to recurring conflicts and environmental pressures. Historically, these ecosystems supported rich biodiversity and provided critical resources for the livelihoods of coastal communities. However, the coastal zone of Gaza has suffered from issues such as severe erosion (Abualtayef et al. 2021) and pollution in seawater and sediments in the Gaza fishing harbor area (Ubeid et al. 2018; Zaqoot et al. 2018). Repeated Israeli assaults have further exacerbated these vulnerabilities, disrupting the ecological balance and threatening the sustainability of marine life due to weapons-related debris and physical damage caused by explosive impacts (Ali 2002; Shaheen et al. 2024; UNEP 2024). The recent war has magnified these challenges, deepening the threats to marine ecosystems and the communities that depend on them.

Sustainable Development Goals (SDGs) serve as a global framework for addressing sustainable development challenges, engaging governments, civil society, the private sector, and the scientific community to strengthen state capacities and drive impactful actions. Among the SDGs, the goal 14 stands out for its focus on the conservation and sustainable use of oceans, seas, and marine resources. This goal addresses the urgent need to safeguard marine biodiversity, reduce pollution, and promote responsible fishing practices. It emphasizes the global effort to enhance the resilience of marine ecosystems, which are vital for maintaining ecological balance and supporting economic stability. By fostering international collaboration and innovative solutions, SDG 14 provides a pathway for addressing the degradation of marine environments and ensuring the sustainable use of these critical resources (Recuero Virto 2018; Ntona and Morgera 2018; Arora et al. 2023; Haas 2023).

Although recent research extensively explores the impacts of wars and conflicts on marine ecosystems and progress toward SDG 14 (Linner et al. 2023; Tahmid et al. 2023; Graham 2024), studies on the recent war on Gaza remain notably scarce. This is despite the unprecedented scale of destruction and environmental degradation caused by the war, which demands much greater attention. The war on Gaza poses significant threats to marine biodiversity, coastal habitats, and the sustainability of vital marine resources. These impacts call for concerted efforts from governments, researchers, and international organizations to assess, address, and mitigate the damage, ensuring the resilience of Gaza's marine environment.

This chapter examines the environmental and socioeconomic impacts of the war on Gaza, with a specific focus on the vulnerabilities of its marine ecosystems and their relevance to achieving SDG 14. It delves into how the war exacerbates environmental degradation, emphasizing the urgent need to address habitat destruction, pollution, and socioeconomic challenges faced by the population in this besieged enclave. Furthermore, it proposes actionable solutions, such as deploying comprehensive environmental monitoring systems, initiating targeted habitat restoration projects, and fostering international partnerships to secure expertise and resources for long-term recovery. By highlighting the interconnectedness of marine conservation, socioeconomic resilience, and peacebuilding, this chapter underscores SDG 14 as a crucial pathway for achieving sustainable development in post-conflict regions.

12.2 Overview of Sustainable Development Goal 14 (SDG 14)—Life Below Water

SDG 14 of the United Nations' 2030 Agenda aims to conserve and sustainably use the oceans, seas, and marine resources for sustainable development. Oceans are integral to global ecological health, climate regulation, and economic systems. Covering over 70% of the Earth's surface, they generate half of the planet's oxygen, absorb 25% of global carbon dioxide emissions, and host rich biodiversity critical for food security and livelihoods. However, marine ecosystems face growing threats from pollution, overexploitation, and habitat degradation, challenges that are further exacerbated by conflicts and wars (Talukder et al. 2022; Haas 2023).

Among the specific targets of SDG 14, several are particularly relevant in the context of conflicts and wars. Target 14.1 aims to prevent and significantly reduce marine pollution, especially from land-based activities, by 2025. Armed conflicts often lead to extensive pollution of marine environments through oil spills, the discharge of hazardous materials, and debris from destroyed infrastructure. Military activities in coastal areas can release toxic substances, such as heavy metals and chemicals, which accumulate in marine sediments and harm biodiversity (Shukla et al. 2023; Meaza et al. 2024; Hryhorczuk et al. 2024). Target 14.2 emphasizes the sustainable management and protection of marine and coastal ecosystems to strengthen their resilience. Coastal ecosystems are highly vulnerable to physical damage caused by warfare, including bombing and the deployment of naval forces. In conflict zones, coral reefs, seagrass beds, and other habitats have been degraded due to both direct impacts and secondary effects such as reduced environmental governance and oversight (Arora et al. 2023; Tahmid et al. 2023).

Target 14.3 focuses on minimizing and addressing the impacts of ocean acidification, often exacerbated by local stressors linked to conflicts. For instance, the destruction of wastewater treatment facilities in war zones leads to untreated sewage entering marine environments, contributing to nutrient imbalances and ecosystem degradation. Target 14.5 highlights the importance of conserving at least 10% of coastal and marine areas in alignment with international law. However, conflicts frequently hinder the establishment and maintenance of Marine Protected Areas (MPAs), as seen in the Eastern Mediterranean and other regions affected by war. Limited resources and disrupted governance significantly reduce the effectiveness of conservation measures. Finally, Targets 14.b and 14.c of SDG 14 aim to support small-scale artisanal fishers by providing access to marine resources and markets, while also enhancing ocean conservation and sustainable use through the implementation of international law, including the United Nations Convention on the Law of the Sea (Okonkwo 2017; Recuero Virto 2018; Campbell and Gray 2019; Gómez et al. 2021; Arora et al. 2023; Rist et al. 2024).

As a result, it becomes evident that the relationship between SDG 14 and conflicts is complex and multifaceted. Wars and armed conflicts intensify the challenges of achieving SDG 14 by inflicting direct damage on marine and coastal ecosystems and disrupting governance needed for effective conservation. They often lead to pollution and habitat destruction, including oil spills, hazardous chemical releases, and widespread physical devastation of critical habitats. Achieving SDG 14 in conflict-affected regions requires a multifaceted approach that integrates environmental restoration with peacebuilding efforts. Enhanced international cooperation is essential to strengthen global commitments to marine conservation through frameworks like the United Nations Convention on the Law of the Sea (UNCLOS) and to promote cross-border initiatives addressing transboundary pollution and habitat protection. Restoration projects, such as coral reef rehabilitation and seagrass replanting programs, can help rebuild ecosystems severely affected by conflict. Capacity building is also crucial, supporting local governments and communities with the technical and financial resources needed to rebuild governance systems and implement sustainable practices (Linner et al. 2023; Tahmid et al. 2023; Graham 2024; Qafisheh 2024). By addressing the unique challenges posed by conflicts, the international community can make significant progress toward the targets of SDG 14 and ensure the resilience of marine ecosystems even in the most fragile regions.

12.3 Importance of Coastal and Marine Ecosystems in Gaza

The coastal and marine ecosystems in Gaza are rich in biodiversity, featuring a variety of habitats and species that hold significant ecological and socioeconomic importance. These ecosystems contribute to regional biodiversity and sustain local livelihoods, forming an integral part of the Mediterranean ecosystem. The diversity of habitats in Gaza's marine environment supports ecological balance in the Eastern Mediterranean and provides essential resources for the local population.

Gaza, covering an area of 365 km^2, is situated along the eastern coast of the Mediterranean Sea. It is characterized by a semi-arid Mediterranean climate, with hot summers, cold winters, and an average annual rainfall ranging from 230 to 400 mm (Al-Najjar et al. 2021). Despite its small size, Gaza's geographical location enhances biodiversity, supporting a variety of marine and terrestrial species, including fish, amphibians, reptiles, birds, and mammals (Abd Rabou 2020). Palestinian waters are home to over 400 marine fish species, with 201 of these species specifically recorded in the waters off Gaza. Bony fishes constitute 81% of the species, with 56% classified as common, 36% as rare, and 8% as very rare (Ali 2002). Moreover, the proximity of the Gaza Strip to the Suez Canal facilitates the migration of Lessepsian species, enriching the marine ecosystem with both native and non-native species, such as the reef stonefish and sohal surgeonfish (Bariche et al. 2019; Rabou et al. 2023). This unique biodiversity underscores the ecological and socioeconomic importance of Gaza's coastal environment.

The marine environment in Gaza hosts a wide variety of species, including fish such as sardines, mackerel, and anchovies, which are essential to the local fishing industry. Coastal waters also support crustaceans, mollusks, and seabirds that depend on the rich biodiversity of this ecosystem (Abd Rabou et al. 2023; Hussein et al. 2022). A long-term study conducted from 2003 to 2022 verified the presence of at least 26 marine organisms in coastal waters, including whales, sea turtles, and both bony and cartilaginous fish (Abd Rabou et al. 2023).

Despite its ecological importance, Gaza's marine environment and wetlands face numerous challenges, including pollution, overfishing, and habitat destruction, leading to the alteration of the coastal ecosystem (Zaqoot et al. 2018; Bariche et al. 2019; Abualtayef et al. 2021). For instance, Wadi Gaza has become heavily contaminated with heavy metals such as lead and cadmium due to untreated waste and agricultural runoff, posing risks to both biodiversity and public health (Ubeid et al. 2018). Moreover, the fisheries sector in Gaza has faced increasing restrictions on access to fishing zones imposed by Israel over the years. While the 1994 Oslo Accords allowed fishing within 20 nautical miles (nm), this zone has been systematically reduced. Following the Second Intifada, it was reduced to 12 nm and later to 6, with further restrictions during the blockade, reducing it to 3 nm. These limitations confine fishermen to overexploited and less biodiverse shallow waters, significantly impacting fish catches and livelihoods. The fishing boundaries have fluctuated between 3 and 12 nm, often influenced by the security situation, creating uncertainty and exacerbating socioeconomic challenges for Gaza's population (Ali 2002; Hussein et al. 2022).

Despite significant challenges, the fisheries sector remains a crucial provider of protein-rich food and a key driver of economic activity, with fish being a substantial part of the diet in the Gaza Strip (Elnabris et al. 2013; Hussein et al. 2022). However, the recent war has exacerbated these challenges by further restricting fishing zones, damaging infrastructure, and increasing pollution in coastal waters. These issues have led to various environmental and socioeconomic consequences, as will be elaborated in the next section.

12.4 Environmental Effects on Gaza's Coastal Areas and Implications for SDG 14

The recent war on Gaza has caused extensive environmental damage to its coastal regions, severely impacting marine life, fisheries, and ecosystems. Military activities have led to the destruction of habitats, destabilization of coastal zones, and significant biodiversity loss, with far-reaching consequences for ecological stability and human livelihoods. These immediate impacts are exacerbated by pollution from untreated sewage, debris, oil spills, and hazardous chemicals, which further disrupt ecological balance and weaken the resilience of marine ecosystems. The cumulative effects of these environmental challenges not only pose immediate threats but also have long-term implications, including the collapse of local fisheries, the extinction of vulnerable species, and reduced capacity for ecosystems to recover. Together, these factors undermine sustainable development efforts and compromise several targets of SDG 14, highlighting the urgent need for targeted restoration and conservation measures. The scale and persistence of the environmental destruction have prompted discussions of ecocide, as the widespread and deliberate damage threatens the survival of entire ecosystems and violates principles of environmental justice (Ahmed et al. 2024; Al Mezan 2024; Gillett 2024; Lambert 2024; Pearce 2024).

12.4.1 Destruction of Coastal Habitats and Loss of Biodiversity

Military activity along the coastline of Gaza has caused extensive destruction to natural habitats, critically damaging fragile ecosystems in the region. Explosions from airstrikes and shelling have led to physical destruction of coastal landscapes, including beaches and wetlands. Furthermore, military operations have destabilized coastal zones, increasing their vulnerability to erosion and further degradation. The fragile coastal areas, which are home to many endangered species (Qumsiyeh 2024), have become increasingly threatened due to this destruction. These species rely on intact and functional habitats for survival, and the loss of these environments further accelerates the decline in biodiversity.

Even before the recent war, the Israeli restriction of fishing to a small area near the coast has led to overfishing and has had a negative impact on breeding grounds. The Israeli restrictions confine Gaza's fishermen to areas within 3 nautical miles, where only small sardines and limited pelagic fish are available, leading to overfishing of these species (Fig. 12.1). This overfishing of shallow waters depletes stocks and undermines long-term sustainability. Inaccessible deeper waters beyond 3 nm hold higher-value species like tuna and mackerel, but restrictions prevent access, cutting off essential economic and ecological resources. These constraints threaten Gaza's food security, livelihoods, and marine biodiversity, highlighting the urgent need for sustainable fishing practices and expanded access.

In the absence of specific studies on Gaza's coastal habitats, insights can be drawn from similar conflicts elsewhere. For instance, during the Ukraine conflict, military activities have damaged coastal and marine habitats, placing marine ecosystems at risk. The degradation of these ecosystems not only diminishes their ecological functions but also has long-term consequences for livelihoods in coastal areas (UNEP and GRID-Arendal 2022; Mammadov et al. 2024; Hryhorczuk et al. 2024). The depletion of natural resources and reduced resilience to future envi-

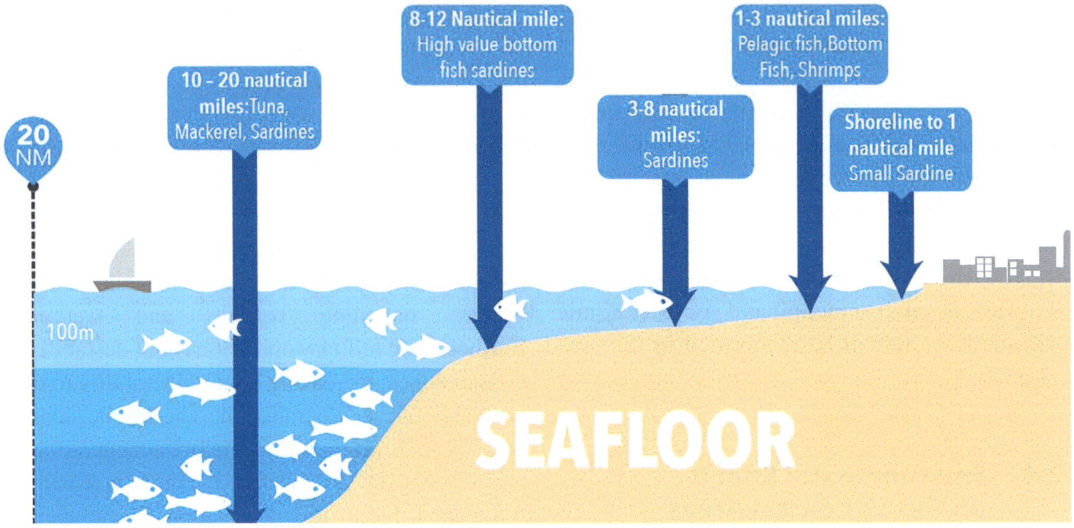

Fig. 12.1 Type of fish caught by sea depth and nautical miles (NM). (Source: [OCHA 2018])

ronmental challenges highlight the critical need for preservation and restoration efforts.

The destruction of coastal habitats due to military activities in Gaza directly impacts several targets of SDG 14. For example, Target 14.2 is undermined as the physical destruction of beaches and wetlands disrupts ecological balance and critical habitats, threatening the sustainable management and protection of coastal ecosystems. The damage also compromises Target 14.4, as the degradation of habitats essential for marine life reduces fish populations, increasing the risk of overexploitation of the remaining stocks. Furthermore, Target 14.5 is negatively impacted as the destruction of vital habitats hinders efforts to conserve and protect coastal and marine areas. Collectively, these impacts weaken the resilience of Gaza's coastal ecosystems, accelerate biodiversity loss, and pose significant challenges for sustainable development and long-term marine conservation.

12.4.2 Pollution from Debris, Oil Spills, and Chemicals

Coastal marine environment in Gaza faces significant threats from wartime pollution. The war has generated approximately 50 million tons of debris, contaminating soil and water with toxic materials such as asbestos and heavy metals (OCHA 2024; Shaheen et al. 2024). Toxic debris from weapons often infiltrates fragile coastal ecosystems, home to endangered species, exacerbating marine pollution (Qumsiyeh 2024). Oil spills pose another critical threat, as petroleum hydrocarbons harm aquatic life and ecosystems, disrupt coral reefs, and cause economic losses in coastal activities (Saadoun 2016). During the recent war, chemical weapons such as white phosphorus have been used extensively, with thousands of bombs being dropped on Gaza. These hazardous materials significantly harm biodiversity in water bodies and adjacent ecosystems (Browne 2024; El-Shewy et al. 2024; Khafaei et al. 2024).

The environmental pollution caused by debris, oil spills, and chemicals in Gaza severely undermines SDG 14. It directly contradicts Target 14.1 by introducing toxic materials, such as heavy metals and asbestos, into coastal waters. The degradation of fragile ecosystems that support endangered species further violates Target 14.2. Moreover, the extensive use of hazardous chemicals exacerbates environmental damage, indirectly impacting Target 14.3. Together, these issues significantly hinder progress toward achieving SDG 14.

12.4.3 Degradation of Water Quality

The war has severely degraded water quality in Gaza. Power outages have shut down all sewage treatment plants, resulting in the daily discharge of approximately 130,000 m^3 of untreated sewage into the Mediterranean Sea (Masria et al. 2024; Rangel-Buitrago et al. 2024; Shaheen et al. 2024). This sewage contains toxic heavy metals, nitrates, and chlorine, posing significant threats to marine life and human health (Wafi 2015; Chakraborty and Krishnani 2022). The resulting pollution disrupts fragile habitats, reduces fish populations, and exacerbates the already unstable environmental conditions along Gaza's coast. Heavy metal contamination has been detected in fish from Gaza's fishing port and other eastern Mediterranean areas, further highlighting the severe risks to biodiversity and human consumption (Shukla et al. 2023; Ramon et al. 2023).

The degradation of water quality and loss of biodiversity in Gaza severely undermine several targets of SDG 14. The daily discharge of large amounts of untreated sewage into the Mediterranean Sea, laden with toxic heavy metals, nitrates, and chlorine, significantly disrupts marine ecosystems, directly violating Target 14.1. This pollution, coupled with the destruction of fragile habitats and the decline in fish populations, hampers progress on Target 14.2. Additionally, the contamination of fish with heavy metals, which poses serious risks to biodiversity and human health, undermines Target 14.4. These impacts underscore the urgent need for interventions to restore water quality and safeguard biodiversity along Gaza's coast.

12.5 Socioeconomic Effects on Gaza's Coastal Areas and Implications for SDG 14

The war on Gaza has devastated the livelihoods of coastal communities, where fishing has historically been a primary source of income and cultural heritage for many Gazans. The destruction of fishing infrastructure and restrictions on access to fishing zones have unleashed a cascade of socioeconomic challenges, including escalating poverty, widespread loss of income, soaring unemployment, and worsening food insecurity. These impacts have severely undermined progress toward achieving SDG 14, which emphasizes sustainable use of marine resources and the support of small-scale fisheries for economic and social resilience. Below is a breakdown of these critical issues.

12.5.1 Increased Poverty

The war has devastated Gaza's fisheries, causing a near collapse of this vital sector and severely exacerbating poverty. Once a cornerstone of Gaza's economy and a primary source of livelihood for thousands of families, the fishing industry has been brought to a standstill. Restrictions on fishing zones, widespread destruction of infrastructure, and the loss of boats and equipment have rendered fishing activities impossible. The port of Gaza City has suffered extensive damage, and the majority of fishing vessels have become inoperable due to military operations and the lack of maintenance resources caused by prolonged blockades and resource shortages. Additionally, the severe lack of fuel has paralyzed remaining operations, leaving fishermen unable to sustain their livelihoods despite the urgent need for food security (FAO 2024; Qafisheh 2024).

Even before the recent war, poverty was already widespread in Gaza, with a significant portion of households living below the poverty line. For fishermen, the situation was particularly dire, as approximately 90% of the fishing community struggled to survive on incomes far below the poverty threshold (Hussein et al. 2022; World Bank Group 2023).

This situation directly undermines Target 14.b of SDG 14, as extreme poverty among Gaza's fishermen, compounded by restrictive policies and limited development initiatives, severely restricts access to resources and markets. The widespread marginalization of the fishing community highlights systemic barriers to equitable resource distribution, further exacerbated by the war, making it nearly impossible to achieve this target or support small-scale fisheries in fostering sustainable livelihoods and resilience.

12.5.2 Loss of Income

The war has devastated Gaza's fishing value chain, leading to a widespread loss of income for fishermen and those dependent on marine resources. Before the recent war, the fisheries sector in Gaza supported 3982 registered fishers, directly sustaining approximately 21,000 individuals. Additionally, about 110,000 people, including family members, relied on this sector for their livelihoods (The World Bank 2020; Hussein et al. 2022; Première Urgence Internationale 2023; FAO 2024).

However, the war has had a catastrophic impact on Gaza's fishing sector. Between October 2023 and April 2024, the average daily fish catch plummeted to a mere 7.3% of 2022 levels, resulting in an estimated $17.5 million loss in production (FAO 2024). Since October 2023, fishermen who once depended on the sea for sustenance now face constant danger, as it has turned into a battleground of suffering and peril, where Israeli war boats relentlessly target their boats, equipment, and nets, forcing their families into poverty and deprivation. According to local authorities, 90% of Gaza's fishing boats were destroyed within the first 2 months of this brutal war (Business Insider 2024; The Palestinian Information Center 2024). Furthermore, restrictions on fishing zones (Fig. 12.2) have further exacerbated the crisis, with the allowed fishing distance progressively reduced from 82 km (44 nm) to 20 nm, then 12, 6, and finally just 3 nm (Milton et al. 2024; Qafisheh 2024; Salah et al. 2024).

This severe limitation drastically reduces access to fishing grounds, depleting income opportunities for fishermen and undermining their ability to sustain livelihoods.

This loss of income directly undermines Target 14.b of SDG 14 (i.e., providing small-scale artisanal fishers access to marine resources and markets). Furthermore, the severe restrictions on fishing zones contradict Target 14.c (i.e., enhancing the conservation and sustainable use of oceans by implementing international law, such as the UN Convention on the Law of the Sea). The progressive narrowing of Gaza's fishing zones violates the principle of equitable access to marine resources, further marginalizing fishing communities and hindering their ability to contribute to sustainable development and economic resilience.

12.5.3 Soaring Unemployment

The Palestinian economy is on the brink of collapse, marked by unprecedented unemployment rates and a 250% surge in the cost of many essential goods (ILO 2024; UNICEF 2024). Even before the recent war, it was reported that in Gaza, the unemployment rate is beyond 50% (Hussein et al. 2022). The Israeli aggression has devastated Gaza's fishing sector, rendering nearly 4000 fishermen unemployed (Al Mayadeen 2024). This collapse has not only stripped them of their livelihoods but also disrupted the primary source of income for thousands of families who depend on the fishing industry. The loss of employment in this critical sector exacerbates poverty and food insecurity, further destabilizing already vulnerable coastal communities.

The soaring unemployment in Gaza, particularly in the fishing sector, directly undermines SDG 14's Target 14.b. The collapse of the fishing industry has displaced thousands of workers, stripping them of livelihoods and opportunities to engage in sustainable fishing practices. With destroyed infrastructure and rising job insecurity, coastal communities face significant barriers to market access and economic stability, making progress toward this target nearly impossible.

12.5.4 Worsening Food Insecurity

The damage to Gaza's fishing sector has exacerbated an already dire food security crisis due to the massive destruction of other food production facilities, including the destruction of 70% of croplands. Moreover, the collapse of Gaza's fishing industry has eliminated a critical source of protein, directly impacting food security and leading to acute malnutrition. According to latest projections in the Integrated Food Security Phase

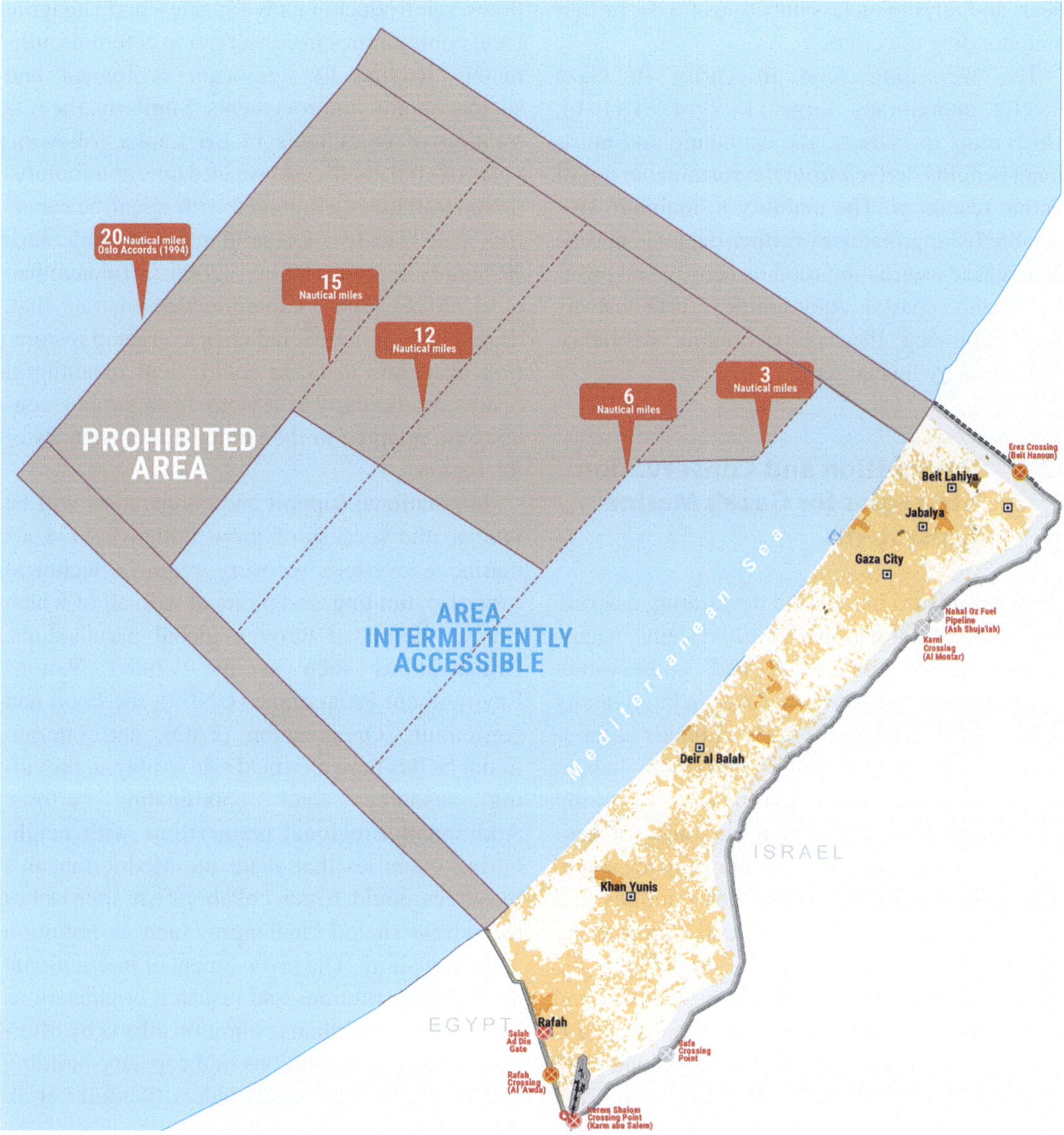

Fig. 12.2 Gaza Strip fishing limits. (Source: [OCHA 2019])

Classification (IPC), almost the entire Gaza Strip is classed as experiencing an emergency (phase 4), and that some areas are in danger of famine (phase 5) (IPC 2024). Many reports have demonstrated high levels of weight loss and marked food insecurity, especially in the Northern Gaza Strip during this war (Alnabih et al. 2024; Hassoun et al. 2024).

Gaza's fisheries sector was once a vital source of high-protein food for the population (Hussein et al. 2022), but the collapse of sustainable fishing capacity, coupled with widespread agricultural destruction, has pushed nearly the entire population into food insecurity and malnutrition, especially children. A UNICEF report from late 2024 revealed that the number of acutely malnourished children had quadrupled since May 2024, with many children tragically losing their lives (UNICEF 2024). Households are resorting to severe coping mechanisms, such as skipping meals and relying on less nutritious foods, further compounding the crisis.

The worsening food insecurity in Gaza directly undermines Target 14.7 of SDG 14, which aims to increase the economic and nutritional benefits derived from the sustainable use of marine resources. The inability to maintain sustainable fishing practices further depletes marine resources, exacerbating food insecurity and pushing Gaza's coastal communities further away from achieving the socioeconomic resilience envisioned by this target.

12.6 Restoration and Conservation Strategies for Gaza's Marine Ecosystems

The war on Gaza has caused devastating destruction to its coastal and marine ecosystems, necessitating urgent restoration and conservation efforts. Gaza's coastline, once rich in biodiversity and a critical resource for the livelihoods of thousands, has been degraded by pollution, habitat destruction, and overexploitation. Restoring these ecosystems is essential not only for ecological recovery but also for supporting the socioeconomic stability of coastal communities and achieving SDG 14. Environmental monitoring and targeted restoration projects should be at the forefront of these efforts, providing the data and framework needed to rehabilitate marine habitats, replenish fish stocks, and address long-term environmental challenges. Robust monitoring can help identify the most critical areas for intervention, track progress, and ensure accountability in the use of resources allocated for recovery (Geist and Hawkins 2016; Manea et al. 2023).

Case studies from other conflict-affected regions highlight the potential for success when restoration efforts are informed by scientific research and bolstered by international cooperation. For example, in post-conflict Iraq, large-scale restoration projects were implemented to rehabilitate the Mesopotamian Marshes, a UNESCO World Heritage Site that had suffered extensive damage due to war and deliberate drainage. These efforts involved restoring water flows, reintroducing native species, and engaging local communities in conservation activities, ultimately leading to significant ecological and socioeconomic improvements. Similarly, the restoration of coral reefs in Sri Lanka following years of civil conflict showcased how community-driven initiatives, combined with scientific expertise, can lead to successful recovery outcomes (Richardson and Hussain 2006; Arulananthan et al. 2021). These examples demonstrate that, despite the immense challenges, targeted restoration programs in Gaza could yield meaningful results, particularly if lessons from similar contexts are adapted to the unique circumstances of the region.

International support and cooperation will be vital to the success of these initiatives. Gaza's marine ecosystem recovery requires technical expertise, funding, and political will, all of which can be facilitated through global partnerships. Organizations such as the United Nations Environment Programme (UNEP), the Food and Agriculture Organization (FAO), and international NGOs have a critical role to play in providing resources and coordinating efforts. Additionally, regional partnerships with neighboring countries that share the Mediterranean's resources could foster collaborative approaches to address shared challenges, such as pollution and overfishing. The involvement of international academic institutions and research organizations can further strengthen restoration efforts by offering cutting-edge solutions and capacity-building initiatives for local stakeholders (Abdulla et al. 2009; Giakoumi et al. 2012).

To align restoration efforts with SDG 14, sustainable management practices must be prioritized. These include implementing regulated fishing practices, protecting critical habitats, and enforcing pollution control measures. The development of MPAs in Gaza's coastal waters could serve as a cornerstone of conservation efforts, providing safe havens for biodiversity while allowing depleted fish stocks to recover. Community engagement will also be critical; empowering local fishers and coastal residents to participate in sustainable practices ensures both economic benefits and long-term buy-in for conservation initiatives. Moreover, incorporating traditional ecological knowledge into modern management strategies can enhance the effectiveness of these practices, fostering a sense of ownership among local communities (Butler et al. 2012; Recuero Virto 2018; Nilsson et al. 2018; Hamelin et al. 2024).

Peacebuilding and environmental diplomacy are equally critical for the success of marine restoration in Gaza. The war has not only destroyed ecosystems but also eroded the socioeconomic fabric that supports conservation efforts. Establishing a stable and collaborative political environment is essential to ensure that restoration programs can be implemented without interruptions. Environmental diplomacy, which frames ecosystem restoration as a shared goal transcending political divides, can create opportunities for dialogue and cooperation even in conflict settings. The success of transboundary conservation projects in other politically unstable regions demonstrates how shared environmental goals can build bridges between conflicting parties and maintain sustainable peace. For Gaza, integrating peacebuilding efforts with environmental recovery could foster not only ecological resilience but also sociopolitical stability (Hwang 2022; UNEP 2024).

12.7 Conclusion

The degradation of coastal and marine ecosystems in Gaza poses significant challenges to achieving SDG 14. The recent brutal war has devastated biodiversity, fisheries, and critical habitats, deepening socioeconomic and environmental crises. Addressing these issues requires a comprehensive approach integrating ecological restoration, socioeconomic recovery, and peacebuilding. Environmental and socioeconomic consequences of this war are catastrophic. The war has destroyed infrastructure, reduced fish stocks, and worsened food insecurity for Gaza's fishing communities. Rebuilding the fisheries sector requires lifting restrictions, restoring infrastructure, and implementing sustainable practices to balance immediate needs with long-term conservation. Peacebuilding is essential for sustainable restoration, as conservation efforts are vulnerable to interruptions without political stability. Environmental peacebuilding can foster synergies for biodiversity and regional stability. Engaging local communities ensures interventions align with their needs and priorities. Finally, it is clear that achieving SDG 14 in Gaza remains unfeasible in the aftermath of the recent genocidal war. The first step must be the establishment of lasting stability and addressing the devastating impacts of the conflict. Only then can efforts focus on the restoration of marine ecosystems, the advancement of socioeconomic recovery, and the promotion of sustainable peace. The global community must prioritize support for recovery efforts and the creation of conditions necessary for meaningful progress toward the SDGs.

References

Abd Rabou AF (2020) The Palestinian marine and terrestrial vertebrate Fauna preserved at the biology exhibition, Islamic University of Gaza, bombarded by the Israeli Army in December, 2008. Israa Univ J Appl Sci 4:9–51. https://doi.org/10.52865/hfrs6095

Abd Rabou AFN, Elkahlout KE, Elnabris KJ et al (2023) An inventory of some relatively large marine mammals, reptiles, and fishes sighted, caught, by-caught, or stranded in the Mediterranean coast of the Gaza strip-Palestine. Open J Ecol 13:119–153. https://doi.org/10.4236/oje.2023.132010

Abdulla A, Gomei M, Hyrenbach D et al (2009) Challenges facing a network of representative marine protected areas in the Mediterranean: prioritizing the protection of underrepresented habitats. ICES J Mar Sci 66:22–28. https://doi.org/10.1093/icesjms/fsn164

Abualtayef M, Rabou MA, Afifi S et al (2021) Change detection of Gaza coastal zone using GIS and remote sensing techniques. J Coast Conserv 25. https://doi.org/10.1007/s11852-021-00825-4

Ahmed K, Gayle D, Mousa A (2024) 'Ecocide in Gaza': does scale of environmental destruction amount to a war crime? Guard

Al Mayadeen (2024) "Israel" kills 200 fishermen in Gaza; fishing sector faces $20mln loss. https://english.almayadeen.net/news/politics/-israel%2D%2Dkills-200-fishermen-in-gaza%2D%2Dfishing-sector-faces. Accessed 12 Jan 2025

Al Mezan (2024) Ecocide: Israel's deliberate and systematic environmental destruction in Gaza

Ali M (2002) The coastal zone of Gaza strip-Palestine management and problems. Present MAMA first kick-off meet, 1–30

Alnabih A, Alnabeh N, Aljeesh Y, Aldabbour B (2024) Food insecurity and weight loss during wartime: a mixed-design study from the Gaza strip. J Health Popul Nutr 43:222

Al-Najjar H, Ceribasi G, Ceyhunlu AI (2021) Effect of unconventional water resources interventions on the management of Gaza coastal aquifer in Palestine. Water Supply 21:4205–4218. https://doi.org/10.2166/WS.2021.170

Arora NK, Mishra I, Arora P (2023) SDG 14: life below water- viable oceans necessary for a sustainable planet. Environ Sustain 6:433–439. https://doi.org/10.1007/s42398-023-00299-0

Arulananthan A, Herath V, Kuganathan S et al (2021) The status of the coral reefs of the Jaffna peninsula (Northern Sri Lanka), with 36 coral species new to Sri Lanka confirmed by DNA Bar-coding. Oceans 2:509–529. https://doi.org/10.3390/oceans2030029

Bariche M, Sayar N, Balistreri P (2019) Records of two non-indigenous fish species Synanceia verrucosa Bloch and Schneider, 1801 and Acanthurus sohal (Forsskål, 1775) from the Gaza strip (eastern Mediterranean Sea). BioInvasions Rec 8:699–705. https://doi.org/10.3391/bir.2019.8.3.27

Botterell ZLR, Lindeque PK, Thompson RC, Beaumont NJ (2023) An assessment of the ecosystem services of marine zooplankton and the key threats to their provision. Ecosyst Serv 63:101542. https://doi.org/10.1016/J.ECOSER.2023.101542

Browne BC (2024) The Coloniality of enforced starvation: Reading famine in Gaza through an Gorta Mór. J Palest Stud. https://doi.org/10.1080/0377919X.2024.2379709

Business Insider (2024) 4 Gazans show how war devastated their coastline. https://www.businessinsider.com/gazans-show-how-war-devastated-their-coastline-2024-9?utm_source=chatgpt.com. Accessed 12 Jan 2025

Butler JRA, Tawake A, Skewes T et al (2012) Integrating traditional ecological knowledge and fisheries management in the torres strait, Australia: the catalytic role of turtles and dugong as cultural keystone species. Ecol Soc 17:34. https://doi.org/10.5751/ES-05165-170434

Campbell LM, Gray NJ (2019) Area expansion versus effective and equitable management in international marine protected areas goals and targets. Mar Policy 100:192–199. https://doi.org/10.1016/j.marpol.2018.11.030

Chakraborty P, Krishnani KK (2022) Emerging bioanalytical sensors for rapid and close-to-real-time detection of priority abiotic and biotic stressors in aquaculture and culture-based fisheries. Sci Total Environ 838:156128. https://doi.org/10.1016/j.scitotenv.2022.156128

Cheung WWL, Pauly D, Sumaila UR (2024) Hope or despair revisited: assessing progress and new challenges in global fisheries. Fish Fish 1–13. https://doi.org/10.1111/faf.12877

Elnabris KJ, Muzyed SK, El-Ashgar NM (2013) Heavy metal concentrations in some commercially important fishes and their contribution to heavy metals exposure in Palestinian people of Gaza strip (Palestine). J Assoc Arab Univ Basic Appl Sci 13:44–51. https://doi.org/10.1016/j.jaubas.2012.06.001

El-Shewy M, Griffiths M, Jones C (2024) Israel's war on Gaza in a global frame. Antipode 1–21. https://doi.org/10.1111/ANTI.13094

FAO (2024) Reviving Gaza's fishing sector hinges on restoring peace and safe access to the sea. https://www.fao.org/newsroom/detail/reviving-gaza-s-fishing-sector-hinges-on-restoring-peace-and-safe-access-to-the-sea/en. Accessed 11 Jan 2025

Geist J, Hawkins SJ (2016) Habitat recovery and restoration in aquatic ecosystems: current progress and future challenges. Aquat Conserv Mar Freshw Ecosyst 26:942–962. https://doi.org/10.1002/aqc.2702

Giakoumi S, Mazor T, Fraschetti S et al (2012) Advancing marine conservation planning in the Mediterranean Sea. Rev Fish Biol Fish 22:943–949. https://doi.org/10.1007/s11160-012-9272-8

Gillett M (2024) Ecocide, environmental harm and framework integration at the international criminal court. Int J Hum Rights:1–37. https://doi.org/10.1080/13642987.2024.2433660

Gómez S, Carreño A, Lloret J (2021) Cultural heritage and environmental ethical values in governance models: conflicts between recreational fisheries and other maritime activities in Mediterranean marine protected areas. Mar Policy 129:104529. https://doi.org/10.1016/J.MARPOL.2021.104529

Graham KJ (2024) Turning the tide on remnants of war at sea? Toward the principles for protection of the environment in relation to armed conflicts. Int J Mar Coast Law 39:768–787. https://doi.org/10.1163/15718085-bja10190

Haas B (2023) Achieving SDG 14 in an equitable and just way. Int Environ Agreements Polit Law Econ 23:199–205. https://doi.org/10.1007/s10784-023-09603-z

Hamelin KM, Charles AT, Bailey M (2024) Community knowledge as a cornerstone for fisheries management. Ecol Soc 29:26. https://doi.org/10.5751/ES-14552-290126

Hassoun A, Al-Muhannadi K, Hassan HF et al (2024) From acute food insecurity to famine: how the 2023/2024 war on Gaza has dramatically set back sustainable development goal 2 to end hunger. Front Sustain Food Syst 8:1402150. https://doi.org/10.3389/fsufs.2024.1402150

Hryhorczuk D, Levy BS, Prodanchuk M et al (2024) The environmental health impacts of Russia's war on Ukraine. J Occup Med Toxicol 19:1–14. https://doi.org/10.1186/S12995-023-00398-Y/FIGURES/2

Hussein DMA, Lucchetti MC, Zaqoot HA et al (2022) Status of fisheries in Gaza strip: past trends and challenges. Int J Euro Mediterranean Stud 15:179–216

Hwang J (2022) Building sustainable peace through environmental cooperation in the island of Ireland: modelling transboundary conservation. Irish Polit Stud 37:548–570. https://doi.org/10.1080/07907184.2021.2011230

ILO (2024) A year of war in Gaza: impacts on employment and livelihoods in the West Bank and Gaza strip. Bulletin No 5

IPC (2024) Gaza Strip: acute food insecurity situation for September–October 2024 and projection for November 2024–April 2025

Khafaei M, Panahi Y, Abolghasemi H et al (2024) Injuries caused by recurrent white phosphorus bombing as a crisis in Gaza strip. Trauma Mon 29:1052–1057. https://doi.org/10.30491/tm.2024.452290.1719

Lambert LA (2024) Ecocide as genocide in Gaza: a human security approach to "Utter Annihilation"

Linner A, Sharifi A, Simangan D et al (2023) Interactions between SDG 14 (life below water) and SDG 16 (peace, justice, and strong institutions): a review of co-benefits, synergies, conflicts, and trade-offs. In: Sharifi A, Simangan D, Kaneko S (eds) Bridging peace and sustainability amidst global transformations. Springer, pp 31–51

Mammadov S, Luhovyi S, Starodubets O et al (2024) Collateral ecocide. The impact of war on Ukrainian flora and fauna. Int J Environ Stud 81:446–454. https://doi.org/10.1080/00207233.2024.2314851

Manea E, Bergami C, Ptacnik R (2023) Editorial: advances in marine and freshwater monitoring to support aquatic ecosystem conservation and restoration. Front Environ Sci 11:1304545. https://doi.org/10.3389/fenvs.2023.1304545

Masria A, Elejla K, Abualtayef M et al (2024) Modeling the dispersion of wastewater pollutants in Gaza's coastal waters. Mar Pollut Bull 208:117071. https://doi.org/10.1016/j.marpolbul.2024.117071

Meaza H, Ghebreyohannes T, Nyssen J et al (2024) Managing the environmental impacts of war: what can be learned from conflict-vulnerable communities? Sci Total Environ 927:171974. https://doi.org/10.1016/j.scitotenv.2024.171974

Milton S, Elkahlout G, Attallah S (2024) Shrinking reconstruction space in the Gaza strip: rebuilding after the 2021 and 2022 wars. Conflict Secur Dev 24:49–78. https://doi.org/10.1080/14678802.2024.2314034

Nilsson M, Chisholm E, Griggs D et al (2018) Mapping interactions between the sustainable development goals: lessons learned and ways forward. Sustain Sci 13:1489–1503. https://doi.org/10.1007/S11625-018-0604-Z/TABLES/2

Ntona M, Morgera E (2018) Connecting SDG 14 with the other Sustainable Development Goals through marine spatial planning. Mar Policy 93:214–222. https://doi.org/10.1016/J.MARPOL.2017.06.020

OCHA (2018) Gaza fisheries: fishing catch Increases amid ongoing protection concerns. https://www.ochaopt.org/content/gaza-fisheries-fishing-catch-increases-amid-ongoing-protection-concerns?utm_source=chatgpt.com. Accessed 12 Jan 2025

OCHA (2019) Gaza's fisheries: record expansion of fishing limit and relative increase in fish catch; shooting and detention incidents at sea continue. https://www.ochaopt.org/content/gaza-s-fisheries-record-expansion-fishing-limit-and-relative-increase-fish-catch-shooting. Accessed 12 Jan 2025

OCHA (2024) Gaza Humanitarian Response Update | 24 November—7 December 2024. https://www.ochaopt.org/content/gaza-humanitarian-response-update-24-november-7-december-2024?_gl=1*1egyngz*_ga*NDE3Nzg4ODEuMTczMzQ2NTkzMQ..*_ga_E60ZNX2F68*MTczNTE0MjExMi4xMC4xLjE3MzUxNDIxMTguNTQuMC4w. Accessed 25 Dec 2024

Okonkwo T (2017) Maritime boundaries delimitation and dispute resolution in Africa. Beijing Law Rev 08:55–78. https://doi.org/10.4236/blr.2017.81005

Pawar PR (2016) Anthropogenic threats to coastal and marine biodiversity: a review. IJMBR 4:35–45

Pearce E (2024) Israel's ecocide is contributing to the forcible displacement of Gazans. Open Glob. Rights. https://www.openglobalrights.org/israel-ecocide-contributing-to-forcible-displacement-gazans/. Accessed 27 Apr 2024

Première Urgence Internationale (2023) The impacts of the enforcement of the access restricted areas at land and sea in the Gaza Strip

Qafisheh MM (2024) From Constantinople to Oslo to Gaza: developments of Palestine's maritime legislation under colonization. Mar Policy 160:105954. https://doi.org/10.1016/j.marpol.2023.105954

Qumsiyeh MB (2024) Impact of the Israeli military activities on the environment. Int J Environ Stud 81:977–992. https://doi.org/10.1080/00207233.2024.2323365

Rabou DAFNA, Elkahlout KE, Elnabris KJ et al (2023) Rare records of the exotic reef stonefish (Synanceia verrucosa Bloch and Schneider, 1801) in the Mediterranean waters of the Gaza strip, Palestine. Int J Fauna Biol Stud 10:24–28. https://doi.org/10.22271/23940522.2023.v10.i3a.963

Ramon D, Britzi M, Davidovich N et al (2023) Accumulation of Lead, mercury and cadmium in coastal sediments in the Eastern Mediterranean Sea. In: Lead, mercury and cadmium in the aquatic environment: worldwide occurrence, fate and toxicity. CRC Press, pp 92–125

Rangel-Buitrago N, Galgani F, Neal WJ (2024) Addressing the global challenge of coastal sewage pollution. Mar Pollut Bull 201:116232. https://doi.org/10.1016/j.marpolbul.2024.116232

Recuero Virto L (2018) A preliminary assessment of the indicators for Sustainable Development Goal (SDG) 14 "Conserve and sustainably use the oceans, seas and marine resources for sustainable development". Mar Policy 98:47–57. https://doi.org/10.1016/j.marpol.2018.08.036

Richardson CJ, Hussain NA (2006) Restoring the garden of eden: an ecological assessment of the marshes of Iraq. Bioscience 56:477–489. https://doi.org/10.1641/0006-3568(2006)56[477:RTGOEA]2.0.CO;2

Rist L, Norström A, Queiroz C (2024) Biodiversity, peace and conflict: understanding the connections. Curr Opin

Environ Sustain 68:101431. https://doi.org/10.1016/J.COSUST.2024.101431
Saadoun IMK (2016) Impact of oil spills on marine life. In: Larramendy ML, Soloneski S (eds) Emerging pollutants in the environment—current and further implications. IntechOpen, p 13
Salah J, Aboutair M, Zava B, Di Natale A (2024) Bluefin tuna (Thunnus thynnus l.) catches in the Palestinian area (southeastern mediterranean sea). Collect Vol Sci Pap Int Comm Conserv Atl Tunas 81:1–13
Shaheen A, Dajani R, Zinszer K et al (2024) The war on the Gaza strip and its consequences on global warming. Front Hum Dyn 6:1463902
Shukla S, Mbingwa G, Khanna S et al (2023) Environment and health hazards due to military metal pollution: a review. Environ Nanotechnol Monit Manag 20:100857. https://doi.org/10.1016/j.enmm.2023.100857
Tahmid A, Khanam S, Rashid MM, Ibnat F (2023) Reviewing the impact of military activities on marine biodiversity and conservation: a study of the Ukraine-Russia conflict within the framework of international law. Grassroots J Nat Resour 6:15–31. https://doi.org/10.33002/nr2581.6853.060302
Talukder B, Ganguli N, Matthew R et al (2022) Climate change-accelerated ocean biodiversity loss & associated planetary health impacts. J Clim Chang Heal 6:100114. https://doi.org/10.1016/J.JOCLIM.2022.100114
The Palestinian Information Center (2024) Gaza fishermen: a struggle for life under Israeli fire. https://english.palinfo.com/reports/2024/12/21/330827/. Accessed 11 Jan 2025
The World Bank (2020) Rapid fishery and aquaculture sector diagnosis using fishery performance indicators in the Gaza Strip
Ubeid KF, Al-Agha MR, El-Turk MF (2018) Heavy metals distribution and pollution in the sediments of the Wadi Gaza mouth, eastern Mediterranean coast, Palestine. Ser Correl Geol 34:71–88
UNEP (2024) Environmental impact of the conflict in Gaza: preliminary assessment of environmental impacts. Nairobi
UNEP, GRID-Arendal (2022) The environmental impact of the conflict in Ukraine: a preliminary review
UNICEF (2024) Palestinian children Palestinian children under siege children
Wafi HN (2015) Assessment of heavy metals contamination in the Mediterranean Sea along Gaza coast—a case. Study of Gaza Fishing Harbor
World Bank Group (2023) West Bank and Gaza: Country climate and development report
Zaqoot HA, Aish AM, Wafi HN (2018) Study on heavy metal pollution in Gaza fishing harbour along the Mediterranean Sea-Gaza beach. Palestine J Aquat Sci Mar Biol 1:24

Life on Land in Peril: How the Recent War on Gaza Has Devastated Terrestrial Ecosystems

13

Abdo Hassoun, Heba Jarrar, Lina Goudali, Silvia Lisciani, Khawla Al-Muhannadi, Mohamed Buheji, and Alaa El-Din A. Bekhit

Abstract

War leaves a devastating imprint on both people and the environment, and nowhere is this more evident than in Gaza. This chapter explores the profound environmental and socioeconomic consequences of the recent war (October 2023–January 2025) on Gaza, focusing on its implications for Sustainable Development Goal 15 (SDG 15). The war has caused extensive damage to terrestrial ecosystems, including soil contamination, biodiversity loss, and the destruction of agricultural land, undermining efforts to conserve ecosystems, combat desertification, and halt biodiversity loss. These challenges are further compounded by food insecurity and socioeconomic vulnerabilities. The chapter underscores the importance of ecological restoration, sustainable land management, and international collaboration to address these critical issues. By highlighting the urgency of recovery efforts, it emphasizes the need for global action to mitigate the long-term environmental impacts of conflict and support sustainable development in Gaza.

Keywords

Agricultural land · Biodiversity · Deforestation · Flora and fauna · Food insecurity · Land rehabilitation · Pollution · SDG 15 · Soil degradation · The Gaza Strip

A. Hassoun
Sustainable AgriFoodtech Innovation & Research (SAFIR), Arras, France

Faculty of Agricultural Engineering, University of Aleppo, Aleppo, Syria
e-mail: a.hassoun@saf-ir.com

H. Jarrar
Earth Charter International, Dubai, UAE

L. Goudali
Higher Normal School of Technical Education of Mohammedia Hassan II, University of Casablanca, Casablanca, Morocco
e-mail: lina.goudali-etu@etu.univh2c.ma

S. Lisciani
Research Centre for Food and Nutrition, Council for Agricultural Research and Economics, Rome, Italy
e-mail: silvia.lisciani@crea.gov.it

K. Al-Muhannadi
Environment Friends Society & Reem Program for Young Environmental Leaders, Manama, Bahrain

M. Buheji
Socioeconomic Institute for Advanced Studies (SIAS), Kigali, Rwanda

Inspiration Economy Institute, Manama, Bahrain

A. E.-D. A. Bekhit (✉)
Department of Food Science, University of Otago, PO, Dunedin, New Zealand
e-mail: aladin.bekhit@otago.ac.nz

A. Hassoun (ed.), *War on Gaza*, Sustainable Development Goals Series,
https://doi.org/10.1007/978-3-031-88500-6_13

13.1 Introduction

The recent war on Gaza has brought catastrophic consequences for both people and the environment, making it one of the most challenging contexts for achieving Sustainable Development Goals (SDGs) stated by the United Nations 2030 Agenda, particularly SDG 15, which emphasizes life on land. Reports highlight that the deliberate destruction of critical infrastructure (e.g., healthcare, food, education facilities) and ecosystems in Gaza aims to render the region uninhabitable (Farhat et al. 2023; PAX 2023; Shaheen et al. 2024; London et al. 2024; Iriqat et al. 2025). The scale of devastation has disrupted natural ecosystems, obliterated biodiversity, and polluted vital resources like soil and water, essential for sustaining life and enabling development (Buheji and Al-Muhannadi 2023; Al Mezan 2024; UNEP 2024).

Achieving the objectives of SDG 15 in conflict zones is inherently difficult. Wars and armed conflicts exacerbate environmental degradation through the destruction of ecosystems, contamination of natural resources, and disruption of conservation efforts. Military activities often lead to deforestation, soil erosion, biodiversity loss, and water pollution, making sustainable development nearly impossible in such contexts. The long-term impacts of these activities often persist for decades, leaving affected regions with degraded ecosystems that are challenging to restore (Solokha et al. 2023, 2024; Leal Filho et al. 2024). In Gaza, these challenges are intensified by its unique environmental and ecological conditions. Limited land availability, high population density, and fragile ecosystems make the region particularly vulnerable to environmental stress and degradation (UN and State of Palestine 2018; Qumsiyeh and Abusarhan 2021; Elkahlout 2024). This chapter examines the profound effects of the war on terrestrial ecosystems in Gaza, with specific attention to soil degradation, biodiversity loss, water scarcity, and waste management. Based on gray literature, official reports, and peer-reviewed publications, the chapter provides actionable insights and recommendations to address the environmental crisis and move toward sustainable development of land and resources in Gaza, even amidst conflict.

13.2 Overview of Sustainable Development Goal 15 (SDG 15)—Life on Land—And Its Relation to Armed Conflicts

SDG 15, a part of the United Nations 2030 Agenda for sustainable development, aims to protect, restore, and promote the sustainable use of terrestrial ecosystems, combat desertification, halt and reverse land degradation, and halt biodiversity loss. Key targets under SDG 15 include the sustainable management of forests, combating desertification, restoring degraded land, reducing habitat destruction, and conserving biodiversity. This goal is critical for ensuring the health of ecosystems that sustain human life, regulate climate, and provide essential services such as clean water, food security, and livelihoods (Krauss 2022; Li et al. 2023; Kumar and Isha 2024).

However, wars and conflicts pose significant barriers to achieving SDG 15. It is well-documented that nature is a silent victim of war, as armed conflicts disrupt ecosystems, degrade soil and water quality, and lead to deforestation, habitat destruction, and loss of biodiversity. Conflict-related activities, such as the use of heavy weaponry, destruction of infrastructure, and contamination of land with toxic substances, contribute to long-term ecological degradation. For instance, military operations often involve the deployment of explosives and chemicals that seep into soil and water systems, rendering land unsuitable for agriculture and reducing its ecological functionality (Hulme 2022; Hsiao et al. 2023; Weir et al. 2024).

Conflicts and wars negatively impact the achievement of all SDGs (Wang et al. 2024), with SDG 15, which includes several particularly vulnerable targets, being among the most affected. The destruction caused by armed conflicts reverses progress made in these areas and exacerbates the environmental challenges, further delaying the achievement of SDG 15 targets. For instance, Target 15.1, which aims to ensure the conservation, restoration, and sustainable use of terrestrial and freshwater ecosystems, is often undermined as conflicts damage natural habitats, disrupt water resources, and lead to the overexploitation of limited environmental assets. Similarly, Target 15.3, focused on combating desertification and achieving land degradation neutrality, is severely affected when military activities cause soil erosion, compaction, and contamination, making land unproductive and difficult to restore. Additionally, Target 15.5, which emphasizes the urgent need to reduce habitat destruction and prevent species extinction, is directly impacted as wars and conflicts destroy biodiversity hotspots, displace wildlife, and reduce the capacity for effective conservation efforts (Fontaine 2024; Meaza et al. 2024; Rist et al. 2024).

Therefore, achieving SDG 15 amidst wars and conflicts requires global cooperation, conflict-sensitive environmental restoration, and adherence to international laws that prevent environmental harm. Continued monitoring and investment in post-conflict environmental recovery are essential to mitigating long-term ecological and socioeconomic impacts.

13.3 Environmental Impacts in Gaza Due to the War

The recent war on Gaza has caused catastrophic environmental and agricultural devastation, exacerbating food insecurity and undermining progress toward SDG 15. Agricultural systems have suffered extensive damage, with most of arable land destroyed and orchard areas reduced drastically, crippling local food production. By early 2024, over 25,000 tons of explosives—equivalent to two nuclear bombs—had been dropped on Gaza (OHCHR 2024). Soil degradation from military activities, such as compaction and contamination by heavy metals, has further compromised land fertility and ecosystem health. Biodiversity has declined sharply, with the loss of forests and vegetation due to deforestation and destructive warfare, endangering wildlife and disrupting ecological balance. Additionally, water scarcity has intensified due to the targeting of water infrastructure and blockade restrictions, leaving millions without clean water. The collapse of waste management systems has worsened the crisis, with untreated sewage and waste polluting water bodies and agricultural land (Fig. 13.1).

13.3.1 Agricultural Devastation: Impact on Food Production and Food Security

The war on Gaza has exacerbated the chronic food insecurity in the region, undermining progress toward achieving SDG 15. The war has disrupted agricultural systems, biodiversity, and environmental stability, critically impacting targets such as sustainable ecosystem management (Target 15.1), combating desertification (Target 15.3), and halting biodiversity loss (Target 15.5). Other conflicts, such as the war in Ukraine, have similarly demonstrated the profound impact of armed violence on sustainability and global food chains. These disruptions create cascading effects on international food systems, driving up food prices, exacerbating hunger, and intensifying economic instability worldwide (Jagtap et al. 2022; Ben Hassen and El Bilali 2022). Small-scale agriculture and livestock farming, which are essential for sustaining subsistence economies, are particularly vulnerable to the devastating effects of wars and armed conflicts (Kemmerling et al. 2022).

In Gaza, the agricultural sector, which occupies 42% of its 36,410 ha, has faced catastrophic losses. Orchards and other trees in Gaza, representing 59% of arable land, have been reduced from 2115 ha to just 293 ha between October 2023 and September 2024, according to FAO and UNOSAT assessments. Agricultural infrastructure, including domestic barns, broiler farms, and sheep farms, has also suffered extensive damage,

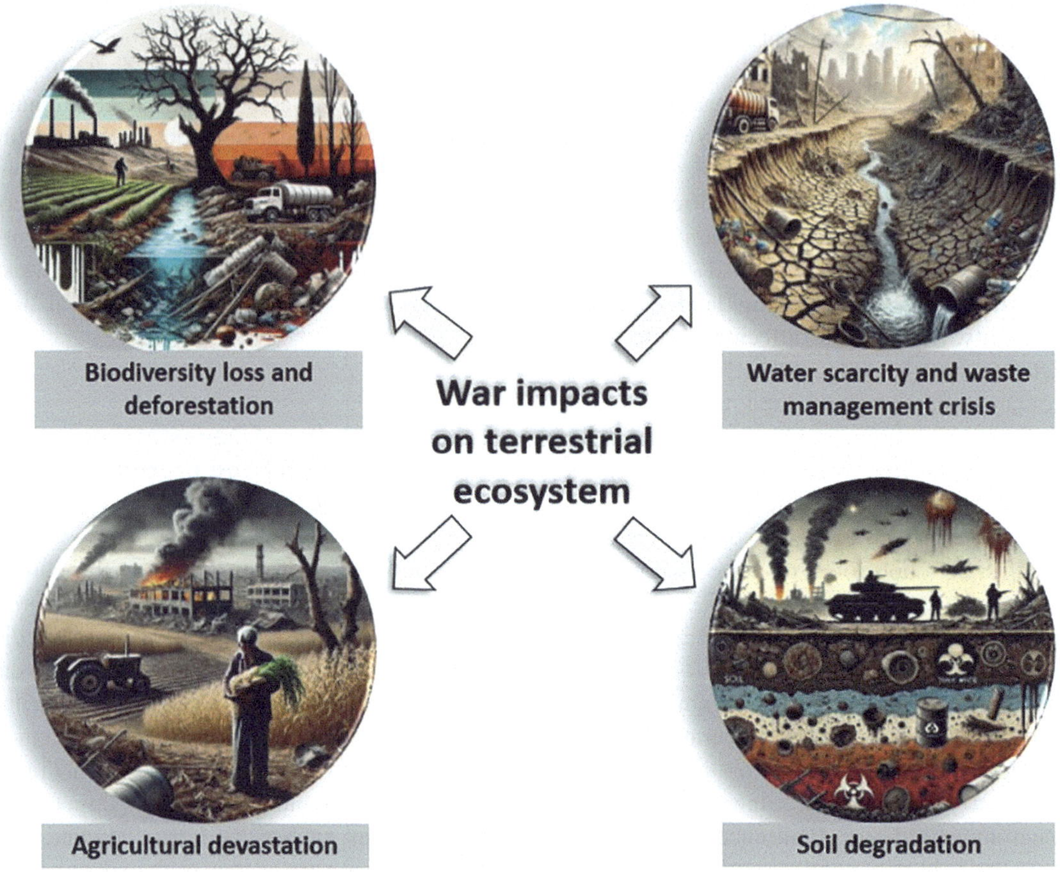

Fig. 13.1 Environmental impacts of the recent war on Gaza

crippling local food production capacity. Nearly 70% of agricultural land, which previously provided up to one-third of daily food needs, has been damaged or destroyed since the intensification of hostilities (FAO/UN 2024; UN 2024).

This current catastrophic situation in Gaza underscores the urgent need for international cooperation to address food insecurity and related issues, including food price volatility, in Gaza. Rebuilding the agricultural sector and stabilizing food prices are essential not only for mitigating the immediate crisis but also for ensuring long-term food security and achieving SDG 15. By focusing on sustainable practices and systemic reforms, it is possible to build a more resilient food economy for Gaza, reducing vulnerability to current environmental challenges and future disruptions.

13.3.2 Soil Degradation and Pollution

War exerts significant and often lasting impacts on soil health, contributing to physical degradation, and chemical and biological contamination that undermine land and soil productivity, disrupt the balance of ecosystems, and pose long-term risks to human and animal health. Military actions such as the deployment of heavy machinery, the destruction of infrastructure, and the use of explosive ordnance inflict severe and lasting physical damage on the soil. The movement of tanks, trucks, and other heavy machinery compacts soil layers, reducing porosity and impeding root growth and water infiltration. This compaction restricts essential soil processes, such as gas exchange and microbial activity, which are necessary for maintaining soil fertility and resilience. Furthermore, the creation of trenches, bomb craters, and destroyed infrastructure accelerates erosion, leading to the rapid loss of nutrient-rich topsoil (Broomandi et al. 2020; Filho et al. 2024; Leal Filho et al. 2024; Hryhorczuk et al. 2024).

In addition to physical degradation, war often introduces harmful chemicals into the soil. Explosives, heavy metals, and toxic residues from munitions, such as lead, mercury, and nitrates, seep into the soil, resulting in chemical pollution that compromises soil quality. These contaminants can persist in the environment, adversely affecting soil biology, reducing crop yields, and entering food chains, where they pose health risks to humans and animals through the bioaccumulation process. Some heavy metals, found naturally in the earth, are essential for various physiological and biochemical activities in the body. However, in high doses, they can be toxic and cause acute and chronic harmful impacts. Lead, in particular, is highly toxic, and prolonged exposure can severely affect human health, impacting the nervous system, gastrointestinal tract, kidneys, and posing a fatal risk, especially to children. Copper at high levels can damage the liver and kidneys and cause anemia, while nickel is carcinogenic. However, the toxicity of these metals in soil varies with factors like soil texture, organic matter content, and pH levels. Health risks from soil contamination depend on exposure route, intensity, and duration, making it challenging to define contamination levels that guarantee immediate health effects (Engwa et al. 2019; Mitra et al. 2022; Solokha et al. 2024).

Studies conducted on multiple war sites highlight the prevalence of heavy metal contamination in conflict zones. For instance, the shelling zones of the Russian–Ukrainian war in the village of Mala Rohan were found to contain significant amounts of cadmium, cobalt, chromium, copper, iron, nickel, and lead (Solokha et al. 2023). Similar findings have been reported in other areas exposed to military activities, such as during the Gulf War of 1991 in Saudi Arabia (Hussain and Gondal 2008), in Kuwait (Sadiq et al. 1992), and in areas affected by military activities during WWI in Pas-de-Calais, France (Williams and Rintoul-Hynes 2022). The development of modern weapons of mass destruction exacerbates these impacts, spreading toxins deeper into soil layers, killing beneficial microorganisms, reducing soil organic matter, and disrupting nutrient cycles essential for plant growth. Regions affected by radiation from nuclear weaponry or damaged nuclear facilities face even graver consequences, as radioactive contamination can render land unsuitable for agriculture and habitation for decades, further compounding ecological and socioeconomic challenges. The threat of unexploded ordnance, including landmines, adds another dimension to soil degradation in conflict and post-conflict zones. Landmines contaminate large areas with explosives and heavy metals that slowly leach into the soil. Even when cleared, mined lands often require extensive monitoring and remediation to restore soil health. This restricted access to arable land leads to the overexploitation of safer, often marginal lands, accelerating soil degradation through unsustainable farming practices and overuse (Griffiths and Rubaii 2024; Rodríguez-Seijo et al. 2024).

The war on Gaza has exacerbated soil degradation and land pollution, presenting unprecedented environmental challenges. An estimated 50 million tons of debris have been generated during this war (ReliefWeb 2024), significantly contaminating the land. The destruction is unparalleled in scale, with debris containing asbestos, unexploded ordnance, and other hazardous substances, which leach into the soil and water systems. Gaza's unique geographical and political context amplifies these issues due to its limited land area, high population density, and fragile environmental governance structures. The collapse of waste management systems, combined with the contamination from munitions and destroyed infrastructure, disrupts soil ecosystems and accelerates land degradation (Elkahlout 2024; Qumsiyeh 2024; UNEP 2024). These issues particularly hinder progress on SDG 15.3, which targets combating desertification, restoring degraded land and soil, and achieving a land degradation-neutral world, as well as SDG 15.1, which emphasizes the conservation and sustainable use of terrestrial ecosystems.

13.3.3 Biodiversity Loss and Deforestation

Severe biodiversity loss is a significant obstacle to achieving SDG 15. Plant and animal species are essential to our survival, contributing to crop pollination, the provision of medicines and economic activity. However, the rate of species extinction in the twenty-first century is currently tens to hundreds of times higher than the natural rate over the last ten million years, driven by habitat loss, invasive species, overexploitation of natural resources, illegal wildlife trade, pollution, and climate change (Isbell et al. 2023; Kumar and Isha 2024; Rist et al. 2024). This situation is further complicated by ongoing geopolitical conflicts, which not only create humanitarian crises but also result in the destruction of natural habitats (Lawrence et al. 2015; Fontaine 2024).

War and related military activities have both direct and indirect impacts on biodiversity. Military operations cause environmental pollution and habitat destruction through deforestation and contamination, putting wildlife in danger and disrupting their natural behaviors (Fontaine 2024; Rist et al. 2024). Many of the warfare products, such as chemical agents with highly toxic and noxious properties, intended for human targets can also have negative impacts on other species. Bombings lead to significant ecological disruption, exacerbating land degradation and biodiversity loss resulting from both acute and chronic effects in ecosystems. The degree to which warfare can impact on ecosystems and its populations depends on the nature of the disturbance, the sensitivity of the biological system, and the timescale of the impacts (Lawrence et al. 2015; Fontaine 2024).

Repeated conflicts in Palestine, particularly in Gaza, have profoundly undermined the "Life on Land" goal of the 2030 Agenda by intensifying environmental degradation and fostering instability through the widespread destruction of ecosystems. Despite its small size, Palestine has diverse climates and environments that have given rise to a rich biodiversity of fauna and flora. This richness is due to the geographical position of Palestine, with its varied topography, geology, and ecoregions. The total number of living organisms in Palestine has been estimated at around 51,000 species, which constitutes 3% of global biodiversity. There are an estimated 30,848 animal species, consisting of an estimated 30,000 invertebrates, 373 birds, 297 fish, 92 mammals, 81 reptiles, and 5 amphibians. However, biodiversity has dramatically declined over the last years, with 24 species now facing extinction. This decline is driven by the expansion of human activities, the impacts of climate change, the establishment of Israeli settlements, and the construction of the Wall (UN and State of Palestine 2018; Abd Rabou 2019; Palestinian Central Bureau of Statistics 2020; Qumsiyeh and Abusarhan 2021; Ali-Shtayeh et al. 2022). According to FAO's rapid assessments, the war has devastated Gaza's livestock sector, with nearly 95% of cattle, amounting to almost 15,000 animals, perishing, and nearly all calves being slaughtered. The surviving livestock population is alarmingly low, with fewer than 25,000 sheep (approximately 43%) and just 3000 goats (around 37%) still alive. The poultry sector has suffered even more catastrophic losses, leaving only 34,000 birds—merely 1% of the original population (FAO 2024).

Palestine, prior to the recent war, had a total forest area of 101.4 km^2, of which 98.2 km^2 was in the West Bank and 3.2 km^2 in Gaza, significantly reduced from 350 km^2 in 1970 due to illegal Israeli settlements and logging practices. The Israeli occupation has been a major obstacle to conservation efforts, with illegal land expropriation for nature reserves, which comprised 13% of the West Bank, and over 1.5 million Palestinian olive trees having been uprooted. Due to Israeli land confiscation and the construction of the Wall, Palestinian green lands decreased from 4.3% to less than 1.5% between 1998 and 2004, causing severe ecological impacts (UN and State of Palestine 2018; Palestinian Central Bureau of Statistics 2020).

The war in Gaza has precipitated this environmental crisis, threatening natural resources and long-term sustainability. Destruction of land, coupled with the massive use of munitions, has led to contamination of soil and groundwater with hazardous materials, posing immediate dangers to human health and jeopardizing biodiversity and agricultural productivity. These ecological impacts are recognized as irreversible, exacerbating the vulnerability of Gaza to climate change and creating a toxic legacy that threatens human health and the ecosystems of future generations (Evereux 2024; FAO/UN 2024; Fontaine 2024; Hassoun et al. 2024). Similar patterns of environmental degradation are evident in other conflicts, such as the war in Ukraine. The term "ecocide" is used to describe the widespread environmental devastation caused by this war, which has caused significant damage using chemical weapons, including white phosphorus and other hazardous substances that cause air, soil, and water pollution, as well as loss of biodiversity of flora and fauna (Gardashuk 2022; Sousa et al. 2022; El Bilali and Ben Hassen 2024; Fontaine 2024). To combat this crisis, it is essential to preserve and restore natural habitats, promote sustainable land use practices, and implement biodiversity conservation measures to restore ecological balance and protect endangered species. This approach not only mitigates the environmental damage caused by deforestation, habitat destruction, and soil degradation but also highlights the interconnectedness of biodiversity, conflict, and peace (Shevchuk et al. 2023; Weir et al. 2024).

Mass removal of trees, deforestation, and destruction of vegetation due to wars and conflicts contribute significantly to desertification and environmental degradation. These activities destabilize soil, increase erosion rates, and reduce biodiversity. Vegetative cover, including trees and root systems, plays a critical role in anchoring soil, retaining moisture, and adding organic material through decomposition. The loss of this cover depletes organic matter in the soil, making it more vulnerable to erosion, while also disrupting essential ecosystem services required for maintaining soil fertility and ecological balance (Aldridge Mugadza 2022; Meaza et al. 2024; Qu et al. 2024). An example of this phenomenon can be observed during the conflict in Syria between 2011 and 2017. Satellite observations documented extensive environmental damage, including deforestation and reduced vegetative cover, which led to significant soil degradation and loss of biodiversity. These findings demonstrate the profound and lasting ecological impacts of war, where vegetation removal disrupts ecosystems and diminishes the availability of arable land (Abdo 2018).

In Gaza, the recent war has intensified these issues, with widespread destruction of vegetation, agricultural lands, and forests. Satellite imagery indicates that Israel has removed large numbers of trees, primarily in orchards and farmland, reportedly for military purposes. Moreover, acute fuel shortages caused by the war have compelled civilians to cut down trees for firewood, further depleting limited natural resources. In addition, extensive use of munitions and intentional clearing of land have stripped the area of vegetation, accelerating soil erosion and desertification (Ahmed et al. 2024; bellingcat 2024). In recent months, the deliberate destruction of old olive trees and other vegetation has been widely documented and shared across social media platforms. Reports indicate that over 1 million olive trees have been uprooted since October 2023 (OCHA 2024a), accompanied by the destruction of forested areas and crop plantations. This devastation exacerbates food insecurity for the people of Gaza, depriving them of critical agricultural resources, while simultaneously threatening the survival of native fauna and flora.

13.3.4 Water Scarcity and Waste Management Crisis

The war on Gaza has severely intensified water scarcity, with the deliberate targeting of water infrastructure and restrictions on access serving as a form of weaponization. These actions have disrupted essential water supply systems, leaving the population without reliable access to clean water and undermining progress toward critical targets under SDG 15. The deliberate destruction of water resources, combined with the collapse of waste management systems, has created a multifaceted humanitarian and environmental crisis, threatening the fragile ecosystems in Gaza.

The deliberate restriction of water access in Gaza, through the destruction of pipelines, desalination plants, and other critical infrastructure, has weaponized water against the civilian population. The blockade further exacerbates the crisis by severely limiting the import of drinking water and essential materials needed to repair damaged infrastructure or develop alternative water sources. These actions have left millions of people without reliable access to clean water, forcing dependence on contaminated or insufficient supplies to meet even basic needs. Water weaponization has become a tool of collective punishment, undermining human dignity and survival. Supply cuts, blockade, and infrastructure destruction have transformed water scarcity from a natural resource issue into a deliberate strategy of deprivation. Families are often left with only a fraction of the water required for drinking, hygiene, and cooking, resulting in severe public health consequences and widespread suffering (ESCWA 2023; Abdul Samad et al. 2024; Cole 2024).

The crisis extends beyond human impact. The lack of water availability disrupts agricultural activity, exacerbating food insecurity and increasing reliance on external aid. Reduced water supply and contamination of groundwater further destabilize the environmental balance in this region, contributing to the degradation of ecosystems that depend on freshwater availability. Wetlands, aquifers, and coastal systems face collapse, undermining their ability to sustain biodiversity and support livelihoods. The use of water as a weapon not only violates basic human rights but also escalates environmental and humanitarian crises. The denial of access to this essential resource deepens vulnerabilities, leaving civilians trapped in a cycle of scarcity, disease, and ecological devastation (Gleick 2019; Schillinger et al. 2020).

In parallel, the collapse of waste management systems has significantly worsened the situation, with untreated sewage and solid waste accumulating in open areas. On October 06, 2024, only one desalination plant out of three in Gaza, which together produce just 7% of the water supply, was operational. Even this plant functioned at reduced capacity due to the blockade on electricity and fuel. Frequent power outages and fuel shortages have critically impeded waste treatment operations, creating a chain reaction of environmental degradation. All five wastewater treatment plants in Gaza and the majority of the 65 sewage pumping stations have been forced to shut down, even at a very early stage of this war, resulting in untreated sewage being discharged directly into the sea and solid waste piling up in the streets. Facilities for waste treatment, which depend on a steady supply of energy, are often unable to operate at full capacity or are completely shut down under these circumstances. This has caused untreated wastewater and solid waste to accumulate at alarming rates. The disruption directly pollutes nearby water bodies and contributes to wider environmental crises. Without functioning infrastructure, sewage is discharged into other water sources and coastal waters without filtration or detoxification. Waste seeps into groundwater and surface water supplies, introducing hazardous substances such as heavy metals, pathogens, and organic pollutants. Therefore, water sources critical for human consumption and agriculture become dangerously unsafe (Khalidi 2023; Dardona et al. 2024; UNDP/PAPP 2024; UNEP 2024).

The consequences extend far beyond human health. Contaminated water infiltrates agricultural land, degrades soil quality, and reduces crop yields, creating a threat to food security. The inability to properly manage and treat waste perpetuates a cycle of environmental degradation that disrupts ecosystems, harms public health, and undermines progress toward sustainable development.

13.4 Mitigation Measures and Recommendations for Achieving SDG 15 in Gaza

Achieving the "Life on Land" goal in Gaza faces immense challenges due to a combination of severe environmental degradation, political instability, scarce resources, and the devastating impact of the recent war. The key targets of SDG 15 are nearly unattainable under the conditions created by occupying forces wielding overwhelming power. Between October 2023 and November 2024, over 85,000 tons of bombs were dropped on Gaza (Middle East Monitor 2024), an amount exceeding the explosives used during World War II. This strategy of large-scale carpet bombing has not only destroyed vital ecosystems but also exacerbated the humanitarian crisis. If the international political landscape shifts toward recognizing Palestinians as entitled to human rights and ensuring their protection through an international mandate, there is hope for progress.

The war on Gaza has severe implications for SDG 15, directly damaging terrestrial ecosystems and intensifying socioeconomic vulnerabilities. The destruction of agricultural lands, infrastructure, and habitats undermines targets such as ecosystem conservation (Target 15.1), combating desertification (Target 15.3), and halting biodiversity loss (Target 15.5). An estimated 50 million tons of hazardous debris contaminate the soil, making large areas unsuitable for agriculture or habitation, and accelerating land degradation (OCHA 2024b; UNEP 2024).

This environmental damage worsens food insecurity in a region heavily dependent on international aid, reduces natural resource availability, and limits economic opportunities tied to ecosystem services like farming and fishing (Pereira et al. 2022; Shevchuk et al. 2023; ESCWA/UNDP 2024). Socially, marginalized groups, especially women, bear the brunt of resource scarcity and environmental degradation (Kotsis 2024). Displacement from habitat destruction exacerbates overcrowding, health risks, and unemployment, further entrenching poverty and delaying recovery (UNCTAD 2023, 2024; ESCWA/UNDP 2024). This cycle of environmental degradation and socioeconomic instability undermines sustainable development and recovery efforts. The following section outlines key mitigation strategies and recommendations to help achieve SDG 15 in Gaza, should such transformative changes occur.

13.4.1 Support a National Plan for Reforestation and Afforestation

Implementing large-scale tree planting, forest restoration programs, and farm re-establishment initiatives is critical to restoring green spaces, improving air quality, and enhancing temperature regulation in Gaza. These efforts should prioritize reforesting degraded areas and creating new green spaces to combat desertification and promote biodiversity. Given Gaza's arid climate and limited water resources, emphasis should be placed on drought-resistant species that can thrive under these challenging conditions. Collaboration with local and international NGOs will be essential to access the expertise, funding, and logistical support needed for these projects. In particular, planting olive trees, which are culturally significant and well-adapted to the region, can help to replace those intentionally destroyed by occupying forces, restore the degraded lands, and provide a sustainable livelihood.

13.4.2 Implement Strategy for Soil Conservation and Land Rehabilitation

Implementing programs to detoxify the land from hazardous chemical weapons used during assaults on Gaza, remove mines and booby traps installed by occupying forces, and adopt soil conservation techniques is essential to prevent further land degradation. These efforts will help restore the environment, improve public safety, and ensure sustainable land use for agriculture and development. To achieve this, collaboration with international and national organizations is vital. These partnerships can support Gaza authorities in developing comprehensive demining strategies and policies, conducting mine clearance operations, and building local capacity through training and education.

Providing risk education and advocating for the affected communities will ensure the safe and effective use of cleared land for rebuilding efforts. Soil conservation techniques must also be prioritized to stabilize the land and combat erosion. Practices such as constructing terraces and check dams can help prevent soil erosion while mulching and planting windbreaks are effective in areas vulnerable to desertification. These methods restore soil fertility and contribute to long-term agricultural resilience and environmental sustainability.

Mitigating the impact of war on Gaza and building resilience requires a holistic approach to land rehabilitation that addresses both environmental and socioeconomic dimensions. Table 13.1 outlines key methodologies for prioritizing rehabilitation efforts, detailing their objectives, expected outputs, strengths, limitations, and practical examples.

Selecting an appropriate land rehabilitation method for post-war Gaza depends on factors such as scope, speed, stakeholder involvement, data requirements, and focus areas (Fig. 13.2).

The Rapid Damage and Needs Assessment (RDNA) is ideal for urgent crises due to its efficiency in short-term recovery. For sustainable, long-term planning, Environmental Impact Assessment (EIA) offers in-depth, environment-focused analysis but requires more time and data. Community-based prioritization ensures alignment with social and cultural priorities through stakeholder engagement, despite relying on qualitative data. Cost-Benefit Analysis (CBA) evaluates economic efficiency but may overlook non-economic factors. Multi-Criteria Decision Making (MCDM) balances environmental, social, and economic objectives, making it suitable for complex scenarios like Gaza's, though it demands extensive data and collaboration. Combining these methods often yields a comprehensive strategy addressing both immediate needs and long-term goals.

13.4.3 Design Programs for Water Management and Conservation

The 2017 Amnesty International report highlighted how Israeli military authorities have consolidated control over all water resources and infrastructure in the occupied Palestinian territories, even before the recent war (Amnesty International 2018). Military Order 158 mandates that Palestinians obtain Israeli military permits for any new water installations. Since its enforcement, Palestinians have faced "devastating" consequences, as described by Amnesty, including restricted access to water resources and the inability to drill new wells, deepen existing ones, or install pumps. Access to the Jordan River and freshwater springs is also denied. Israel controls rainwater collection throughout much of the West Bank, with Palestinian rainwater cisterns frequently destroyed by the Israeli army. Additionally, large areas of the West Bank are off-limits to Palestinians, designated as "closed military areas" due to their proximity to Israeli settlements, settler roads, military training zones, or protected nature reserves. Meanwhile, Israeli settlers living mere meters away from Palestinians face no such restrictions, and they enjoy abundant water access, enabling irrigated farmlands and swimming pools, starkly contrasting the severe water shortages experienced by their Palestinian neighbors. Due to water restrictions imposed on Gazans, Oxfam concluded that the Israeli Government indiscriminately weaponized water (Abdul Samad et al. 2024).

Therefore, improved water use efficiency and management practices need to be implemented as a strategy to eliminate the unfair water control imposed on the population in Gaza and occupied territories. As the conflict in Gaza remains unresolved at the time of writing this chapter, a complete understanding of the impact caused by military actions on groundwater resources is yet to be fully established. Concerns have also been raised about the potential contamination and destruction of water resources, further exacerbating the humanitarian crisis and long-term challenges for the population. In July 2024 alone, the occupying military forces blew up more than 30 water wells in Gaza, according to Reuters (Reuters 2024). Sustainable water management practices, such as rainwater harvesting, water recycling, and the use of efficient irrigation systems (e.g., drip irrigation) should be promoted for agriculture and domestic use. Implementing

Table 13.1 Key methodologies for prioritizing sustainable land rehabilitation

Approach	Objective	Methodology	Output	Strength	Limitations
Rapid damage and needs assessment (RDNA)	Provides a quick evaluation of damage and the urgent needs for immediate recovery.	–focuses on collecting and analyzing data rapidly to guide emergency response and prioritize urgent rehabilitation actions. –often involves a qualitative assessment, with basic quantitative estimates of damage.	A high-level assessment used for mobilizing resources and guiding immediate, short-term rehabilitation priorities.	Fast and efficient for urgent situations, suitable for early recovery planning	Lacks in-depth analysis; not comprehensive for long-term planning or sustainable solutions.
Environmental impact assessment (EIA)	Assesses the potential environmental consequences of rehabilitation projects before they are implemented.	–involves a detailed evaluation of how proposed projects might affect the environment, including biodiversity, water resources, and air quality. –includes baseline studies, public consultations, and alternatives analysis.	Comprehensive reports that guide the planning and design of environmentally sustainable projects.	Ensures environmental considerations are integrated into project planning, focusing on minimizing negative impacts.	Time-consuming and resource-intensive; may not be suitable for urgent or large-scale rapid response efforts.
Community-based prioritization (CBP)	Involves local communities in decision-making to ensure projects align with their needs and values.	–uses participatory approaches like workshops, surveys, and focus groups to gather input from residents. –communities identify priorities based on social, cultural, and economic considerations.	A list of projects prioritized based on community needs and preferences.	Builds local ownership, ensures social and cultural relevance, and enhances project acceptance.	Subjective and may lack technical rigor; could lead to bias if not well-facilitated or if diverse community voices are not adequately represented.
Cost-benefit analysis (CBA)	Evaluates the economic efficiency of projects by comparing the expected costs with the anticipated benefits.	–involves assigning monetary values to both the costs (e.g., project expenses) and the benefits (e.g., economic gains from restored land). –calculates net present value (NPV), benefit-cost ratio (BCR), or internal rate of return (IRR) to determine project viability.	A financial assessment that ranks projects based on economic returns.	Quantitative and objective; helps in making decisions where budget constraints are critical.	Difficult to monetize all environmental and social benefits, especially for intangible or long-term impacts; may not fully capture non-economic priorities.
Multicriteria decision analysis (MCDA)	Integrates multiple criteria, including environmental, social, economic, and technical factors, to evaluate and rank projects.	–uses structured frameworks (like AHP, TOPSIS, or SAW) to assign weights to criteria and score alternatives. –balances trade-offs between conflicting objectives, using both quantitative and qualitative data.	A comprehensive ranking of projects based on overall performance across all criteria.	Flexible and comprehensive; considers a wide range of factors, suitable for complex and multidimensional problems.	Requires extensive data and stakeholder input; more complex to implement and analyze.

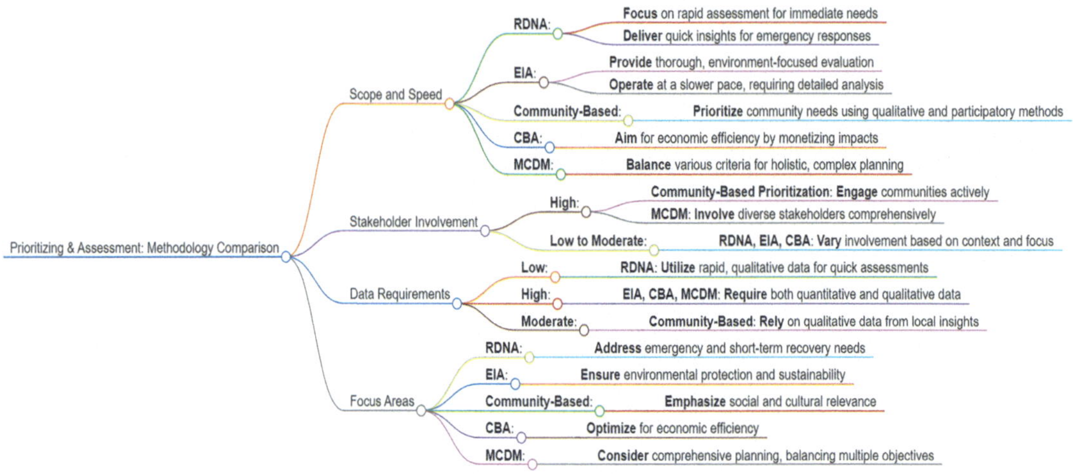

Fig. 13.2 Prioritizing and assessment methodologies comparison, emphasizing key attributes, including scope, speed, stakeholder involvement, data requirements, and focus areas

community-based rainwater harvesting systems can provide a buffer against water scarcity. Promoting drought-resistant crops and low-water-use irrigation systems in agriculture could significantly reduce pressure on water resources.

13.4.4 Planning Biodiversity Protection and Habitat Restoration

The environmental impact of intense military operations is profound and far-reaching. Over 110,000 bombing sorties were carried out in the first 6 weeks of the assault on Gaza, leading to the destruction of more than 70% of residential structures (Qumsiyeh 2024). Half of the aerial bombs used were indiscriminate, causing extensive collateral damage. Similarly, the deployment of around 2500 tanks and 5000 armored personnel carriers has exacerbated the destruction. Throughout the ground operations, the military contributed to an adapted scorched-earth policy as a military strategy, which significantly led to environmental degradation and the killing of native plants, birds, and wildlife. The release of toxic fumes, fuel, heavy metals like lead into the soil and water, and the use of white phosphorus munition, all contaminate ecosystems and harm biodiversity. Protecting and restoring biodiversity, particularly in coastal and agricultural ecosystems, is a vital goal for rebuilding Gaza. This effort involves restoring natural habitats such as wetlands, coastal areas, and forests, which play a crucial role in supporting biodiversity. Establishing protected areas or nature reserves and raising awareness within local communities about the importance of biodiversity conservation are key steps in this process. For instance, initiatives to safeguard coastal ecosystems in Gaza, such as creating marine reserves and implementing mangrove restoration projects, can enhance biodiversity while also providing critical protection against coastal erosion. These measures contribute to both ecological resilience and long-term environmental sustainability.

13.4.5 Implementation of Sustainable Agriculture and Agroforestry Strategy

The extent of soil contamination, both intentional and unintentional, resulting from actions by occupying forces remains unclear. Testing agricultural land for illegal and long-term contaminants is crucial to identify potential health risks and ensure the safety of farming practices. Promoting sustainable agriculture and agroforestry practices is essential to enhance soil health and reduce pressure on natural ecosystems. Palestinians, known for their resourcefulness and deep agricultural heritage, have long utilized techniques such as agroecology, organic farming, and integrated pest management to minimize environmental impacts. There is significant potential to further develop agroforestry systems in the region. These systems, which integrate trees with crops, can improve biodiversity, enhance soil fertility, and increase water retention. Expanding such practices can strengthen the resilience of agricultural systems while contributing to environmental sustainability in Gaza.

13.4.6 Implementation of Waste Management and Recycling Strategies

A comprehensive strategic plan for waste management and recycling is essential to reduce land contamination and support recovery efforts in Gaza. This plan should focus on implementing solid waste recycling programs and composting initiatives to minimize reliance on landfills and promote sustainable waste disposal. Gaza could benefit from the expertise of nations such as Vietnam, Korea, and Japan, which have successfully recovered from similar challenges and developed efficient waste management systems after experiencing significant devastation. Community involvement is critical to the success of such initiatives. Engaging local populations in sorting and recycling waste can help mitigate environmental damage while fostering a sense of stewardship over the land. Community-based composting programs, in particular, offer a sustainable solution for reducing organic waste, which would otherwise contribute to pollution and soil degradation. By converting organic waste into nutrient-rich compost, these programs can enhance soil health and contribute to long-term environmental restoration and agricultural resilience.

13.4.7 International Collaboration and Funding

International collaboration and funding are essential to scaling up environmental restoration and sustainable development initiatives in Gaza. Support from organizations such as the United Nations, the World Bank, and various NGOs can provide technical expertise and financial resources to implement impactful environmental projects. For example, partnerships with international environmental organizations can help fund reforestation efforts, water conservation initiatives, and sustainable land management programs to combat land degradation and restore ecosystems. However, the repeated destruction of homes and infrastructure in Gaza, occurring in cycles of approximately 3 years as part of the so-called "lawn mowing" strategy, poses a significant challenge. These cycles of devastation often undo progress made through international funding and rebuilding efforts, creating a frustrating and unsustainable pattern. The global community bears a responsibility not only to support the restoration of Gaza but also to hold accountable those responsible for the destruction. It is imperative that international efforts address both the immediate needs of environmental restoration and the systemic issues driving recurring devastation, ensuring long-term progress and stability for the people and ecosystems of Gaza.

13.5 Conclusion

The destruction caused by the recent war on Gaza represents a monumental challenge to achieving SDG 15, "Life on Land." Key targets, including the sustainable management of ecosystems, combating desertification, and halting biodiversity loss, are far from attainable under current conditions. Military actions have devastated agricultural land, forests, water resources, and biodiversity, leaving a toxic legacy that will require decades to remediate. This environmental degradation has not only caused immediate humanitarian crises but also long-term ecological damage that threatens the resilience and sustainability of ecosystems in the Gaza Strip.

In response to these challenges, this chapter underscores the urgent need for international collaboration, sustainable restoration efforts, and conflict-sensitive environmental policies. Immediate measures, such as fair access to water, food, waste services and other essential resources, the repair of critical infrastructure, decontamination of land, reforestation, and community-based conservation programs, must be prioritized. International stakeholders, including governments, NGOs, and environmental organizations, have a responsibility to provide technical and financial support to rebuild ecosystems in Gaza and hold accountable those responsible for environmental destruction. Ultimately, addressing the environmental crises in Gaza requires not only technical solutions but also a commitment to peace and justice. Recognizing the deep connection between ecological health and human wellbeing, achieving SDG 15 in Gaza depends on establishing stability, ensuring fair access to resources, and mobilizing global efforts to restore life on land in this war-affected region.

References

Abd Rabou AFN (2019) The mammalian, reptilian and amphibian fauna of Al-Mawasi ecosystem, South-Western Gaza strip-Palestine. Res Technol Open Access J. https://doi.org/10.19080/ARTOAJ.2019.23.556233

Abdo HG (2018) Impacts of war in Syria on vegetation dynamics and erosion risks in Safita area, Tartous, Syria. Reg Environ Chang 18:1707–1719. https://doi.org/10.1007/s10113-018-1280-3

Abdul Samad L, Butcher M, Khalidi B (2024) Water war crimes: how Israel has weaponised water in its military campaign in Gaza. Oxfam Int 6

Ahmed K, Gayle D, Mousa A (2024) ‘Ecocide in Gaza’: does scale of environmental destruction amount to a war crime? Guard

Al Mezan (2024) Ecocide: Israel’s deliberate and systematic environmental destruction in Gaza

Aldridge Mugadza A (2022) The disastrous effects of deforestation and Forest degradation in the climate vulnerability era. CIFILE J Int Law 3:12–40. https://doi.org/10.30489/CIFJ.2022.324393.1049

Ali-Shtayeh MS, Jamous RM, Abuzaitoun SY (2022) Analysis of floristic composition and species diversity of vascular plants native to the State of Palestine (West Bank and Gaza strip). Biodivers Data J 10:e80427. https://doi.org/10.3897/BDJ.10.e80427

Amnesty International (2018) Amnesty International Report 2017/18: The state of the world’s human rights

bellingcat (2024) Gaza’s trees disappear, showing a humanitarian crisis. https://www.bellingcat.com/news/2024/03/15/gazas-trees-disappear-showing-a-humanitarian-crisis/. Accessed 28 Dec 2024

Ben Hassen T, El Bilali H (2022) Impacts of the Russia-Ukraine war on global food security: towards more sustainable and resilient food systems? Food Secur 11:2301. https://doi.org/10.3390/FOODS11152301

Broomandi P, Guney M, Kim JR, Karaca F (2020) Soil contamination in areas impacted by military activities: a critical review

Buheji M, Al-Muhannadi K (2023) Mitigating risks of environmental impacts on Gaza - review of precautions & solutions post (2023 war). Int J Adv Res Eng Technol 14:15–47

Cole J (2024) Weaponizing a water crisis: the destruction of water supplies in Gaza. New Lines Inst:1–7

Dardona Z, Amane M, Dardona A, Boussaa S (2024) Health and environmental impacts of Gaza conflict (2023–2024): a review. One Heal Bull. https://doi.org/10.4103/ohbl.ohbl

El Bilali H, Ben Hassen T (2024) Disrupted harvests: how Ukraine – Russia war influences global food systems—a systematic review. Policy Stud 1–26. https://doi.org/10.1080/01442872.2024.2329587

Elkahlout G (2024) Sisyphus but the Boulder gets a little heavier every time. In: Elkahlout G (ed) Gaza’s cycle of destruction and rebuilding. Palgrave Macmillan, pp 1–16

Engwa GA, Ferdinand PU et al (2019) Mechanism and health effects of heavy metal toxicity in humans. In: Karcioglu O, Arslan B (eds) Poisoning in the modern World—new tricks for an old dog? IntechOpen, p 13

ESCWA (2023) War on Gaza: weaponizing access to water, energy and food

ESCWA/UNDP (2024) Gaza war: expected socioeconomic impacts on the State of Palestine—October 2024

Evereux VD (2024) Some insights on application of the international law to the damaged environment during the war in Gaza. ACTA Univ CAROLINAE—Iurid 4:177–194. https://doi.org/10.14712/23366478.2024.167

FAO (2024) Protecting livelihoods and lives in Gaza. https://www.fao.org/newsroom/story/protecting-livelihoods-and-lives-in-gaza/en. Accessed 10 Jan 2025

FAO/UN (2024) Agricultural Damage Assessment in the Gaza Strip from October 7th 2023 to September 1st 2024

Farhat T, Ibrahim S, Abdul-Sater Z, Abu-Sittah G (2023) Responding to the humanitarian crisis in Gaza: damned if you do … damned if you don’t! Ann Glob Heal 89:1–6. https://doi.org/10.5334/AOGH.3975

Filho WL, Fedoruk M, Paulino Pires Eustachio JH et al (2024) The environment as the first victim: the impacts of the war on the preservation areas in Ukraine. J Environ Manag 364:121399. https://doi.org/10.1016/j.jenvman.2024.121399

Fontaine R (2024) Strengthening environmental resilience in conflict zones: analysis of UNEA-6 resolution and the PERAC principles

Gardashuk T (2022) Is Russian aggression in Ukraine ecocide? Envigogika 17:1–6. https://doi.org/10.14712/18023061.642

Gleick PH (2019) Water as a weapon and casualty of conflict: freshwater and international humanitarian law. Water Resour Manag 33:1737–1751. https://doi.org/10.1007/s11269-019-02212-z

Griffiths M, Rubaii K (2024) Late modern war and the geos: the ecological 'beforemaths' of advanced military technologies. Secur Dialogue 1–20. https://doi.org/10.1177/09670106241265636

Hassoun A, Al-Muhannadi K, Hassan HF et al (2024) From acute food insecurity to famine: how the 2023/2024 war on Gaza has dramatically set back sustainable development goal 2 to end hunger. Front Sustain Food Syst 8:1402150. https://doi.org/10.3389/fsufs.2024.1402150

Hryhorczuk D, Levy BS, Prodanchuk M et al (2024) The environmental health impacts of Russia's war on Ukraine. J Occup Med Toxicol 19:1–14. https://doi.org/10.1186/S12995-023-00398-Y/FIGURES/2

Hsiao ELY, Garside A, Weir D, Plumptre AJ (2023) Protected zones in context: exploring the complexity of armed conflicts and their impacts on the protection of biodiversity. Int Rev Red Cross 105:1412–1440. https://doi.org/10.1017/S1816383123000450

Hulme K (2022) Using international environmental law to enhance biodiversity and nature conservation during armed conflict. J Int Crim Justice 20:1155–1190. https://doi.org/10.1093/jicj/mqac060

Hussain T, Gondal MA (2008) Monitoring and assessment of toxic metals in gulf war oil spill contaminated soil using laser-induced breakdown spectroscopy. Environ Monit Assess 136:391–399. https://doi.org/10.1007/s10661-007-9694-2

Iriqat D, Alousi R, Aldahdouh TZ et al (2025) Educide amid conflict: the struggle of the Palestinian education system. Qual Educ All 2:81–99. https://doi.org/10.1108/QEA-10-2024-0120

Isbell F, Balvanera P, Mori AS et al (2023) Expert perspectives on global biodiversity loss and its drivers and impacts on people. Front Ecol Environ 21:94–103. https://doi.org/10.1002/fee.2536

Jagtap S, Trollman H, Trollman F et al (2022) The Russia-Ukraine conflict: its implications for the global food supply chains. Food Secur 11:2098. https://doi.org/10.3390/foods11142098

Kemmerling B, Schetter C, Wirkus L (2022) The logics of war and food (in)security. Glob Food Sec 33:100634. https://doi.org/10.1016/J.GFS.2022.100634

Khalidi B (2023) The unfolding water catastrophe in Gaza. Oxfam https://views-voices.oxfam.org.uk/2023/11/water-catastrophe-gaza/. Accessed 28 Dec 2024

Kotsis KT (2024) The impact of war on the environment. Eur J Ecol Biol Agric 1:89–100. https://doi.org/10.1093/acprof:oso/9780195311181.003.0005

Krauss JE (2022) Unpacking SDG 15, its targets and indicators: tracing ideas of conservation. Globalizations 19:1179–1194. https://doi.org/10.1080/14747731.2022.2035480

Kumar N, Isha A (2024) Life on land: progress, outcomes and future directions to achieve the targets of SDG 15. Environ Sustain. https://doi.org/10.1007/s42398-024-00330-y

Lawrence MJ, Stemberger HLJ, Zolderdo AJ et al (2015) The effects of modern war and military activities on biodiversity and the environment. Environ Rev 23:443–460. https://doi.org/10.1139/er-2015-0039

Leal Filho W, Eustachio JHPP, Fedoruk M, Lisovska T (2024) War in Ukraine: an overview of environmental impacts and consequences for human health. Front Sustain Resour Manag 3:1423444. https://doi.org/10.3389/fsrma.2024.1423444

Li X, Han Q, Wang J, Xu W (2023) Harnessing big data to track progress towards SDG 15: life on land. Int J Digit Earth 16:4597–4600. https://doi.org/10.1080/17538947.2023.2273661

London L, Watterson A, Mergler D et al (2024) A call from 40 public health scientists for an end to the continuing humanitarian and environmental catastrophe in Gaza. Environ Heal A Glob Access Sci Source 23:1–6. https://doi.org/10.1186/S12940-024-01097-9/PEER-REVIEW

Meaza H, Ghebreyohannes T, Nyssen J et al (2024) Managing the environmental impacts of war: what can be learned from conflict-vulnerable communities? Sci Total Environ 927:171974. https://doi.org/10.1016/j.scitotenv.2024.171974

Middle East Monitor (2024) Israel dropped over 85,000 tons of bombs on Gaza. https://www.middleeastmonitor.com/20241107-israel-dropped-over-85000-tons-of-bombs-on-gaza/. Accessed 28 Dec 2024

Mitra S, Chakraborty AJ, Tareq AM et al (2022) Impact of heavy metals on the environment and human health: novel therapeutic insights to counter the toxicity. J King Saud Univ Sci 34:101865. https://doi.org/10.1016/J.JKSUS.2022.101865

OCHA (2024a) Humanitarian Situation Update #224 | Gaza Strip. https://www.unocha.org/publications/report/occupied-palestinian-territory/humanitarian-situation-update-224-gaza-strip-enarhe. Accessed 18 Dec 2024

OCHA (2024b) Gaza Humanitarian Response Update | 24 November - 7 December 2024. https://www.ochaopt.org/content/gaza-humanitarian-response-update-24-november-7-december-2024?_gl=1*1egyngz*_ga*NDE3Nzg4ODEuMTczMzQ2NTkzMQ..*_ga_E60ZNX2F68*MTczNTE0MjE

xMi4xMC4xLjE3MzUxNDIxMTguNTQuMC4w. Accessed 25 Dec 2024
OHCHR (2024) UN Special Committee finds Israel's warfare methods in Gaza consistent with genocide, including use of starvation as weapon of war. https://www.ohchr.org/en/press-releases/2024/11/un-special-committee-finds-israels-warfare-methods-gaza-consistent-genocide. Accessed 10 Jan 2025
Palestinian Central Bureau of Statistics (2020) Press Release by the Palestinian Central Bureau of Statistics (PCBS) and the Environment Quality Authority on World Environment Day (WED) about "Celebrating Biodiversity"
PAX (2023) Uninhabitable? The reverberating public health and environmental risks from the war in Gaza
Pereira P, Zhao W, Symochko L et al (2022) The Russian-Ukrainian armed conflict will push back the sustainable development goals. Geogr Sustain 3:277–287. https://doi.org/10.1016/j.geosus.2022.09.003
Qu X, Li X, Bardgett RD et al (2024) Deforestation impacts soil biodiversity and ecosystem services worldwide. PNAS 121:1–10. https://doi.org/10.1073/pnas
Qumsiyeh MB (2024) Impact of the Israeli military activities on the environment. Int J Environ Stud 81:977–992. https://doi.org/10.1080/00207233.2024.2323365
Qumsiyeh MB, Abusarhan MA (2021) Biodiversity and environmental conservation in palestine. In: Biodiversity, conservation and sustainability in Asia: volume 1: prospects and challenges in West Asia and Caucasus, pp 1–22
ReliefWeb (2024) Gaza humanitarian response update | 24 November—7 December 2024. https://reliefweb.int/report/occupied-palestinian-territory/gaza-humanitarian-response-update-24-november-7-december-2024. Accessed 18 Dec 2024
Reuters (2024) Destruction of Gaza water wells deepens Palestinian misery. https://www.reuters.com/world/middle-east/destruction-gaza-water-wells-deepens-palestinian-misery-2024-07-30/. Accessed 28 Dec 2024
Rist L, Norström A, Queiroz C (2024) Biodiversity, peace and conflict: understanding the connections. Curr Opin Environ Sustain 68:101431. https://doi.org/10.1016/J.COSUST.2024.101431
Rodríguez-Seijo A, Fernández-Calviño D, Arias-Estévez M, Arenas-Lago D (2024) Effects of military training, warfare and civilian ammunition debris on the soil organisms: an ecotoxicological review. Springer, Berlin Heidelberg
Sadiq M, AlThagafi KM, Mian AA (1992) Preliminary evaluation of metal contamination of soils from the Gulf war activities. Bull Environ Contam Toxicol 49:633–639. https://doi.org/10.1007/BF00200774
Schillinger J, Özerol G, Güven-Griemert Ş, Heldeweg M (2020) Water in war: understanding the impacts of armed conflict on water resources and their management. Wiley Interdiscip Rev Water 7:e1480. https://doi.org/10.1002/WAT2.1480
Shaheen A, Dajani R, Zinszer K et al (2024) The war on the Gaza strip and its consequences on global warming. Front Hum Dyn 6:1463902
Shevchuk O, Rochshyna N, Lazarenko I, Stets O (2023) Towards a sustainable future: overcoming the challenges of post-war ecosystem reconstruction in Ukraine. IOP Conf Ser Earth Environ Sci 1269:1–13. https://doi.org/10.1088/1755-1315/1269/1/012018
Solokha M, Pereira P, Symochko L et al (2023) Russian-Ukrainian war impacts on the environment. Evidence from the field on soil properties and remote sensing. Sci Total Environ 902:166122. https://doi.org/10.1016/j.scitotenv.2023.166122
Solokha M, Demyanyuk O, Symochko L et al (2024) Soil degradation and contamination due to armed conflict in Ukraine soil degradation and contamination due to armed conflict in Ukraine. Land 13:1614
Sousa R, da Silva JP, Douda K, Mammola S (2022) The cost of war for biodiversity: a potential ecocide in Ukraine. Front Ecol Environ 20:394–396. https://doi.org/10.1002/fee.2548
UN (2024) Over 1.8 million in Gaza face extreme hunger. https://news.un.org/en/story/2024/10/1155836. Accessed 10 Jan 2025
UN, State of Palestine (2018) Palestinian National Voluntary Review on the Implementation of the 2030 Agenda
UNCTAD (2023) Developments in the economy of the Occupied Palestinian Territory
UNCTAD (2024) Preliminary assessment of the economic impact of the destruction in Gaza and prospects for economic recovery
UNDP/PAPP (2024) Impact of the War on Solid Waste Management in the Gaza Strip
UNEP (2024) Environmental impact of the conflict in Gaza: preliminary assessment of environmental impacts. Nairobi
Wang D, Hao M, Li N, Jiang D (2024) Assessing the impact of armed conflict on the progress of achieving 17 sustainable development goals. iScience 27:111331. https://doi.org/10.1016/j.isci.2024.111331
Weir D, Durant SM, Schulte H et al (2024) Conservationists must address how war affects species and ecosystems. Nature 634:538–541
Williams OH, Rintoul-Hynes NLJ (2022) Legacy of war: Pedogenesis divergence and heavy metal contamination on the WWI front line a century after battle. Eur J Soil Sci 73:e13297. https://doi.org/10.1111/ejss.13297

14 Economic Devastation: The Impact of the Israeli War on Gaza's Work and Economic Growth

Mohammed Migdad, Abdalrahman Migdad, Isra Migdad, and Abdo Hassoun

Abstract

This chapter examines the severe economic consequences of the recent war on Gaza, focusing on Sustainable Development Goal 8 (SDG 8), which aims to achieve inclusive economic growth and decent work for all. This genocidal war has caused extensive job losses, destruction of industries, and widespread poverty, while the devastation of infrastructure has undermined economic stability. Movement restrictions, the blockade, and military operations have disrupted trade and humanitarian aid, further worsening economic conditions. Unemployment has surged, particularly among youth and women, while unsafe working conditions and the rise in child labor reflect dire socioeconomic conditions. The destruction of businesses, along with educational and vocational facilities, has severely hindered skill development and undermined long-term economic prospects. However, amidst the devastation, inspiring examples of resilience and innovation, such as youth-led entrepreneurial ventures and community-driven initiatives, demonstrate the potential for recovery and economic renewal. This chapter highlights the urgent need for sustainable peace and international interventions to rebuild Gaza's economy. Strategies for recovery, including infrastructure reconstruction, job creation, and fostering local entrepreneurship, are explored, underscoring the importance of restoring economic stability and achieving the inclusive and sustainable development envisioned by SDG 8.

Keywords

Economic recovery · Infrastructure destruction · Job loss · Poverty · Resilience · SDG 8 · Sustainable development · Unemployment

14.1 Introduction

Decent work and economic growth are essential for sustainable development, forming the foundation of Sustainable Development Goal 8 (SDG 8). This goal seeks to promote inclusive and sus-

M. Migdad (✉)
Faculty of Economics and Managerial Sciences, Islamic University of Gaza, Gaza, Palestine
e-mail: mmigdad@iugaza.edu.ps

A. Migdad
Sabahattin Zaim University, Istanbul, Turkey
e-mail: abdalrahman.migdad@izu.edu.tr

I. Migdad
Independent Researcher, Gaza, Palestine

A. Hassoun
Sustainable AgriFoodtech Innovation & Research (SAFIR), Arras, France

Faculty of Agricultural Engineering, University of Aleppo, Aleppo, Syria
e-mail: a.hassoun@saf-ir.com

A. Hassoun (ed.), *War on Gaza*, Sustainable Development Goals Series,
https://doi.org/10.1007/978-3-031-88500-6_14

tainable economic growth, employment, and decent work for all. However, in Gaza, these objectives remain out of reach due to decades of violence, an enduring blockade, and systematic restrictions that have dismantled its economic base and left its labor force highly vulnerable (Rai et al. 2019; Alagha and Hussein 2024; ESCWA/UNDP 2024a). Conflicts and wars profoundly undermine SDG 8 by destabilizing economies, destroying infrastructure, and creating widespread unemployment. Armed conflicts often result in the destruction of critical industries, public services, and labor markets, leaving populations without stable income sources or opportunities for decent work. Prolonged instability disrupts education and vocational training, leading to skill shortages and a generational loss of economic potential. Women, youth, and other vulnerable groups are disproportionately affected, facing heightened barriers to employment and economic participation. The ripple effects of conflict extend beyond immediate economic loss, exacerbating poverty, inequality, and social exclusion, and hindering post-conflict recovery efforts. In many conflict-affected regions, achieving SDG 8 remains an aspirational goal, as wars deepen socioeconomic disparities and dismantle the very structures necessary for sustainable growth (Thies and Baum 2020; Maphiri et al. 2021; Mhlanga and Ndhlovu 2024; van Tulder 2024).

Gaza, situated on the eastern Mediterranean coast and bordered by Egypt to the south and Israel to the north and east, has endured immense suffering under a blockade imposed since 2007. This blockade, compounded by repeated military assaults in 2008–2009, 2012, 2014, 2021, and the most recent war, has severely restricted the movement of people and goods. These measures have shattered Gaza's economy, dismantling industries, crippling livelihoods, and creating chronic shortages of essential supplies. Even before the recent war, Gaza faced profound economic and humanitarian challenges. Israeli restrictions on the flow of goods and humanitarian aid amplified these difficulties, undermining any prospects for stability. Deliberate policies and repeated destruction have reversed progress, dismantling infrastructure and eroding economic systems, leaving the region trapped in a state of de-development. Geographic isolation and systematic economic suppression have further deprived Gazans of opportunities for growth or self-sufficiency, intensifying the humanitarian crisis and perpetuating cycles of dependency (Jebril 2023; Alagha and Hussein 2024; Choonara 2024; Elkahlout 2024).

The military assault on Gaza, characterized by its unprecedented scale of destruction and human toll—about 43,000 people as of November 2024 (Samuel 2024)—has exacerbated these long-standing challenges. The repeated targeting of infrastructure, including factories, schools, hospitals, and public institutions, has crippled economic activity and disrupted daily life. Entire industries have collapsed, further eroding Gaza's economic base. Women and youth, already among the most vulnerable, face even greater barriers to employment, education, and vocational training. This generational loss of skills and opportunities deepens the economic stagnation and social inequalities in Gaza (Agha et al. 2024; Butter 2024; ILO 2024a; UN Women 2024; Abed Alah 2024a; Alser et al. 2024).

This chapter aims to examine the impact of the war (from October 2023 to January 2025) on Gaza's economic landscape and its alignment with SDG 8, which focuses on promoting decent work and sustainable economic growth. By analyzing the interplay of social, economic, and political factors, the chapter seeks to highlight the urgent need for international intervention, sustainable peace efforts, and targeted policies that can restore economic rights and promote inclusive growth in Gaza. In the following sections, readers will find a detailed exploration of the challenges and opportunities related to SDG 8 in Gaza. Section 14.2 outlines the methodology used in this study, including data collection, comparative analysis, and case studies. Section 14.3 provides an estimation of the economic impact of the war, focusing on key indicators such as poverty, unemployment, and Gross Domestic Product (GDP) contraction. Section 14.4 discusses the implications of the war on decent work and economic growth, particularly in relation to SDG 8. Section 14.5 highlights surviving jobs and work opportunities during the war, showcasing examples of resilience and innovation. Section 14.6 examines the specialization and division of labor among displaced families, illustrating how communities adapt to extreme conditions. Section 14.7 offers recommendations for mitigation measures, emphasizing the need for sustainable recovery strategies and international support to rebuild Gaza's economy and achieve the targets of SDG 8. Finally, Sect. 14.8 concludes the chapter by summarizing key findings and reiterating the importance of international cooperation and long-term strategies to achieve sustainable development in Gaza.

14.2 Methodology

14.2.1 Data Collection and Analysis

Qualitative data: This study utilizes secondary data sources, including reports from international organizations such as the United Nations (UN), the World Bank, and local economic agencies. Data was also gathered from the Palestinian Central Bureau of Statistics (PCBS) and annual trade reports from various official organizations. Primary data was collected through interviews with individuals living in tents, along with direct observations made by one of the authors, who resides in a tent at the time of writing this chapter. This data highlights trends in unemployment rates, underlying conditions, and common causes contributing to the increase in unemployment in Gaza. Open interviews were analyzed to explore factors affecting work conditions in Gaza, the influence of key stakeholders on employment, and the obstacles and barriers limiting work opportunities.

Comparative analysis: The authors illustrate the realities of work in Gaza through detailed examples and case studies of individuals living in Gaza or displaced to other countries following the October assault on Gaza. These case studies offer critical insights into how unemployment has disproportionately affected youth and women, highlighting vulnerabilities and inequities in the current labor market.

Case studies: The authors explain the reality of work in Gaza through examples and case studies of people in Gaza or displaced to other countries after the October assault in Gaza. Through the case studies it will be clear to understand how the youth and women were impacted by unemployment by highlighting vulnerabilities in the current labor market.

14.2.2 Field Observations and Interviews

Conducting fieldwork with a diverse range of stakeholders, including business owners, employees, community leaders, and NGOs, offers valuable insights into the ground realities of Gaza's economic landscape. Observations can shed light on unsafe working conditions, informal labor practices, and the evolving dynamics of the informal economy, providing a deeper understanding of the challenges faced by workers and businesses alike.

This methodology will provide a robust foundation for addressing the impact of conflict on decent work and economic growth, directly aligned with SDG 8's targets in Gaza. It allows for a nuanced understanding of both the statistical and human dimensions of the issue, ensuring that the research is comprehensive and impactful.

14.3 Estimation of the Economic Impact of the War

Even before this recent genocidal war, many economic indicators showed degraded economic situation in Gaza due to the long-standing blockade, political division since 2007, and repeated military assaults on the Strip (Tannira 2021). The war on Gaza has driven its economy into a state of near-total collapse, marked by a convergence of devastating economic indicators, resulting in an almost complete halt of economic activity and the deepest recession ever recorded in this region

(World Bank Group 2024). This section explores critical indicators, including poverty, unemployment, labor force participation, wages, GDP contraction, and the rise in child labor.

Before the recent war, poverty levels in Gaza were already alarming, with over 53% of the population living below the poverty line, and approximately 80% of residents depending on international aid for basic survival (OCHA 2018; The World Bank 2023). The war has exacerbated this crisis, with close to 100% of the Gaza population lives in poverty (World Bank 2024). International aid delivery has been disrupted by border closures and blockades, leaving many without access to essential resources. Disruption of aid delivery, coupled with economic collapse and food insecurity, has left children particularly vulnerable, as nearly nine out of ten children in Gaza now face severe food poverty, consuming diets that fail to meet even minimal nutritional diversity (Abu-Ras and Mohamed 2018; UNICEF 2024).

Unemployment, which was already one of the highest globally at 45% prior to the recent war, has skyrocketed to nearly 80% amid the recent war (The World Bank 2023; ILO 2024a). This war has caused extensive destruction to productive assets, including agricultural land, crops, livestock shelters, greenhouses, orchards, storage facilities, fishing boats, and equipment. Critical infrastructure such as agribusinesses, irrigation systems, water-pumping facilities, electricity grids, Internet networks, factories, office spaces, housing units, schools, and healthcare centers has also suffered significant damage, further crippling the region's socioeconomic foundation (The World Bank 2024; UNCTAD 2024; UNITAR-UNOSAT 2024).

The collapse of wages and household incomes has deepened Gaza's economic crisis. Workers who remain employed have faced significant pay cuts as employers struggle with resource shortages and restricted market access. Simultaneously, GDP of Gaza contracted 86% in the last quarter of 2023, due to the destruction of critical infrastructure and the cessation of most economic activities. Key sectors such as agriculture, manufacturing, and trade have been devastated, while supply chain disruptions have prevented any meaningful recovery. This contraction has left Gaza's economy on the brink of total collapse, with minimal prospects for immediate stabilization (Alagha and Hussein 2024; The World Bank 2024; UNCTAD 2024).

Child labor in Palestine has been a deeply rooted issue even prior to the recent war, fueled by economic instability, widespread poverty, and the sociopolitical consequences of prolonged conflict (Thabet et al. 2011; Di Maio and Nandi 2013; Jildeh et al. 2014). The recent war has exacerbated the situation, pushing even more children into labor. Children who are abandoned, separated from their families, or orphaned are among the most vulnerable, facing heightened risks of exploitation, neglect, and severe mental health challenges. With limited access to essential resources for survival, these children often endure hazardous working conditions that expose them to physical harm and exploitation. Such circumstances deprive them of education and perpetuate cycles of poverty. The psychological burden on these children is profound, as they navigate extreme economic and emotional hardships during critical developmental years (Jabr and Berger 2022; Buheji and Buheji 2024; Wells et al. 2024).

The widespread destruction has led to estimates suggesting it could take until 2040 to rebuild the destroyed housing units alone, based on an optimistic scenario (ESCWA/UNDP 2024b). Other assessments from ESCWA, the UN, and related organizations estimate that Gaza's development has been set back by up to 69 years (UNDP 2024). Some reports, drawing on UN data, predict that under current conditions, it could take as long as 350 years for Gaza to fully rebuild (UN 2024).

14.4 Decent Work and Economic Growth (SDG 8) in the Context of Gaza

SDG 8 focuses on promoting sustained, inclusive, and sustainable economic growth, full and productive employment, and decent work for all.

Its primary targets include fostering steady economic growth, enhancing global productivity, reducing unemployment (especially among youth), ensuring fair wages, and creating safe, supportive working environments. This goal emphasizes the importance of equal access to job opportunities, protection of workers' rights, and elimination of exploitative practices across all sectors (Rai et al. 2019; Kreinin and Aigner 2022).

However, the destruction of Gaza's critical infrastructure and economic hubs has resulted in a sharp contraction of the local economy, with unemployment surging to nearly 80% and GDP contracts by almost 85% in Gaza (ILO 2024b). This economic collapse contradicts the target of achieving at least 7% annual growth in developing regions and leaves Gaza far from achieving sustained development.

Labor productivity and economic diversification, both integral components of SDG 8, have also been severely impacted. Gaza's key economic sectors, such as agriculture, manufacturing, and tourism, have been devastated by hostilities, leaving the economy heavily reliant on low-productivity informal jobs. Many workers are employed in precarious and unsafe conditions, such as hazardous construction sites, which provide little opportunity for upward mobility or skill development. The loss of access to global markets, coupled with restrictions on resources and movement, further stifles the potential for economic innovation and diversification, leaving Gaza unable to rebuild its economic foundation (Alagha and Hussein 2024; Elkahlout 2024; ESCWA/UNDP 2024a; Morrar et al. 2024).

The promotion of decent work and employment opportunities for youth is another critical aspect of SDG 8 that has been severely undermined in Gaza. Youth unemployment rates in Gaza were already among the highest in the world, reaching approximately 70% even before the recent war (Al Jazeera 2024a), The war has further exacerbated this crisis, pushing unemployment rates even higher as economic opportunities continue to vanish. The destruction of schools, vocational training centers, and other educational facilities has further compounded this issue, depriving young people of the skills and knowledge necessary for productive employment. These challenges are closely tied to the increase in child labor, as many children are forced to work in hazardous and exploitative conditions to support their families (Hamad et al. 2024; Repo 2024). This directly contradicts SDG 8's targets, including the aim to eradicate forced labor and child labor in all its forms (Target 8.7), as well as the commitment to create safe, secure, and productive working environments for all workers (Target 8.8). The absence of stable and secure work opportunities perpetuates cycles of poverty, fosters social instability, and deprives Gaza's youth of their potential, making it nearly impossible to achieve the inclusive and productive employment envisioned by SDG 8.

The goal of ensuring safe and secure working environments remains unattainable under the current conditions in Gaza. Intensive bombardments have either destroyed workplaces or left them structurally unsafe, exposing workers to daily risks of injury, exploitation, and even death. In the absence of basic labor protections, individuals are forced into informal sectors where they are highly vulnerable to exploitative practices, further exacerbating their precarious situation (Buheji and Al-Muhannadi 2023; Abed Alah 2024b).

Moreover, this war has severely hindered entrepreneurship and innovation, a key component of SDG 8 (Target 8.3). Small businesses, startups, and entrepreneurial ventures have faced immense challenges due to the destruction of physical assets, loss of financial capital, and restricted access to both local and international markets. Entrepreneurs, particularly youth and women, who play a vital role in driving economic innovation and diversification, have been disproportionately affected. The destruction of Gaza's infrastructure and limited access to resources, such as technology and communication networks, have further stifled creativity and the ability to develop new business ventures. This has reduced income-generation opportunities and curtailed the potential for long-term economic growth and resilience (Buheji and Mushimiyimana 2023; Morrar et al. 2024; PFI/FPCCIA 2024).

14.5 Survived Jobs and Work Opportunities during the War

Despite the severe disruption of business and industry in commerce, manufacturing, and services due to the war on Gaza, several sectors have managed to continue operating, though at minimal capacity. Jobs within UN agencies, international organizations, and newly created roles in these organizations, both temporary and permanent, have persisted. A limited number of positions within private sector companies have also remained, including roles in marketing and solution services, supported by sporadic Internet access. Higher education institutions have demonstrated remarkable resilience by continuing their services, even at reduced capacity. Many educators, in the absence of colleagues lost to the war, have stepped into voluntary teaching roles, enabling almost the fall semester of 2024 to proceed online. Additionally, an increasing number of higher education students have been able to defend their master's and doctoral theses, showcasing the unwavering commitment to education even amid crisis (Faculty of Education University of Cambridge/Centre for Lebanese Studies & UNRWA 2024; Jebril 2024; Iriqat et al. 2025).

The private sector, along with broader social and community-driven initiatives, has highlighted creative resilience in addressing the urgent needs of daily life. In response to the near-total cessation of traditional job opportunities, individuals have developed innovative, necessity-driven ventures that provide employment and help sustain families. For instance, street vendors have adapted to selling a variety of goods, while temporary stalls for fruits and vegetables have emerged based on produce availability. However, food shortages and the risk of famine persist, with the entire population of the Gaza Strip (about 2.2 million people) is experiencing acute food insecurity at Crisis level (IPC Phase 3) or worse (The Integrated Food Security Phase Classification (IPC) 2024; Alnabih et al. 2024; Hassoun et al. 2024; IPC 2024; Khoodoruth and Khoodoruth 2024).

Water scarcity, exacerbated by damaged infrastructure and limited humanitarian access, has given rise to water distribution jobs. Carts and vehicles deliver drinking water and other essential supplies to areas facing acute shortages, as only 10% of Gaza's water supply is safe for drinking (Abdul Samad et al. 2024; Farajallah and Farajallah 2024; Razum et al. 2024). The use of water as a weapon of war has been highlighted, with claims that access to this essential resource has been made contingent on meeting certain objectives in the war (Human Rights Watch 2024). In the energy sector, mobile phone and battery charging services have become essential due to the absence of a stable power supply, ensuring communication and lighting within displaced communities (Migdad et al. 2024a).

The transportation sector has adapted creatively to fuel shortages, with alternative modes such as horse- or donkey-drawn carts and vehicles modified to run on cooking oil, ensuring the movement of goods and passengers. Clay ovens fueled by wood have become increasingly common, offering bread for sale or baking services to meet the needs of the community. These initiatives address the critical demand for affordable food production, given the skyrocketing costs of essential goods (Al Jazeera 2024b). Numerous small-scale ventures, including currency exchange and money transfer services, have created additional employment opportunities and played a vital role in addressing the needs of the displaced population. The construction of tents and temporary shelters, along with related facilities such as latrines and wells, has provided temporary jobs, especially for those living in makeshift accommodations. These examples show alternative models of economic development and a resistance economy in practice (McGahern et al. 2024).

14.6 Specialization and Division of Labor Among Displaced Families

In displacement settings, the structured division of labor among family members in Gaza's tent communities has become a critical survival mechanism for the at least 1.9 million people displaced by the recent war (Hanbali et al. 2024;

Kerbaj et al. 2024). Each family member assumes specific responsibilities to address the household's immediate needs, reflecting an adaptive system that balances efficiency and equity. For example, some members focus on water collection, often waiting in long lines at distribution points, while others sell surplus goods or engage in small trade to earn modest amounts of money for purchasing necessities. Tasks such as charging batteries for night lights and cell phones are essential for maintaining communication and safety in the absence of reliable electricity. Women and daughters manage cooking and cleaning, while men frequently gather firewood and build fires for meal preparation. This division of labor ensures the family's survival and supports vulnerable members by distributing tasks based on individual capacities, fostering collaboration and resilience in the face of adversity.

Unpaid household labor within these displaced families represents a significant yet undervalued economic contribution, particularly when considered through the lens of opportunity cost. Activities such as cooking, water fetching, and resource gathering could command financial value if outsourced. Women take on roles such as kneading dough and cooking, daughters assist with baking, and boys contribute by collecting firewood and fetching water. In addition to these essential tasks, families often pool their efforts to develop self-employment initiatives, generating income to meet their needs. Such unpaid labor underscores the resilience and resourcefulness of displaced families, offering a vital informal economic buffer in a time of crisis. The following real-life examples of self-employment illustrate how individuals have leveraged their skills and resourcefulness to overcome the severe challenges posed by displacement and war:

Safa, a 19-year-old university student majoring in Fine Arts at Al-Aqsa University, illustrates the power of creativity and entrepreneurial spirit amid displacement. Driven by the urgent need to support her family, Safa leveraged her artistic skills to create and sell paintings internationally (Fig. 14.1).

She also encouraged her mother and sisters to produce handwoven textiles, combining cultural heritage with economic necessity. Safa markets these woven pieces with bilingual descriptions that highlight their unique stories. Despite the war, she participated in an art exhibition in Dublin, Ireland, displaying her paintings and crafts. This initiative has provided essential

Fig. 14.1 Examples of Safa's work (Migdad et al. 2024b)

income for her family while illustrating the potential of creative industries to sustain livelihoods during crises.

Ahmed, a 21-year-old university student in Turkey, represents another example of adaptability and innovation. Although he was not directly displaced, Ahmed faced the financial strain imposed on his family in Gaza by the war. Determined to reduce his family's burden, he used his wood carving and drawing skills to establish a small workshop, producing artistic pieces such as carved maps of Palestine (Fig. 14.2). Ahmed's efforts have not only enabled him to support himself but also positioned him to expand his business. He is preparing for a specialized exhibition and actively pursuing partnerships with business incubators to secure marketing and production support, aiming to develop a sustainable enterprise.

Similarly, Mr. Nabeel, a displaced entrepreneur, founded a bakery specializing in pastries and baked goods. Initially operating with limited resources, he gradually expanded his business to employ over 20 people, including family members and skilled displaced bakers. Despite challenges such as ingredient scarcity and high costs, Mr. Nabeel upgraded the bakery's tools and equipment, transforming it into a thriving enterprise. The bakery not only provides income for many families but also demonstrates how small businesses can adapt and grow even in displacement settings.

Fig. 14.2 Examples of Ahmed's work (Migdad et al. 2024b)

These examples highlight the resilience and ingenuity of Gaza's displaced population, who have adapted to extreme conditions with remarkable creativity and resourcefulness. The structured division of labor within families, the economic value of unpaid contributions, and innovative self-employment initiatives collectively offer a glimpse of how individuals and communities can overcome immense hardships. By sustaining livelihoods and fostering resilience, these efforts provide a critical lifeline for families navigating the challenges of displacement and war.

14.7 Recommendations for Mitigation Measures

Achieving the targets of SDG 8 in Gaza requires immediate and sustained international intervention to address the devastating economic and labor impacts of this unprecedented war. The international community must exert pressure on Israel to comply with international law, including an immediate ceasefire and an end to the blockade. Reopening all commercial crossings, restoring access to electricity, water, and communication networks, and initiating large-scale reconstruction efforts are critical first steps. Reconstruction should prioritize local labor and businesses to create immediate employment opportunities and stimulate economic recovery. International donors must support the rebuilding of residential, industrial, agricultural, and service infrastructure, including schools, hospitals, and factories, to lay the foundation for long-term economic resilience.

Economic empowerment and job creation must be central to recovery strategy of Gaza. Financial assistance, grants, and low-interest loans should be provided to small businesses and startups, with a focus on supporting women and youth entrepreneurs. Business incubators and innovation hubs can help drive economic diversification and access to global markets. Key sectors such as agriculture, manufacturing, and tourism, which have been destroyed by the war, must be revived through targeted investments in infrastructure, tools, and market access. Public works programs should be launched to provide temporary employment in reconstruction projects, while ensuring fair wages, safe working conditions, and the elimination of exploitative practices such as child labor.

Strengthening governance and financial systems is essential to ensure sustainable economic recovery. Introducing a digital Palestinian currency would facilitate internal trade and reduce dependency on Israeli-controlled financial systems. Palestinian authorities must be empowered to have full control over their finances, including clearance funds and natural resources, to promote economic self-sufficiency. Legal and institutional frameworks should be enhanced to enforce labor protections, ensure accountability for violations, and promote transparency in the allocation of reconstruction funds. Training programs for judges, lawmakers, and labor inspectors can help address the illegal conduct of the Israeli occupation and safeguard workers' rights.

International advocacy and awareness campaigns are crucial to mobilizing global support for the recovery of Gaza. Social media campaigns should highlight the economic and humanitarian crisis, emphasizing the impact of the war on unemployment, poverty, and labor rights. The international community, including the UN, the Arab League, and the European Union, must pressure Israel to end the blockade and allow unrestricted access to Gaza for humanitarian and economic purposes. Supporting political leaders and movements that advocate for Palestinian human and development rights, such as the Boycott, Divestment, and Sanctions (BDS) movement, can help hold Israel accountable for its violations of international law.

Long-term strategies for sustainable development must focus on education, vocational training, and regional partnerships. Rebuilding schools and vocational training centers destroyed during the war is critical to addressing the generational loss of skills and opportunities. Partnerships with international universities and organizations can provide scholarships and training programs for Gazan youth, particularly in fields such as technology, engineering, and healthcare. Regional and international economic partnerships should be fostered to create trade opportunities and attract foreign investment. Establishing free trade zones and special economic zones can stimulate economic growth and integrate Gaza into the global economy. By addressing immediate humanitarian needs and implementing long-term strategies, Gaza can begin to rebuild its economy, create decent work opportunities, and achieve the targets of SDG 8.

14.8 Conclusion

Sustainable development is a universal aspiration, yet it remains an unattainable goal for many conflict-affected regions, including Gaza, where decades of violence, occupation, and systematic oppression have crippled economic and social progress. Achieving the targets of SDG 8 in Gaza is a significant challenge due to the genocide, movement restrictions, and the closure of the main crossings. The persistent violence and violations of international law and human rights by the Israeli occupation have had a severe impact on Gaza's labor market, infrastructure, and land, halting progress toward sustainable development. Unemployment becomes extremely high, with many men and women losing jobs since October 2023, leaving job opportunities scarce and forcing most people to rely on informal income sources.

The economy and society have been devastated by the unprecedented military assault and over 17 years of blockade, increasing dependence on foreign aid—aid that barely enters Gaza due to punitive measures imposed by Israel, part of its strategy of ethnic cleansing. Moreover, the absence of a safe environment is a primary barrier to achieving SDG 8 targets. Ending the genocide and the humanitarian crisis is fundamental to any hope for sustainable development in Gaza. This includes creating a framework that fosters long-term economic empowerment and job opportunities for all.

The genocide in Gaza serves as a stark test of the commitment of democratic and advanced nations that claim to uphold the principles of humanity and human rights. The international community, including the United Nations, the Arab League, the European Union, and other global actors, must unite to demand an immediate end to the war. Beyond this, they must work collectively to pave the way for sustainable development in Gaza and Palestine as a whole. This includes addressing chronic poverty, unemployment, and the systemic barriers that hinder economic growth and social progress. Only through sustained international cooperation and a commitment to justice can Gaza begin to rebuild and achieve the inclusive and sustainable development envisioned by SDG 8.

References

Abdul Samad L, Butcher M, Khalidi B (2024) Water war crimes: how Israel has weaponised water in its military campaign in Gaza. Oxfam Int 6

Abed Alah M (2024a) Shattered hierarchy: how the Gaza conflict demolished Maslow's pyramid of needs. Curr Psychol:1–7. https://doi.org/10.1007/S12144-024-05784-1/FIGURES/1

Abed Alah M (2024b) Echoes of conflict: the enduring mental health struggle of Gaza's healthcare workers. Confl Heal 18:1–6. https://doi.org/10.1186/S13031-024-00577-6/METRICS

Abu-Ras W, Mohamed RA (2018) Child poverty and youth unemployment in Palestine. Poverty Public Policy 10:354–370. https://doi.org/10.1002/pop4.227

Agha Z, Esson J, Griffiths M, Joronen M (2024) Gaza: A decolonial geography. Trans Inst Br Geogr 49:e12675. https://doi.org/10.1111/tran.12675

Al Jazeera (2024a) Gaza graduates demand UNRWA solutions for high unemployment rate. https://www.aljazeera.com/news/2023/9/27/gaza-graduates-demand-unrwa-solutions-for-high-unemployment-rate?utm_source=chatgpt.com. Accessed 5 Jan 2025

Al Jazeera (2024b) As people run out of fuel in Gaza, clay ovens are back in demand. https://www.aljazeera.com/gallery/2023/11/21/traditional-clay-ovens-in-demand-as-fuel-shortages-continue-in-gaza?utm_source=chatgpt.com. Accessed 4 Jan 2025

Alagha M, Hussein N (2024) The political economy of Israeli economic warfare on the Gaza strip. In: Elkahlout G (ed) Gaza's cycle of destruction and rebuilding. Palgrave Macmillan, Cham, pp 125–145

Alnabih A, Alnabeh N, Aljeesh Y, Aldabbour B (2024) Food insecurity and weight loss during wartime: a mixed-design study from the Gaza strip. J Health Popul Nutr 43:222

Alser K, Mallah SI, El-Oun YRA et al (2024) Trauma care supported through a global telemedicine initiative during the 2023–24 military assault on the Gaza strip, occupied Palestinian territory: a case series. Lancet 404:874–886. https://doi.org/10.1016/S0140-6736(24)01170-X

Buheji M, Al-Muhannadi K (2023) Mitigating risks of environmental impacts on Gaza - review of precautions & solutions post (2023 war). Int J Adv Res Eng Technol 14:15–47

Buheji M, Buheji B (2024) Mitigating risks of slow children development due to war on Gaza 2023. Int J Psychol Behav Sci 14:11–21. https://doi.org/10.5923/j.ijpbs.20241401.02

Buheji M, Mushimiyimana E (2023) "Gaza"—towards an agile resilience. Int J Manag 14:120–136

Butter D (2024) Economic impact of the Gaza war. IEMed Mediterr Yearb:273–275

Choonara J (2024) A second Nakba? Int Social

Di Maio M, Nandi TK (2013) The effect of the Israeli-Palestinian conflict on child labor and school attendance in the West Bank. J Dev Econ 100:107–116. https://doi.org/10.1016/j.jdeveco.2012.08.007

Elkahlout G (2024) Gaza's cycle of destruction and rebuilding. Understanding the actors, dynamics, and responses. Palgrave Macmillan

ESCWA/UNDP (2024a) Gaza war: Expected socioeconomic impacts on the State of Palestine—October 2024

ESCWA/UNDP (2024b) Gaza war: expected socioeconomic impacts on the State of Palestine Update—May 2024

Faculty of Education University of Cambridge/Centre for Lebanese Studies & UNRWA (2024) Palestinian education under attack in Gaza: restoration, recovery, rights and responsibilities in and through education

Farajallah I, Farajallah H (2024) Challenges in access to safe drinking water and its impact on maternal and child health in Gaza. Anatol Clin J Med Sci:52–65. https://doi.org/10.21673/anadoluklin.1575390

Hamad BA, Vintges J, Diab R, et al (2024) Shattered lives and dreams: the toll of the war on Gaza on young people

Hanbali L, Kwong EJL, Neilson A et al (2024) Israeli necropolitics and the pursuit of health justice in Palestine. BMJ Glob Heal 9:e014942. https://doi.org/10.1136/BMJGH-2023-014942

Hassoun A, Al-Muhannadi K, Hassan HF et al (2024) From acute food insecurity to famine: how the 2023/2024 war on Gaza has dramatically set back sustainable development goal 2 to end hunger. Front Sustain Food Syst 8:1402150. https://doi.org/10.3389/fsufs.2024.1402150

Human Rights Watch (2024) Extermination and acts of Genocide Israel deliberately depriving Palestinians in Gaza of water

ILO (2024a) A year of war in Gaza: impacts on employment and livelihoods in the West Bank and Gaza strip. Bulletin No 5

ILO (2024b) A year of war: Unemployment surges to nearly 80 per cent and GDP contracts by almost 85 per cent in Gaza. https://www.ilo.org/resource/news/year-war-unemployment-surges-nearly-80-cent-and-gdp-contracts-almost-85. Accessed 19 Dec 2024

IPC (2024) Gaza Strip: acute food insecurity situation for September–October 2024 and projection for November 2024–April 2025

Iriqat D, Alousi R, Aldahdouh TZ et al (2025) Educide amid conflict: the struggle of the Palestinian education system. Qual Educ All 2:81–99. https://doi.org/10.1108/QEA-10-2024-0120

Jabr S, Berger E (2022) The children of Palestine: struggle and survival under occupation. In: Tiliouine H, Benatuil D, Lau MKW (eds) Handbook of Children's risk, vulnerability and quality of life. International handbooks of quality-of-life. Springer, Cham, pp 279–296

Jebril M (2023) Between construction and destruction: the experience of educationalists at Gaza's universities. Compare 53:986–1004. https://doi.org/10.1080/03057925.2021.1987190

Jebril M (2024) War, higher education and development: the experience for educationalists at Gaza's universities. High Educ. https://doi.org/10.1007/s10734-024-01353-4

Jildeh C, Abdeen Z, Al Sabbah H et al (2014) Labor and related injuries among schoolchildren in palestine: findings from the national study of Palestinian schoolchildren (HBSC-WBG2006). ISRN Pediatr 2014:729573. https://doi.org/10.1155/2014/729573

Kerbaj M, Traboulsi J, Maasri Z, Toukan H (2024) Gaza in plain sight: witnessing in solidarity. J Vis Cult 22:1–19. https://doi.org/10.1177/14704129241227847

Khoodoruth MAS, Khoodoruth WNC kai (2024) From past famines to present crises: the epigenetic impact of maternal malnutrition on offspring health in Gaza. Asian J Psychiatr 95:103999. https://doi.org/10.1016/j.ajp.2024.103999

Kreinin H, Aigner E (2022) From "decent work and economic growth" to "sustainable work and economic degrowth": a new framework for SDG 8. Springer

Maphiri M, Matasane MA, Mudimu G (2021) Challenges to the effective implementation of SDG 8 in creating decent work and economic growth in the southern African hemisphere: perspectives from South Africa, Lesotho and Zimbabwe. In: Vertigans S, Idowu SO (eds) Global challenges to CSR and sustainable development. CSR, sustainability, ethics & governance. Springer, Cham, pp 39–63

McGahern U, Iriqat D, Al-Haj A (2024) Building a border-city university in occupied Palestine: developing a cross-border "resistance economy" in practice. Polit Geogr 108:103030. https://doi.org/10.1016/J.POLGEO.2023.103030

Mhlanga D, Ndhlovu E (2024) A critical analysis of the sustainable development goals framework in the wake of the Russia-Ukraine conflict. In: Mhlanga D, Ndhlovu E (eds) The Russia-Ukraine conflict and development in Africa. Contributions to political science. Springer, Cham, pp 51–67

Migdad M, Buheji M, Migdad IM (2024a) Reviving resilience from the tents—Gaza (2024). Int J Manag 15:93–108
Migdad M, Migdad SM, Migdad AM, Buheji M (2024b) Waves of inspiration despite genocide—cases from youth and children of Gaza. J Res Adm 6:508–529
Morrar R, Yerousis G, Iwidat Q, Shahwan R (2024) Realities and possibilities of Palestinian entrepreneurial ventures: a comprehensive study on youth entrepreneurship in the State of Palestine. In: Musleh Al-Sartawi AMA, Nour AI (eds) Artificial intelligence and economic sustainability in the era of industrial revolution 5.0. Studies in systems, decision and control. Springer, Cham, pp 925–938
OCHA (2018) 53 per cent of Palestinians in Gaza live in poverty, despite humanitarian assistance. https://www.ochaopt.org/content/53-cent-palestinians-gaza-live-poverty-despite-humanitarian-assistance. Accessed 4 Jan 2025
PFI/FPCCIA (2024) Impact of the Gaza war on private sector and pathways for recovery
Rai SM, Brown BD, Ruwanpura KN (2019) SDG 8: decent work and economic growth—a gendered analysis. World Dev 113:368–380. https://doi.org/10.1016/J.WORLDDEV.2018.09.006
Razum O, Agha H, Davidovitch N et al (2024) Gaza ceasefire: improve WASH, promote cooperation. Int J Public Health 69:1607412. https://doi.org/10.3389/IJPH.2024.1607412/BIBTEX
Repo J (2024) Genocide and the destruction of the means of social reproduction in Gaza. Eur J Polit Gend:1–8. https://doi.org/10.1332/25151088Y2024D000000061
Samuel MT (2024) The Israel-Hamas war one year later: mass violence and Palestinian dispossession. Middle East Policy 31:3–16. https://doi.org/10.1111/mepo.12792
Tannira A (2021) The political economy ofthe Gaza strip under Hamas. In: Tartir A, Dana T, Seidel T (eds) Political economy of Palestine. Palgrave Macmillan, pp 135–154
Thabet AA, Matar S, Carpintero A et al (2011) Mental health problems among labour children in the Gaza strip. Child Care Health Dev 37:89–95. https://doi.org/10.1111/j.1365-2214.2010.01122.x
The World Bank (2023) Impacts of the conflict in the Middle East on the Palestinian Economy
The World Bank (2024) Gaza Strip—interim damage assessment
Thies CF, Baum CF (2020) The effect of war on economic growth. Cato J 40:199–212. https://doi.org/10.36009/CJ40.1.10
UN (2024) Economic costs of the Israeli occupation for the Palestinian people: the economic impact of the Israeli military operation in Gaza from October 2023 to May 2024 - UNCTAD Report (A/79/343) . https://www.un.org/unispal/document/unctad-report-10sep24/. Accessed 4 Jan 2025
UN Women (2024) Gender alert: the gendered impact of the crisis in Gaza
UNCTAD (2024) Preliminary assessment of the economic impact of the destruction in Gaza and prospects for economic recovery
UNDP (2024) New UN report: Impacts of war have set back development in Gaza by as much as 69 years. https://www.undp.org/press-releases/new-un-report-impacts-war-have-set-back-development-gaza-much-69-years. Accessed 4 Jan 2025
UNICEF (2024) 1 in 4 children globally live in severe child food poverty due to inequity, conflict, and climate crises. https://www.unicef.org/press-releases/1-4-children-globally-live-severe-child-food-poverty-due-inequity-conflict-and. Accessed 5 Dec 2024
UNITAR-UNOSAT (2024) UNOSAT Gaza Strip comprehensive damage assessment—December 2024
van Tulder R (2024) International business and SDG8: are we on course? Business and SDG8: are we on course? In: Sinkovics N, Sinkovics RR, Boussebaa M, Fletcher M (eds) International business and SDG 8. The academy of international business. Palgrave Macmillan, Cham, p 44
Wells K, Cortés-Morales S, Esson J, et al (2024) Israel's war on Gaza and the violation of children's rights. Child Geogr 1–4. https://doi.org/10.1080/14733285.2024.2316752
World Bank (2024) Note on the impacts of the conflict in the Middle East on the Palestinian economy—September 2024. Washington, DC
World Bank Group (2024) Macro poverty outlook: country-by-country analysis and projections for the developing world

Shattered Foundations: How the War on Gaza Has Undermined Industry, Innovation, and Infrastructure

15

Samer Abuzerr, Najla Trabelsi, Toufik Ouatmani, Zuhaib Bhat, Mohammed Nijim, Abderahman Rejeb, and Abdo Hassoun

Abstract

This chapter explores the devastating impact of repeated Israeli hostilities on Gaza, especially the recent war; October 2023–January 2025, on industry, innovation, and infrastructure in this region, focusing on the significant setbacks to achieving Sustainable Development Goal 9 (SDG 9). This brutal war has led to widespread destruction of critical infrastructure, including power plants, health facilities, and communication networks, crippling industrial activities and halting technological advancements. The destruction of manufacturing facilities and the scarcity of essential resources have severely limited production capabilities, leading to massive job losses and economic decline. Furthermore, the war has severely hindered the development of new technologies and innovations by destroying educational institutions and research centers, while restrictions on the import of technological equipment and resources have further exacerbated these challenges, limiting access to tools necessary for advancement and growth. Through detailed analysis of reports and scientific publications, this chapter highlights the urgent need for international intervention and support to rebuild industry and infrastructure in Gaza. This

S. Abuzerr (✉)
Department of Medical Sciences, University College of Science and Technology,
Khan Younis, Gaza, Gaza Strip, Palestine

N. Trabelsi
Laboratory of Olive Biotechnology, Center of Biotechnology of Borj-Cédria, Hammam-Lif, Tunisia
e-mail: najla.trabelsi@cbbc.rnrt.tn

T. Ouatmani
Laboratoire de Biomathématique, Biophysique, Biochimie et Scientometrie, Faculté des Sciences de la Nature et de la Vie, Université de Bejaia,
Bejaia, Algeria
e-mail: toufik.ouatmani@univ-bejaia.dz

Z. Bhat
Division of Livestock Products Technology,
SKUAST-J, Jammu, India

M. Nijim
Department of Sociology and Anthropology, Carleton University, Ottawa, ON, Canada
e-mail: mohammednijim@cmail.carleton.ca

A. Rejeb
Faculty of Business and Economics, Széchenyi István University, Győr, Hungary
e-mail: rejeb.abderahman@sze.hu

A. Hassoun
Sustainable AgriFoodtech Innovation & Research (SAFIR), Arras, France

Faculty of Agricultural Engineering, University of Aleppo, Aleppo, Syria
e-mail: a.hassoun@saf-ir.com

A. Hassoun (ed.), *War on Gaza*, Sustainable Development Goals Series,
https://doi.org/10.1007/978-3-031-88500-6_15

comprehensive examination of the impact of the war on Gaza industry, innovation, and critical infrastructure offers valuable insights into the broader challenges of achieving SDG 9 in conflict-affected regions.

Keywords

Resilience · Economy · Education · Energy · Gaza reconstruction · Health · International support · Post-conflict recovery · Resilience · SDG 9 · Sustainable development

15.1 Introduction

Sustainable Development Goal 9 (SDG 9), one of the 17 United Nations Sustainable Development Goals, underscores the necessity of resilient infrastructure, inclusive and sustainable industrialization, and innovation. These components are integral to fostering economic growth, reducing inequalities, and improving well-being (Küfeoğlu 2022; Singh and Ru 2023). However, achieving SDG 9 in conflict zones is fraught with challenges, especially in regions experiencing systematic violence, such as Gaza. Repeated wars on Gaza illustrate the systematic dismantling of infrastructure, industry, and innovation—key pillars of SDG 9—resulting in profound impacts on development and human lives (Roy 2016; Nijim 2023, 2024; Amnesty International 2024; Human Rights Watch 2024).

The Gaza Strip, home to over 2 million people, has endured decades of occupation, blockades, and recurrent Israeli assaults. The war that escalated in October 2023 has led to unprecedented destruction of Gaza's physical infrastructure, economic systems, and social networks. As of November 2024, approximately 43,000 people have been killed, the majority of whom are women and children (Samuel 2024). Recent studies report an even higher mortality rate in the Gaza Strip (Khatib et al. 2024; Jamaluddine et al. 2025), underscoring the devastating human toll of this genocidal war. Decades of systemic neglect, external control, and destruction have left Gaza's infrastructure and economy in a fragile state. The blockade imposed in 2007 has restricted the flow of goods, people, and resources, stifling industrial growth and technological innovation. Even before this recent war, Gaza struggled with chronic shortages of electricity, clean water, and essential materials, limiting basic services and industrial activity (Barakat et al. 2018; Nijim 2023; Agha et al. 2024; Elkahlout 2024a). Sara Roy argues that economy has been systematically eroded by Israel and turned into the state of de-development (Roy 2016). Therefore, Gaza's historical and geopolitical context is critical to understanding the challenges to achieving SDG 9 in its three dimensions, namely infrastructure, industry, and innovation.

Infrastructure is the backbone of development, supporting connectivity, economic activity, and access to essential services. In Gaza, its destruction has severed supply chains, isolated communities, and disrupted humanitarian aid. Damage to power grids and water systems has had cascading effects on public health, education, and the economy, deepening the humanitarian crisis (Elkahlout 2024b; ESCWA/UNDP 2024; The World Bank 2024a; Whyte 2024). The industrial sector in Gaza has long been a cornerstone of its economy, providing employment and income. However, the war has devastated this sector, with aerial bombardments destroying factories and warehouses. Gaza's GDP has contracted by over 84% during this recent war, plunging the region into economic collapse. This destruction undermines not only economic stability but also the social fabric, as countless families are left in poverty and insecurity (Nijim 2023; Alagha and Hussein 2024; ESCWA/UNDP 2024; ILO 2024; UNDP/PAPP 2024). Moreover, innovation, dependent on intellectual capital and collaboration, has been stifled by the destruction of educational institutions and research facilities. Laboratories and research centers, crucial for advancements in renewable energy, agriculture, and healthcare, have been damaged or totally destroyed. Restrictions on Internet access, movement, the import of specialized equipment, and interaction with the external world have further isolated Gaza from global scientific and technological networks, halting progress and adaptation to new challenges (UNCTAD 2019; Hamad et al. 2024; Shellah 2024).

The erosion of industry, innovation, and infrastructure in Gaza highlights the challenges of achieving the 2030 Agenda for sustainable development in conflict zones. It underscores the need for preventive measures, international accountability, and addressing the root causes of conflict. Sustainable peace and development in Gaza require not just humanitarian aid but policies enabling the aspirations of SDG 9 and the broader 2030 Agenda. This chapter explores how the war (October 7, 2023–January 19, 2025) on Gaza has severely undermined SDG 9, focusing on the destruction of industry, innovation, and infrastructure. Drawing on UN reports and academic research, it highlights the broader consequences for humanitarian crises and long-term recovery, emphasizing the need for international accountability and sustainable solutions to address the root causes of conflicts.

15.2 Destruction of Critical Infrastructure in Gaza

The recent war on Gaza has caused unprecedented damage to vital infrastructure and has further exacerbated the challenges faced by its population. The destruction of critical infrastructure in Gaza has had devastating consequences on the economic, social, and environmental stability of Gaza. Repeated military assaults have severely damaged essential infrastructure, including power plants, water supply systems, health and education facilities, and communication networks, leaving the region in a state of disrepair. This widespread destruction has not only disrupted daily life but also crippled the foundations necessary for economic activity and development in the Gaza Strip. The deliberate targeting of civilian vital infrastructure and livelihoods has turned it into one of the most uninhabitable places on Earth (Tawil-Souri 2022; Alkhaldi et al. 2024; UNCTAD 2024; Kerbaj et al. 2024; Holail et al. 2024).

The collapse of Gaza's infrastructure during this war has worsened the already dire living conditions in the region, which have persisted for the last two decades. This destruction has impacted every aspect of daily life, from access to essential utilities such as electricity and clean water to opportunities for economic and social development. These harsh conditions have had far-reaching consequences, particularly during emergencies such as the COVID-19 pandemic (Abuhabib et al. 2020). For example, the challenges faced by the education sector during the pandemic illustrate the broader impact of these difficulties. Students in Gaza struggled to adapt to online learning due to frequent electricity outages, poor Internet infrastructure, and the inability of many families to afford personal computers. Medical students reported difficulties accessing educational materials, unreliable Internet connections, and a lack of quiet spaces for study at home. These obstacles severely limited the effectiveness of e-learning and highlighted the systemic vulnerabilities caused by Gaza's deteriorated infrastructure (Ismail et al. 2023).

Ensuring access to affordable, reliable, and clean energy is a cornerstone of SDGs. Energy is not only vital for powering homes, industries, and public services, but it also plays a critical role in promoting economic growth, social equity, and environmental sustainability (AbuQamar et al. 2023). By providing affordable and clean energy, societies can reduce poverty, improve health outcomes, and enable technological advancements while minimizing environmental impact. This goal underscores the importance of transitioning to renewable energy sources and energy-efficient technologies, ensuring universal access to modern energy systems, and addressing energy inequality, particularly in underserved and vulnerable communities (Pan et al. 2023). However, one of the most pressing issues resulting from the war is the destruction of Gaza's energy infrastructure. Prior to the recent escalation of hostilities, Gaza's power supply was already inadequate, with frequent electricity shortages that left the population with as little as 4–6 h of electricity per day (Weinthal and Sowers 2019). The war has drastically worsened Gaza's energy crisis. Power supply across the region has sharply declined, with satellite imagery showing a 94% reduction in power availability by March 2024 compared to pre-war levels. Fuel deliveries have ceased entirely since the war began in October 2023, leaving Gaza's power plants unable to operate. Since the first day of the genocide, electricity was deliberately cut off under the orders of Israel's defense minister, Yoav Gallant (Browne 2024; Devereux 2024). As a result, prolonged blackouts have affected 95% of buildings, causing severe power losses, while 70% of hospitals and schools have been left without electricity, further exacerbating the humanitarian crisis. These energy shortages have plunged the region into a deeper humanitarian and economic crisis, further compounding the challenges of recovery and resilience (ESCWA 2023; Cao et al. 2025).

Furthermore, the destruction of Gaza's energy infrastructure has far-reaching consequences for other critical sectors, particularly water and sanitation. Gaza's water supply infrastructure has been decimated, resulting in a lack of access to clean water for millions of people. Water treatment plants, desalination facilities, and sewage treatment plants have been severely damaged or destroyed by airstrikes. As a result, access to clean drinking water has become increasingly scarce, forcing residents to rely on unsafe sources or purchase bottled water at inflated prices (ESCWA 2023; Cole 2024; Farajallah and Farajallah 2024). The destruction of wastewater treatment facilities has led to untreated sewage being released into the streets and the Mediterranean Sea, creating severe public health risks. Waterborne diseases, such as cholera and dysentery, have begun to spread, particularly among vulnerable populations such as children and the elderly. The damage to water infrastructure has compounded an already dire humanitarian crisis, as the region's limited water resources are further depleted. Water has long been a critical issue in Gaza, with 97% of the region's aquifer deemed unfit for human consumption, forcing residents to rely heavily on bottled water. However, during the genocide, even this limited option became unavailable, exacerbating the already dire humanitarian crisis (Abuzerr et al. 2024b; Zaqout et al. 2024; Zinszer and Abuzerr 2024; Seeberg 2024).

The healthcare sector has suffered devastating consequences from the war, with hospitals and clinics being directly targeted or severely damaged by Israeli assaults and military operations. The vast majority of healthcare facilities in Gaza have been partially or completely destroyed, with only two hospitals—Al-Awda and Shuhada'a Al-Aqsa in the middle area—remaining operational. This destruction has overwhelmed remaining hospitals, which were already operating at or beyond capacity before the war, with an influx of casualties requiring urgent care. Additionally, the severe fuel shortages caused by the blockade have crippled the functioning of healthcare facilities. Without fuel to power generators, hospitals face interruptions in critical services such as life support, surgeries, and refrigeration of medicines. In many cases, surgeries have been performed under the dim light of mobile phones, highlighting the desperate conditions faced by medical professionals striving to save lives amid extreme resource constraints. This dire situation has pushed the health system in Gaza to the brink of collapse, with many reports and on-ground observations indicating that it has already collapsed. Thousands are left without timely or adequate treatment, further worsening the humanitarian crisis and creating ripple effects across other critical sectors (Ahmed et al. 2024; Mansour et al. 2024; Perugini 2024; Alkhaldi and Alrubaie 2024).

Gaza's educational infrastructure has also suffered significant damage. Schools and universities, which once served as beacons of hope for the younger generation, have been destroyed or repurposed as shelters for displaced families. Many scholars refer to this as Scholasticide or Educide (Milton et al. 2023; Dader et al. 2024; Hajir and Qato 2025; Iriqat et al. 2025). The scale of destruction in Gaza's education sector is staggering, with all 12 universities reduced to ruins, leaving higher education in complete disarray. Additionally, at least 88% of schools across Gaza have been either damaged or completely destroyed, severely disrupting the education of tens of thousands of students (UNRWA 2024;

Ageel 2025). The interruption of education in Gaza has long-term implications, as the loss of educational opportunities limits the prospects for future economic development, innovation, and the creation of a skilled workforce.

The war on Gaza has caused catastrophic damage to the information and communication technology (ICT) sector, a vital component of modern infrastructure. Bombings and military operations have destroyed telecommunications towers, data centers, and Internet facilities, with 75% of Gaza's ICT infrastructure damaged and at least 50% completely destroyed according to 7amleh; the Arab Center for the Advancement of Social Media (7amleh 2024). Mobile phone and Internet services throughout the Gaza Strip have nearly ceased due to the depletion of fuel for backup generators. This total outage has paralyzed communication, leaving residents, businesses, and emergency responders entirely cut off and unable to coordinate, further exacerbating the region's already dire humanitarian crisis (Gritten 2023). Existing challenges, including blockades and restrictions on ICT equipment imports, have further exacerbated the crisis, leaving Gaza technologically isolated (Centivany 2024).

Infrastructure development in Gaza faces critical challenges due to the Israeli blockade and the consequences of the recent devastating war. The combination of these factors has severely disrupted efforts to build and maintain essential infrastructure, including electricity grids, water systems, and healthcare facilities. The war has caused inconceivable destruction to existing infrastructure, while the blockade continues to restrict the import of vital materials and technologies needed for reconstruction. Donor policies, often focused on short-term humanitarian aid, exacerbate the situation by neglecting sustainable infrastructure projects, leading to dependency rather than fostering long-term resilience. Additionally, years of isolation and restrictions on access to training and resources have left local capacity for maintenance and repair critically underdeveloped (Centivany 2024).

15.3 Devastating Effects of War on Industry and Economic Activities

Israel has exercised near-total control over Gaza's economy to enforce a state of dependence, a strategy often described as "Israeli economic warfare on Gaza" (Alagha and Hussein 2024) and de-development (Roy 2016). By controlling imports, exports, labor movements, and access to resources, Israel has stifled economic growth and rendered Gaza reliant on Israeli goods and external aid. The dual-use restrictions and militarization of border crossings block essential materials, crippling industries and halting infrastructure development. Military operations have further destroyed Gaza's industrial base and infrastructure, exacerbating poverty and unemployment, with the settler-colonial context posing fundamental challenges to sustainability efforts, as the ongoing struggle is not merely between local groups but between opposing entities with deeply conflicting interests. These policies trap Gaza in a cycle of destruction, dependency, and economic stagnation, deepening its vulnerability and preventing self-sufficiency, and evolving it into an economically isolated state (Adnan 2022; Al-raee 2024; Alagha and Hussein 2024). This isolation, however, dates back to the initial occupation in 1967 rather than solely the well-known siege that began in 2007, highlighting the chronic nature of economic stagnation and the prolonged state of de-development imposed on the region.

The damage trajectory to key economic private sectors from February to October 2024 has been staggering, with losses estimated to range between 62% and 66% (Fig. 15.1). This collapse has disrupted supply chains, reduced the availability of locally produced goods, and left thousands unemployed, pushing families deeper into poverty.

Agriculture and fishing, vital industries for Gaza's economy and food security, have been devastated. Around 70% of farmlands have been destroyed by bombings, while restrictions on access to agricultural areas near the border prevent farmers from cultivating their land. During the recent genocide, constant aerial surveillance

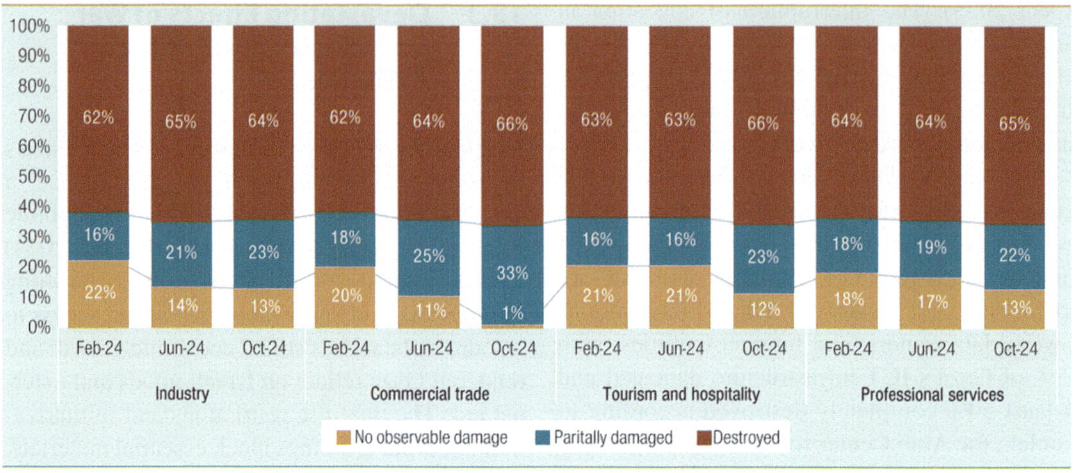

Fig. 15.1 Evolution of private sector damage patterns in Gaza from February to October 2024 (The World Bank 2024b)

and the risk of targeting made it nearly impossible for farmers to work their lands, as any use of equipment was perceived as a threat. The destruction of irrigation systems and water resources has further reduced agricultural productivity. Similarly, the fishing industry has suffered from severe naval restrictions and the destruction of boats and equipment, significantly reducing catches and undermining an essential source of income and nutrition. Prior to the recent war on Gaza, fishing was restricted to within three nautical miles under strict limitations, but during this genocidal war, venturing into the sea meant facing imminent danger, as Israeli naval forces targeted anyone attempting to fish. These losses, combined with the blockade, have exacerbated food insecurity and malnutrition in the Gaza Strip (Abuzerr et al. 2024a; Hassoun et al. 2024).

Indeed, according to the Integrated Food Security Phase Classification (IPC), between November 2024 and April 2025, the entire territory is anticipated to remain classified under emergency food insecurity (IPC Phase 4). During this period, 345,000 individuals are expected to experience severe food shortages, starvation, and the collapse of their livelihoods, while 876,000 people, representing 41% of the population, will face emergency-level conditions (IPC Phase 4). Acute malnutrition is projected to deteriorate, exacerbated by seasonal diseases and overcrowded living conditions (IPC 2024).

The production of construction materials, essential for rebuilding efforts, has been entirely halted. Even before October 2023, restrictions on materials such as cement and iron under dual-use policies, coupled with targeted destruction of production facilities, have crippled this sector. Limited availability of construction materials has delayed reconstruction projects and driven up costs, leaving homes and infrastructure in ruins. This stagnation has also eliminated many jobs, further compounding the economic crisis (UNCTAD 2020; Alagha and Hussein 2024; Buheji and Marouf 2024).

Restrictions on the import and export of goods, compounded by blockades, damaged transportation routes, and the destruction of key industrial zones, have severely undermined sustainable industrialization and exacerbated poverty, unemployment, and economic decline in Gaza (Alagha and Hussein 2024). These factors have severed supply chains, leaving only scarce and heavily restricted aid entering Gaza, and have further isolated what remains of economic activity from essential raw materials and international markets. With the near-total destruction of economic foundations, both small businesses and larger enterprises have been left struggling for survival. The sharp decline in industrial activity has drastically reduced job opportunities, driving unemployment rates to critical levels, with the majority of people unable to work during the initial months. This loss of employment has left many households without a stable source of income, further deepening the economic crisis. Additionally, the lack of security and stability has discouraged investment, further contracting GDP growth and weakening economic resilience. This prolonged stagnation has trapped the region in a cycle of dependency and fragility, creating substantial barriers to recovery and sustainable development (Abu-Rmeileh and Iriqat 2024; Khalidi and Iwidat 2024; PFI/FPCCIA 2024; The World Bank 2024b; UNCTAD 2024).

15.4 Collapse of Innovation and Technological Development

The war on the Gaza Strip has led to the collapse of innovation and technological development, which were already fragile due to years of restrictions and economic isolation. The war has further crippled the Strip's ability to foster technological advancements and industrial innovations that are essential for the long-term development and resilience of its economy. With the destruction of critical infrastructure, restricted access to external markets, and the displacement of skilled workers, Gaza's innovation landscape has been decimated, leaving behind a region that struggles to maintain even basic technological capacities.

Before the recent escalations from October 2023 to January 2025, there were some efforts in Gaza to advance its technological and innovation sectors despite significant constraints. While progress was made through various initiatives and projects, these efforts faced continuous challenges due to long-standing economic and political restrictions. Small startups and tech companies emerged in areas such as software development (ex. Sky Geeks Gaza), mobile applications, and communication technologies. However, numerous challenges persisted as Gaza operated on outdated 2G technology, lacked mobile broadband coverage and reliable electricity, and faced significant restrictions on international connectivity. A recent report further highlighted the sector's struggles, including limited access to advanced technologies and infrastructure, a shortage of skilled labor, and the blockade, which hindered the import of critical components (Bibi and Banat 2024; The World Bank 2024b).

The collapse of innovation in Gaza is not solely due to the destruction of infrastructure but also the severe "brain drain" resulting from the loss of intellectual capital. Many engineers, developers, and other skilled professionals have been displaced by the war or forced to leave Gaza in search of safer and more stable environments (Nijim 2023). This exodus has depleted an already limited pool of talent, exacerbated by long-standing restrictions on education and professional development. The departure of these professionals further undermines Gaza's capacity to develop new technologies, foster innovation in existing sectors, or even rebuild its basic technological infrastructure, compounding the challenges of recovery and progress (Ali Abou-AL-Ross and AlWaheidi 2021; Abusamra 2024).

Moreover, the constant threat of violence, destruction, and insecurity discourages investment in technological research and innovation. Investors and entrepreneurs are less likely to take risks in an environment where the return on investment is uncertain, and the infrastructure required for technological innovation is regularly damaged or destroyed. Therefore, the uncertainty caused by the war has made Gaza an unattractive destination for both local and international investment in the tech sector and other innovative industries. Investors are hesitant to inject capital into a war-ridden region where property is frequently destroyed, and the future of businesses is always uncertain (Fischhendler et al. 2022). For example, fragile regions often face high socioeconomic and political turbulence, which disrupts local systems and further deters investment. Without strong institutions or governance capacity, implementing measures to rebuild or sustain development becomes increasingly challenging (Queiroz et al. 2021).

The war has also led to a decline in education, particularly in the fields of science, technology, engineering, and mathematics. Educational institutions, including universities and technical colleges, have suffered extensive damage, with many buildings destroyed and resources lost, leading to the destruction of the physical environment for learning. Many students, professors, university presidents, researchers, and administrative staff in Gaza have been killed or displaced due to the recent war. The loss of educational opportunities in critical fields directly affects the availability of qualified professionals who could drive innovation and technological development. Without the necessary education and training, the next generation of innovators in Gaza is unable to acquire the skills needed to advance technological capacities in Gaza (Faculty of Education University of Cambridge/Centre for Lebanese Studies & UNRWA 2024; Hussein et al. 2024; Iriqat et al. 2025).

In addition, the destruction of key infrastructure has hindered efforts to develop and deploy innovative technologies in essential sectors such as energy, healthcare, and water management. Prior to this recent war, Gaza had made some progress in adopting renewable energy technologies, such as solar power, to mitigate the region's chronic electricity shortages (Fathi Nassar and Yassin Alsadi 2019; Fischhendler et al. 2022; Younis and Tunç 2023). However, with the destruction of solar panel installations, power grids, and energy infrastructure, these efforts have been severely set back. The limited access to energy resources has undermined the capacity of Gaza to continue its work in the energy sector, particularly in the development and adoption of alternative energy solutions. This setback directly affects Gaza's overall industrial capacity, as many industries rely on a stable energy supply to function (ESCWA 2023; Nassar et al. 2024).

The global healthcare sector has seen transformative advancements in recent years, with innovations like telemedicine, electronic health records, and AI-driven diagnostics revolutionizing patient care. These technologies have improved access, streamlined processes, and enhanced the quality of medical services, especially in resource-constrained settings (Ahmed et al. 2023; Ajaj et al. 2024). In Gaza, the healthcare sector had begun adopting some of these modern solutions, including telemedicine and digital health initiatives, to improve patient care and resource management. However, during the war, the destruction of hospitals and clinics, coupled with disrupted supply chains for medical equipment and medicines, has severely undermined these efforts. The war has stalled the integration of advanced technologies, such as electronic health records and telemedicine services, which could have alleviated pressure on the overstretched system. This has led to a significant regression, with healthcare providers struggling to meet even basic needs under dire circumstances (Parkes et al. 2022; Haimi 2024; Alser et al. 2024).

The digital divide in Gaza has also been exacerbated by this war. Before these hostilities, there was a push to improve digital literacy and Internet access in Gaza, particularly among youth and marginalized communities. Digital literacy programs aimed to empower individuals with the skills necessary for the information economy, thereby providing opportunities for economic advancement. However, the war has further deepened the digital divide, as many people in Gaza now lack access to basic Internet services, digital devices, and modern communication technologies. This has made it even harder for the population to access education, healthcare, and employment opportunities that depend on digital technologies (Traxler 2018; Leake 2024).

Furthermore, international collaborations that could have helped Gaza rebuild and foster technological innovation have been severely hampered by the war. Research institutions, universities, and international organizations that could provide technical support or investment in Gaza's technological infrastructure are reluctant to engage in the region due to the instability and violence. The absence of such partnerships limits Gaza's access to global technological networks, depriving it of opportunities to benefit from international knowledge-sharing, research funding, and technological expertise.

15.5 Implications of the War on Gaza for SDG 9

The war on Gaza has profoundly undermined progress toward SDG 9, particularly its targets on resilient infrastructure (Target 9.1), sustainable industrialization (Target 9.2), and fostering innovation (Target 9.5). The destruction of critical systems has pushed Gaza further into economic isolation and underdevelopment.

15.5.1 Decimation of Infrastructure

Critical infrastructure such as power plants and water systems, healthcare facilities, communication networks, and educational institutions has been destroyed, directly hindering SDG Target 9.1. The destruction of healthcare facilities has severely limited access to medical care, exacerbating public health crises (PHRI 2024). The collapse of communication networks has disrupted emergency responses and access to vital information, while the loss of educational institutions has deprived Gazans of critical opportunities for learning and development (Amnesty International 2024; Buheji and Hasan 2024; Elkahlout 2024a; Iriqat et al. 2025).

15.5.2 Collapse of Industrial Activities

The industrial sector in Gaza, which includes a range of small and medium-sized enterprises across manufacturing, construction, and agriculture, has been severely impacted by repeated assaults and the destruction of key infrastructure, undermining SDG Target 9.2. Many Gazans are employed by international organizations such as the UN, the Palestinian Authority, and local governance structures, but the functionality and persistence of private businesses have been severely hindered by the long-standing blockade and restrictive policies. These challenges have undermined economic activity, limiting industrial growth and sustainability. Restrictions on essential raw materials and the destruction of energy systems and other critical infrastructure, caused by the Israeli military hostilities, have crippled production, increasing unemployment and halting the rebuilding of industrial capacity (Cherqaoui and Hani 2024; The World Bank 2024a; UNCTAD 2024).

15.5.3 Erosion of Innovation and Technological Development

The war has collapsed Gaza's nascent innovation sector, impeding SDG Target 9.5. Startups, tech companies, and educational institutions have ceased operations due to the destruction of facilities and displacement of skilled workers. This has led to a significant "brain drain" stalling advancements in education, innovation, and technical expertise, and further depleting Gaza's intellectual capital. With ongoing challenges and uncertainties, many more professionals are expected to emigrate permanently, exacerbating the region's long-term development prospects (Abusamra 2024; Al-raee 2024; UNCTAD 2024).

15.5.4 Implications for Sustainable Development

The impact of the war on industry, innovation, and infrastructure in Gaza creates cascading effects on multiple SDGs. Economic growth (SDG 8) is severely hindered, leading to critical unemployment levels and worsening poverty (SDG 1). The destruction of critical infrastructure has disrupted access to clean water and sanitation (SDG 6), worsening public health crises (SDG 3) and deepening inequalities (SDG 10). Dependency on humanitarian aid has surged, while the inability to rebuild infrastructure or foster innovation (SDG 9) limits long-term recovery and hinders Gaza's ability to meet emerging challenges, such as climate change adaptation (SDG 13) and sustainable urban development (SDG 11) (ESCWA 2023; Singh and Ru 2023; The World Bank 2024b).

15.6 Strategies for Sustainable Reconstruction and Resilience Building

The extensive destruction caused by the war on Gaza calls for a comprehensive and strategic approach to reconstruction—one that goes beyond merely rebuilding physical infrastructure to fostering long-term resilience and sustainability. However, the success of any reconstruction effort is contingent on the cooperation of Israel, which has historically imposed significant restrictions, and the extent of international pressure to ensure meaningful progress. Given the recurring nature of violence, reconstruction initiatives must integrate sustainable practices while addressing the deep-rooted economic and social vulnerabilities that have long plagued the Gaza Strip. Efforts should prioritize restoring critical infrastructure, promoting economic diversification, and strengthening local industries. Additionally, innovative solutions and technological advancements should be embraced to drive development, while social cohesion, regional collaboration, and robust international support remain essential for achieving lasting recovery.

15.6.1 Restoring Critical Infrastructure

One of the foremost priorities in the reconstruction of Gaza is the restoration of essential infrastructure, including energy, water, sanitation, and communication networks. However, these efforts must go beyond mere rebuilding; they should incorporate resilience and sustainability into their design. This means integrating renewable energy sources, such as solar power, which can reduce Gaza's dependence on external electricity supplies. Harnessing renewable energy solutions is not only a cost-effective response to Gaza's chronic energy shortages but also provides a sustainable path forward, especially given the environmental benefits of reducing fossil fuel consumption (Fathi Nassar and Yassin Alsadi 2019).

Additionally, the restoration of water and sanitation infrastructure in Gaza should prioritize water conservation and efficiency measures. Given the scarcity of freshwater resources, the use of advanced water management technologies, such as desalination plants and wastewater treatment systems, is critical. The use of renewable energy in powering such systems could further reduce the environmental impact while ensuring access to safe drinking water and sanitation for the population. The inclusion of these technologies will be vital in rebuilding Gaza's capacity to meet its water and sanitation needs while reducing its vulnerability to future crises. Addressing the nexus of water and energy security in Gaza is crucial for ensuring sustainable access to clean water and reliable electricity, both of which are essential for improving public health, supporting livelihoods, and fostering long-term resilience in the Gaza Strip (Fathi Nassar and Yassin Alsadi 2019; Zaqout et al. 2024; Rantissi et al. 2024; Abuzerr and Zinszer 2025).

Additionally, efforts must also focus on rebuilding other vital infrastructure, particularly healthcare facilities and educational institutions, which are essential for the well-being and future of the population of Gaza. The restoration of healthcare facilities should prioritize the development of resilient and well-equipped medical centers capable of addressing both immediate health crises and long-term public health needs (Alkhaldi and Alrubaie 2024). Similarly, rebuilding educational infrastructure is critical to providing safe and inclusive learning environments, fostering innovation, and equipping future generations with the necessary skills to rebuild and sustain development in Gaza (Iriqat et al. 2025).

15.6.2 Fostering Economic Diversification and Strengthening Local Industries

The war has wiped out the industrial base of Gaza, leaving its economy in ruins. The war has completely destroyed Gaza's industrial sector, necessitating not only its reconstruction but also strategic efforts to diversify and strengthen local industries. Gaza's economic recovery hinges on rebuilding the foundations of its industrial base and implementing sustainable development initiatives to foster long-term growth. Diversification across the entire supply chain and at various organizational levels is widely acknowledged as a critical strategy for fostering resilience, reducing vulnerabilities, and promoting sustainable economic growth (Hertel et al. 2021). A strategy for sustainable reconstruction must include support for small and medium-sized enterprises (SMEs), which form the backbone of Gaza's economy. Strengthening local industries, particularly in the fields of agriculture, manufacturing, and technology, is essential for fostering economic resilience. Efforts should focus on providing training and resources to help local entrepreneurs scale up their businesses, invest in modern technologies, and create job opportunities that reduce the dependence of Gazans on external aid (Morrar et al. 2024; PFI/FPCCIA 2024).

15.6.3 Fostering Innovative Solutions and Promoting Technological Advancement

Harnessing innovation during crises is vital for rebuilding Gaza's technological and economic capacities. Amid the devastation, innovative solutions such as mobile clinics, telemedicine, and community-based health initiatives have demonstrated the potential to adapt and sustain essential services under extreme conditions. Scaling and institutionalizing such solutions into formal frameworks can ensure resilience and efficiency in responding to future crises (Mansour et al. 2024). Educational institutions play a pivotal role in fostering technological innovation. Projects like TEFL-ePal demonstrate how digital technologies can support local capacity-building by redesigning curricula, introducing open educational resources, and enhancing teacher development. Such initiatives enable adaptive, participatory learning models, building resilience among educators and students in regions facing chronic instability (Scott and Smith 2024).

Furthermore, renewable energy initiatives, such as solar-powered solutions, can mitigate energy shortages and promote sustainability. Studies in Gaza reveal a growing acceptance of solar energy, with its potential to reduce dependency on conventional fuels and enhance socio-economic resilience. However, awareness campaigns and funding mechanisms are essential to overcoming barriers to adoption and expanding renewable energy's role in rebuilding efforts (El-Khozenadar et al. 2022). The reconstruction of Gaza requires a dual focus on advancing digital literacy and creating innovation hubs equipped with resources and mentorship for aspiring entrepreneurs. These centers can foster partnerships with global research institutions, bridging technological gaps and driving knowledge exchange. Reimagining infrastructural development, as seen in proposals like the Internet Pigeon Network, underscores the importance of blending traditional methods with innovative technologies. These creative approaches to communication infrastructure highlight how local contexts and ecological constraints can drive innovative solutions that are both functional and adaptable (Tawil-Souri 2022).

15.6.4 Promoting Social Cohesion, Capacity Building, and Governance Reforms

Sustainable reconstruction in Gaza requires addressing the fractured social fabric, empowering local communities, and strengthening governance. A holistic approach prioritizing social cohesion through community-driven rebuilding efforts can foster a sense of ownership, ensuring long-term sustainability. Capacity-building programs in emerging sectors like renewable energy and technology can further equip the workforce to adapt to evolving economic conditions, aligning with strategies for green economy. Moreover, effective governance is equally essential for sustainable recovery (Al Attar and Brik 2024; Milton 2024; Scott and Smith 2024). Transparent and accountable institutions are needed to ensure efficient resource allocation and project implementation. Strengthening institutional capacity in planning, infrastructure management, and public services, supported by international partnerships, can provide technical assistance, funding, and expertise. These measures are vital for creating resilient, inclusive, and self-sustaining reconstruction frameworks (Hope 2020; Wong 2021).

15.6.5 Fostering Regional Cooperation and International Support

Public-private collaborations are key to driving innovation and sustainability in Gaza. Partnerships between municipalities and private enterprises have shown promise in addressing challenges like waste management and energy efficiency by combining resources and expertise to advance smart urban development (AlAstal 2023).

Regional cooperation with neighboring countries, such as Egypt and Jordan, is equally vital. These partnerships can offer logistical support, market access, and investments in critical sectors like water management, healthcare, and renewable energy, addressing immediate needs while laying the groundwork for long-term sustainability and regional stability. Mobilizing local resources through community-driven initiatives further enhances the efficiency and inclusivity of reconstruction efforts. Integrating appropriate technologies and capacity-building programs ensures that projects are both sustainable and tailored to local needs, fostering resilience and economic recovery in Gaza (Hassan and El-Essy 2014; AlAstal 2023; The World Bank 2024b).

15.7 Conclusions

The recent war on Gaza has caused catastrophic damage to industry, innovation, and infrastructure in the region, significantly hindering progress toward SDG 9. The destruction of critical infrastructure, including energy systems, water networks, healthcare facilities, and educational institutions, has disrupted daily life and undermined the foundations necessary for economic activity and development. The industrial sector has been severely impacted, resulting in widespread job losses, economic decline, and increased poverty. Additionally, the collapse of innovation and technological development has severely constrained the capacity of Gaza to respond to emerging challenges, further hindering its ability to foster growth, resilience, and competitiveness in an interconnected world. The war has worsened existing vulnerabilities, deepening the humanitarian crisis and creating long-term obstacles to recovery. The destruction of educational institutions and research centers has deprived the younger generation in Gaza of opportunities for learning and innovation, restricting prospects for future economic development. The erosion of social cohesion and the displacement of skilled workers have further compounded these challenges, trapping the Gaza Strip in a cycle of dependency and fragility.

Addressing these challenges requires a comprehensive and strategic approach to reconstruction. Efforts should prioritize the restoration of critical infrastructure, the promotion of economic diversification, and the implementation of innovative solutions that enhance resilience and sustainability. International support and regional cooperation will be essential in rebuilding the industrial and technological foundations of Gaza, ensuring access to essential resources, and creating an environment conducive to sustainable development.

Achieving SDG 9 in Gaza demands not only immediate humanitarian aid but also long-term policies that address the root causes of conflicts and promote peace and stability. By integrating sustainable practices, building local capacity, and strengthening governance, Gaza can begin to rebuild its industry, innovation, and infrastructure, laying the groundwork for a more resilient and prosperous future. The international community must take decisive action to support these efforts, ensuring that the population of Gaza can recover from the devastation and achieve aspirations for sustainable development.

References

7amleh (2024) Gaza telecommunications infrastructure: assessment of damages and humanitarian impact

Abuhabib AA, Abu-Aita SN, Procter C, Al-Smeri I (2020) Unique situation of Gaza strip dealing with COVID-19 crisis. Int J Infect Dis 100:149–151. https://doi.org/10.1016/j.ijid.2020.08.070

AbuQamar M, Eltayyan D, Kuznetsova I et al (2023) The impact of access to electricity on mental health in conflict-affected territories: an exploratory study in Gaza. Int J Soc Psychiatry 69:2148–2156. https://doi.org/10.1177/00207640231194479

Abu-Rmeileh Q, Iriqat I (2024) Navigating economic development in Palestine: analysis of key challenges. Int J Sustain Dev Plan 19:591–597. https://doi.org/10.18280/ijsdp.190216

Abusamra A (2024) Brain drain or brain circulation? The impact of Gaza's 7th October war on the displaced academicians and personnel of higher education institutions. J High Educ Theory Pract 24:60–88

Abuzerr S, Zinszer K (2025) Addressing limited access to water sanitation and hygiene in Gaza strip shelters during conflict. Discov Public Heal 22:21–23. https://doi.org/10.1186/s12982-025-00404-0

Abuzerr S, Al-jawaldeh A, Ashour Y et al (2024a) The silent crisis: effet of malnutrition and dehydration on children in Gaza during the war. Front Nutr 11:1395903

Abuzerr S, Hadi M, Zinszer K et al (2024b) Quantitative microbial risk assessment for Escherichia Coli O157: H7 via drinking water in the Gaza strip, Palestine. SAGE Open Med 12:1–11. https://doi.org/10.1177/20503121241258071

Adnan W (2022) From economic integration to near elimination: the economic consequences of isolation. J Dev Stud 58:1160–1180. https://doi.org/10.1080/00220388.2022.2029416

Ageel G (2025) Israel may burn Gaza schools, but Palestinians shall resist. Al Jazeera. https://www.aljazeera.com/opinions/2025/1/13/israel-may-burn-gaza-schools-but-palestinians-shall-resist. Accessed 13 Jan 2025

Agha Z, Esson J, Griffiths M, Joronen M (2024) Gaza: a decolonial geography. Trans Inst Br Geogr 49:e12675. https://doi.org/10.1111/tran.12675

Ahmed SK, Hussein S, Chandran D et al (2023) The role of digital health in revolutionizing healthcare delivery and improving health outcomes in conflict zones. Digit Heal 9:1–3. https://doi.org/10.1177/20552076231218158

Ahmed F, Alhajeeli N, Badheeb M et al (2024) A call for global intervention: the pressing requirement to address the health emergency in Gaza health crisis. Avicenna 2024:1–6. https://doi.org/10.5339/avi.2024.5

Ajaj R, Buheji M, Hassoun A (2024) Optimizing the readiness for industry 4.0 in fulfilling the sustainable development goal 1: focus on poverty elimination in Africa. Front Sustain Food Syst 8:1393935. https://doi.org/10.3389/FSUFS.2024.1393935/BIBTEX

Al Attar HMJ, Brik AB (2024) The historical tragedy: evaluating social protection programs in Gaza. In: Brik AB (ed) Public policy in the Arab World, pp 213–243

Alagha M, Hussein N (2024) The political economy of Israeli economic warfare on the Gaza strip. In: Elkahlout G (ed) Gaza's cycle of destruction and rebuilding. Palgrave Macmillan, Cham, pp 125–145

AlAstal AY (2023) Emerging technological innovation in Gaza strip municipalities: an entrepreneurial approach. J Innov Entrep 12:1–18. https://doi.org/10.1186/S13731-023-00293-0/FIGURES/10

Ali Abou-AL-Ross SN, AlWaheidi T (2021) The impact of living conditions and organizational practices as brain drain cause factors on the human resource assets of the health sector in the Gaza strip—Palestine. Glob J Econ Bus 11:297–314. https://doi.org/10.31559/gjeb2021.11.2.10

Alkhaldi M, Alrubaie M (2024) Roadmap for rebuilding the health system and scenarios of crisis path in Gaza. Int J Health Plann Manag 1–13. https://doi.org/10.1002/HPM.3861

Alkhaldi M, Asi Y, AlBada M, Mansour W (2024) Rethinking and advancing the movement of resistance, activism, and advocacy in health in four central arenas of the Middle East Region. World Med Heal Policy 1–17. https://doi.org/10.1002/WMH3.633

Al-raee M (2024) Gaza's industrial sector: between the hammer of war and the horizon of recovery. Soc Econ Policies Monit "Al Marsad"

Alser K, Mallah SI, El-Oun YRA et al (2024) Trauma care supported through a global telemedicine initiative during the 2023–24 military assault on the Gaza strip, occupied Palestinian territory: a case series. Lancet 404:874–886. https://doi.org/10.1016/S0140-6736(24)01170-X

Amnesty International (2024) 'You feel like you are subhuman'—Israel's genocide against Palestinians in Gaza

Barakat S, Milton S, Elkahlout G (2018) The Gaza reconstruction mechanism: old wine in new bottlenecks. J Interv Statebuilding 12:208–227. https://doi.org/10.1080/17502977.2018.1450336

Bibi A, Banat I (2024) Disrupted futures: how war is crippling Palestinian youth and the tech industry. SPARK. https://spark.ngo/disrupted-futures-how-war-is-crippling-palestinians-youth-and-tech-industry/. Accessed 14 Jan 2025

Browne BC (2024) The Coloniality of enforced starvation: Reading famine in Gaza through an Gorta Mór. J Palest Stud. https://doi.org/10.1080/0377919X.2024.2379709

Buheji M, Hasan A (2024) Beyond famine and chaos—case of Gaza. Int J Manag. https://doi.org/10.17605/OSF.IO/B6SPQ

Buheji M, Marouf M (2024) Mapping reconstruction of GAZA to overcome donors' fatigue. Gradiva 63:116–130. https://doi.org/10.5281/zenodo.13148160

Cao H, Li X, Belabbes S et al (2025) Geo-spatial information science estimates of power supply during the 2023 Gaza humanitarian crisis using night-time light images. Geo-spat Inf Sci. https://doi.org/10.1080/10095020.2024.2439387
Centivany A (2024) Infrastructure development and donation culture in the Gaza strip. Proc Assoc Inf Sci Technol 61:56–67. https://doi.org/10.1002/pra2.1008
Cherqaoui S, Hani M (2024) Gaza: economic collapse and humanitarian catastrophe. Arab Cent Res Policy Stud
Cole J (2024) Weaponizing a water crisis: the destruction of water supplies in Gaza. New Lines Inst:1–7
Dader K, Ghantous W, Masad D et al (2024) Topologies of scholasticide in Gaza: education in spaces of elimination. Fenn Int J Geogr 202:1–12. https://doi.org/10.11143/FENNIA.147002
Devereux S (2024) Was there a famine in Gaza in 2024?
Elkahlout G (2024a) Gaza's cycle of destruction and rebuilding. Understanding the actors, dynamics, and responses. Palgrave Macmillan
Elkahlout G (2024b) War on Gaza: an unceasing humanitarian tragedy. In: Elkahlout G (ed) Gaza's cycle of destruction and rebuilding. Palgrave Macmillan, Cham, pp 209–223
El-Khozenadar HJ, Khatib T, Attaee B, El-Khozondar RJ (2022) Assessment of solar e-cookers social acceptance in Gaza strip. Sci Rep 12:17226. https://doi.org/10.1038/s41598-022-22326-6
ESCWA (2023) War on Gaza: weaponizing access to water, energy and food
ESCWA/UNDP (2024) Gaza war: expected socioeconomic impacts on the State of Palestine—October 2024
Faculty of Education University of Cambridge/Centre for Lebanese Studies & UNRWA (2024) Palestinian education under attack in Gaza: restoration, recovery, rights and responsibilities in and through education
Farajallah I, Farajallah H (2024) Challenges in access to safe drinking water and its impact on maternal and child health in Gaza. Anatol Clin J Med Sci:52–65. https://doi.org/10.21673/anadoluklin.1575390
Fathi Nassar Y, Yassin Alsadi S (2019) Assessment of solar energy potential in Gaza strip-Palestine. Sustain Energy Technol Assess 31:318–328. https://doi.org/10.1016/j.seta.2018.12.010
Fischhendler I, Herman L, David L (2022) Light at the end of the panel: the Gaza strip and the interplay between geopolitical conflict and renewable energy transition. New Polit Econ 27:1–18. https://doi.org/10.1080/13563467.2021.1903850
Gritten D (2023) Gaza faces communications blackout due to lack of fuel. BBC. https://www.bbc.com/news/world-middle-east-67441025. Accessed 14 Jan 2025
Haimi M (2024) Telemedicine in war zones: prospects, barriers, and meeting the needs of special populations. Front Med 11:1417025
Hajir B, Qato M (2025) Academia in a time of genocide: scholasticidal tendencies and continuities. Glob Soc Educ:1–9. https://doi.org/10.1080/14767724.2024.2445855
Hamad BA, Vintges J, Diab R, et al (2024) Shattered lives and dreams: the toll of the war on Gaza on young people
Hassan AM, El-Essy M (2014) Towards sustainable urban livelihoods and poverty reduction in Gaza: the role of partnership and appropriate technology. In: Bolay JC, Hostettler S, Hazboun E (eds) Technologies for sustainable development—a way to reduce poverty. Springer, Cham, pp 171–183
Hassoun A, Al-Muhannadi K, Hassan HF et al (2024) From acute food insecurity to famine: how the 2023/2024 war on Gaza has dramatically set back sustainable development goal 2 to end hunger. Front Sustain Food Syst 8:1402150. https://doi.org/10.3389/fsufs.2024.1402150
Hertel T, Elouafi I, Tanticharoen M, Ewert F (2021) Diversification for enhanced food systems resilience. Nat Food 2:832–834. https://doi.org/10.1038/s43016-021-00403-9
Holail S, Saleh T, Xiao X et al (2024) Time-series satellite remote sensing reveals gradually increasing war damage in the Gaza strip. Natl Sci Rev 11:nwae304. https://doi.org/10.1093/NSR/NWAE304
Hope KR (2020) Peace, justice and inclusive institutions: overcoming challenges to the implementation of sustainable development goal 16. Glob Chang Peace Secur 32:57–77. https://doi.org/10.1080/14781158.2019.1667320
Human Rights Watch (2024) Extermination and acts of Genocide Israel deliberately depriving Palestinians in Gaza of water
Hussein A, Wong S, Bright A (2024) History and impact of Israeli siege and attacks on education in Gaza, Palestine. Oxford Res Encycl Educ. https://doi.org/10.1093/ACREFORE/9780190264093.013.1903
ILO (2024) A year of war in Gaza: impacts on employment and livelihoods in the West Bank and Gaza strip. Bulletin No 5
IPC (2024) IPC global initiative—special brief—Gaza Strip: Projection for 1 November–30 April 2025
Iriqat D, Alousi R, Aldahdouh TZ et al (2025) Educide amid conflict: the struggle of the Palestinian education system. Qual Educ All 2:81–99. https://doi.org/10.1108/QEA-10-2024-0120
Ismail A, Ismail A, Alazar A et al (2023) E-learning medical education in Gaza during COVID-19: students' experiences and policy recommendations. J Med Educat Curri Develop 10:238212052311642. https://doi.org/10.1177/23821205231164228
Jamaluddine Z, Abukmail H, Aly S et al (2025) Traumatic injury mortality in the Gaza strip from Oct 7, 2023, to June 30, 2024: a capture—recapture analysis. Lancet 2024:1–9. https://doi.org/10.1016/S0140-6736(24)02678-3
Kerbaj M, Traboulsi J, Maasri Z, Toukan H (2024) Gaza in plain sight: witnessing in solidarity. J Vis Cult 22:1–19. https://doi.org/10.1177/14704129241227847

Khalidi R, Iwidat Q (2024) Assessing the economic and social impacts of Israel's war on Palestine. Arab Cent Res Policy Stud 7:80–90

Khatib R, McKee M, Yusuf S (2024) Counting the dead in Gaza: difficult but essential. Lancet 404:237–238. https://doi.org/10.1016/S0140-6736(24)01169-3

Küfeoğlu S (2022) SDG-9: industry, innovation and infrastructure. In: Emerging technologies. Sustainable development goals series. Springer, Cham, pp 349–369

Leake M (2024) Palestinians struggle to connect and get news amid digital shutdowns in Gaza: "Without the internet everything stops". Reuters Inst Study Journal https://reutersinstitute.politics.ox.ac.uk/news/palestinians-struggle-connect-and-get-news-amid-digital-shutdowns-gaza-without-internet?utm_source=chatgpt.com. Accessed 14 Jan 2025

Mansour W, Theobald S, Fouad FM et al (2024) The urgent struggle for health justice in Gaza: a crisis of human rights and inequity. Int J Health Plann Manag:1–3. https://doi.org/10.1002/hpm.3882

Milton S (2024) Rethinking the "day after" in the Gaza strip: learning from previous rounds of reconstruction. In: Elkahlout G (ed) Gaza's cycle of destruction and rebuilding. Palgrave Macmillan, Cham, pp 225–240

Milton S, Elkahlout G, Barakat S (2023) Protecting higher education from attack in the Gaza strip. Compare 53:1024–1042. https://doi.org/10.1080/03057925.2021.1987192

Morrar R, Yerousis G, Iwidat Q, Shahwan R (2024) Realities and possibilities of Palestinian entrepreneurial ventures: a comprehensive study on youth entrepreneurship in the State of Palestine. In: Musleh Al-Sartawi AMA, Nour AI (eds) Artificial intelligence and economic sustainability in the era of industrial revolution 5.0. Studies in systems, decision and control. Springer, Cham, pp 925–938

Nassar YF, El-Khozondar HJ, Elnaggar M et al (2024) Renewable energy potential in the State of Palestine: proposals for sustainability. Renew Energy Focus 49:100576. https://doi.org/10.1016/j.ref.2024.100576

Nijim M (2023) Genocide in Palestine: Gaza as a case study. Int J Hum Rights 27:165–200. https://doi.org/10.1080/13642987.2022.2065261

Nijim M (2024) Gazacide: Palestinians from Refugeehood to ontological obliteration. Crit Sociol 1944:1–23. https://doi.org/10.1177/08969205241286530

Pan X, Shao T, Zheng X et al (2023) Energy and sustainable development nexus: a review. Energy Strateg Rev 47:101078. https://doi.org/10.1016/J.ESR.2023.101078

Parkes P, Pillay TD, Bdaiwi Y et al (2022) Telemedicine interventions in six conflict-affected countries in the WHO Eastern Mediterranean region: a systematic review. Confl Heal 16:64. https://doi.org/10.1186/s13031-022-00493-7

Perugini N (2024) Medical Lawfare: the Nakba and Israel's attacks on Palestinian healthcare. J Palest Stud 53:68–91. https://doi.org/10.1080/0377919X.2024.2330366

PFI/FPCCIA (2024) Impact of the Gaza war on private sector and pathways for recovery

PHRI (2024) The destruction of Gaza's healthcare infrastructure

Queiroz C, Norström AV, Downing A et al (2021) Investment in resilient food systems in the most vulnerable and fragile regions is critical. Nat Food 28(2):546–551. https://doi.org/10.1038/s43016-021-00345-2

Rantissi T, Gitis V, Zong Z, Hankins N (2024) Transforming the water-energy nexus in Gaza: a systems approach. Glob Challenges 8:2300304. https://doi.org/10.1002/GCH2.202300304

Roy S (2016) The Gaza strip: the political economy of de-development, 3rd edn. Institute for Palestine Studies, Washington, DC

Samuel MT (2024) The Israel-Hamas war one year later: mass violence and Palestinian dispossession. Middle East Policy 31:3–16. https://doi.org/10.1111/mepo.12792

Scott H, Smith M (2024) Innovation from necessity: digital technologies, teacher development and reciprocity with organisational innovation. Open Learn 39:170–187. https://doi.org/10.1080/02680513.2024.2307627

Seeberg P (2024) Israeli hydro-hegemony and the Gaza war. Middle East Policy 31:33–46. https://doi.org/10.1111/MEPO.12740

Shellah D (2024) War on Gaza: attacks on science are erasing the future. Lancet 404:27–28. https://doi.org/10.1016/S0140-6736(24)00767-0

Singh S, Ru J (2023) Goals of sustainable infrastructure, industry, and innovation: a review and future agenda for research. Environ Sci Pollut Res 30:28446–28458. https://doi.org/10.1007/s11356-023-25281-5

Tawil-Souri H (2022) Speculation on infrastructural ecology: pigeons, Gaza, and internet access. Environ Plan D Soc Sp 40:1064–1081. https://doi.org/10.1177/02637758221139857

The World Bank (2024a) Gaza Strip—interim damage assessment

The World Bank (2024b) Impacts of the conflict in the middle east on the Palestinian economy—December 2024

Traxler J (2018) Digital literacy: a palestinian refugee perspective. Res Learn Technol 26:1–21. https://doi.org/10.25304/rlt.v26.1983

UNCTAD (2019) The economic costs of the Israeli occupation for the Palestinian people: the unrealized oil and Natural gas potential

UNCTAD (2020) The economic costs of the Israeli occupation for the Palestinian people

UNCTAD (2024) Preliminary assessment of the economic impact of the destruction in Gaza and prospects for economic recovery

UNDP/PAPP (2024) Impact of the Gaza war on private sector and pathways for recovery

UNRWA (2024) UNRWA Situation Report #152 on the Humanitarian Crisis in the Gaza Strip and the West Bank, including East Jerusalem. https://www.unrwa.

org/resources/reports/unrwa-situation-report-152-situation-gaza-strip-and-west-bank-including-east-jerusalem. Accessed 31 Dec 2024

Weinthal E, Sowers J (2019) Targeting infrastructure and livelihoods in the West Bank and Gaza. Int Aff 95:319–340. https://doi.org/10.1093/IA/IIZ015

Whyte J (2024) A "tragic humanitarian crisis": Israel's Weaponization of starvation and the question of intent. J Genocide Res:1–15. https://doi.org/10.1080/14623528.2024.2339637

Wong P-H (2021) Economic deprivation: approaches, causes, and consequences for violent conflicts. In: Leal Filho W, Marisa Azul A, Brandli L, Lange Salvia A, Özuyar PG, Wall T (eds) Peace, justice and strong institutions, Encyclopedia of the UN sustainable development goals. Springer, Cham, pp 187–197

Younis SMS, Tunç G (2023) Investigation of electricity distribution, shortages in the Gaza strip, and the role of renewable energy technologies for addressing the electricity crisis. Int J Energy Stud 8:557–579. https://doi.org/10.58559/ijes.1317267

Zaqout M, Fayad M, Barrington DJ et al (2024) Sanitation is political: understanding stakeholders' incentives in funding sanitation for the Gaza strip, Palestine. Third World Q 45:1437–1457. https://doi.org/10.1080/01436597.2024.2318476

Zinszer K, Abuzerr S (2024) Water, sanitation, and hygiene insecurity and infectious disease outbreaks among internally displaced populations in Gaza: implications of conflict-driven displacement on public health. J Water Sanit Hyg Dev 14:1182–1192. https://doi.org/10.2166/washdev.2024.361

16 Deepening Inequalities: Social and Economic Disparities in War-Torn Gaza

Yasmine Fakhry, Dalal Iriqat, and Abdo Hassoun

Abstract

This chapter examines the deepening social and economic inequalities in Gaza resulting from the recent war (October 2023–January 2025) on Gaza, with a particular focus on Sustainable Development Goal 10 (SDG 10), which seeks to reduce inequality within and among countries. The war has exacerbated preexisting disparities, disproportionately impacting the most vulnerable populations. The destruction of infrastructure, businesses, and critical services, compounded by the siege, has led to soaring unemployment, poverty, and a sharp decline in living standards. As families struggle to survive, there has been a rise in child and forced labor, further entrenching social and economic disparities. Access to education and healthcare has severely deteriorated, contributing to the deepening of inequality. Women and children, in particular, face heightened risks of violence, exploitation, and marginalization, while persons with disabilities encounter additional barriers to basic services. The economic consequences of this war are devastating, with hostilities hindering investment and economic development, thus perpetuating a cycle of poverty. This work also reviews the international community's response, highlighting the need for stronger global action to support Gaza's recovery and reducing inequalities. Drawing on data and reports from UN agencies and research studies, the chapter concludes by offering policy recommendations aimed at addressing these inequalities, emphasizing the importance of sustainable peace, international solidarity, and inclusive development strategies. The goal is to contribute to a deeper understanding of the obstacles to social and economic equity in the region and offers insights into potential pathways for a more just and equitable future.

Keywords

Discrimination · Economic development · Economic inequality · Equality · SDG 10 · Social inequality · Sustainable development

Y. Fakhry
School of Social Sciences, University of East London, London, UK
e-mail: u2009750@uel.ac.uk

D. Iriqat (✉)
Arab American University of Palestine, Ramallah, Palestine
e-mail: dalal.s.iriqat@aaup.edu

A. Hassoun
Sustainable AgriFoodtech Innovation & Research (SAFIR), Arras, France

Faculty of Agricultural Engineering, University of Aleppo, Aleppo, Syria
e-mail: a.hassoun@saf-ir.com

A. Hassoun (ed.), *War on Gaza*, Sustainable Development Goals Series,
https://doi.org/10.1007/978-3-031-88500-6_16

16.1 Introduction

Social inequality is a concept used to designate the unequal distribution of resources, social positions and rewards in a certain society. Economic inequality, characterized by the concentration of wealth among a small percentage of the population, leads to significant disparities in living standards (Little 2023; Belle and Bullock 2023). Inequalities have devastating impacts on multiple levels. For instance, economic inequality leads to educational disparities, with educational attainment gaps and lower educational outcomes for disadvantaged groups. Additionally, social identities can exacerbate inequalities, as individuals from marginalized groups face systemic barriers that hinder their life prospects (Yılmaz and Bağçe 2022; Simangunsong et al. 2023).

Inequalities pose a threat to long-term economic and social development, hindering poverty reduction and diminishing people's sense of self-worth and commitment. Extreme inequality can stifle economic growth by preventing marginalized groups from accessing resources, creating social exclusion and perpetuating cycles of poverty. Moreover, inequality can worsen environmental problems, as people with limited resources may overuse them due to necessity, while the wealthy often waste resources, exacerbating environmental strain (Sinha 2019; Gradín et al. 2021).

Inequality is a persistent issue with far-reaching consequences, influenced by factors like income, gender, age, disability, sexual orientation, race, class, nationality, ethnicity, religion, and opportunity (Veenstra 2011). This can lead to crime, disease, and environmental degradation. Addressing inequality is crucial for achieving sustainable development and ensuring a better future for all. Other factors, such as COVID-19 pandemic can exacerbate existing inequalities (Gauthier et al. 2021; Panneer et al. 2022). While the incomes of the poorest 40% of people in many countries were increasing faster than the overall national income before the pandemic, evidence suggests that the pandemic might have slowed down this progress. Furthermore, the pandemic significantly increased inequality between countries, marking the largest rise in three decades (Cojocaru et al. 2022; Remington 2023).

On the level of social cohesion, inequality can fuel social unrest and conflict, destabilizing societies and hindering effective governance. Addressing these issues are essential for achieving sustainable development (Gradín et al. 2021). The most vulnerable are disproportionately affected by inequality, including women and children who lack access to healthcare, older persons, migrants, and refugees who face discrimination and limited opportunities. One in six people globally has experienced discrimination, with women and people with disabilities being particularly affected. These inequalities highlight the urgent need to combat discriminatory practices and hate speech (United Nations 2023).

Countries experiencing colonization or occupation are particularly vulnerable to inequality on multiple levels (Ashiagbor 2021). In Palestine, the Israeli occupation has created significant disparities and inequalities in access to healthcare, education, and other essential services for Palestinians, especially for Gazans (Rosenthal 2021; Tanous et al. 2023). Indeed, in Gaza, the destruction of critical infrastructure, including hospitals, schools, and water sanitation systems, during the recent war has severely exacerbated these challenges, leaving millions without access to basic needs. The blockade further restricts the entry of medical supplies, educational resources, and essential goods, deepening the humanitarian crisis.

Therefore, it is crucial to recognize that the social and economic disparities in Gaza result from systemic barriers compounded by the strategy of collective punishment; a component of genocide aimed at dismantling the foundations of Gazan society. This deliberate destruction targets not only physical infrastructure but also the social fabric, undermining access to healthcare, education, and economic opportunities. Addressing these profound inequalities requires a multifaceted approach that confronts these systemic barriers head-on, including international advocacy, sustainable development initiatives, and ensuring accountability for violations of international law. This chapter explores the systemic inequalities that characterize Gaza under occupation and siege, exacerbated by the war. It focuses on how these disparities have profoundly affected access to healthcare, education, and economic opportunities, while reversing progress toward achieving Sustainable Development Goal 10 (SDG 10). The chapter also examines the broader socioeconomic consequences of the conflict, evaluates the international response to this humanitarian crisis in Gaza, and offers policy recommendations to reduce inequalities and promote sustainable recovery in the region.

16.2 Sustainable Development Goal 10 (SDG 10) and Its Relevance to Gaza

SDG 10 seeks to reduce inequality within and among countries by addressing disparities in income, access to resources, and opportunities. It aims to empower marginalized populations and promote social, economic, and political inclusion. Key targets include ensuring the income growth of the bottom 40% of the population at a rate higher than the national average (Target 10.1), eliminating discriminatory laws and policies that perpetuate inequality (Target 10.3), and fostering the representation of developing countries in global decision-making (Target 10.6). SDG 10 highlights the importance of fair-trade practices, equitable access to essential services, and social protection systems to create a more inclusive and just society. However, there are significant obstacles to achieving these targets, particularly in conflict-affected and politically unstable regions. Key obstacles include economic disparities, social exclusion, political barriers, unequal access to education and healthcare, and the effects of globalization. For instance, economic inequality remains a significant barrier, with wealth disparities impeding equality efforts (Kuhn 2020; Bhandari 2024).

Conflicts and wars significantly hinder progress toward achieving SDG 10 by exacerbating economic disparities, limiting access to education and healthcare, and displacing vulnerable populations. The destruction of infrastructure, loss of livelihoods, and political instability caused by conflict often push marginalized communities deeper into poverty. Wars disrupt social cohesion, marginalize women and children, and create barriers to upward mobility, entrenching cycles of inequality. In many cases, conflict-affected regions see a reversal of progress, with disparities widening and vulnerable groups becoming increasingly reliant on external aid. These challenges highlight the need for targeted strategies to address inequality in conflict zones (Omoeva et al. 2018; Ujunwa et al. 2021; Bin-Nashwan et al. 2024).

In Gaza, the relevance of SDG 10 is particularly acute. The Israeli-imposed siege and recurring military conflicts have devastated economy in Gaza, restricted access to essential services, and entrenched inequality. High unemployment rates, reliance on foreign aid, and limited access to healthcare and education have left Gazans among the most vulnerable populations globally. The destruction of infrastructure and the imposition of trade restrictions have severely undermined Gaza's economic potential, transforming it into a consumptive economy marked by de-development (Roy 2016). Social and political exclusion further marginalize Palestinians, undermining their ability to achieve economic and social mobility (Tartir et al. 2021; Elkahlout 2024). Addressing these challenges through the lens of SDG 10 requires dismantling systemic barriers, fostering inclusive economic recovery, and rebuilding resilient infrastructure to reduce inequalities and support long-term development in Gaza.

16.3 Historical Overview of the Israeli-Palestinian Conflict

Israel was created 75 years ago on the land of Palestine when a settler-colonial project was kick-started and never put to a halt under the framework of an illegal military occupation. In Gaza, civilians have been suffering under a deadly blockade for 17 years, during which Israeli military forces waged four wars, before the recent one. Within its 365 km^2, the Gaza Strip is home to a dense population of 2.3 million people, the most of whom are refugees. The area is severely restricted by an Israeli military occupation that extends across land, air, and sea. Often referred to as the largest open-air prison in the world, this little strip of land is beset by severe humanitarian crises that are made worse by military bombardments and rigorous blockades (Abdelnour 2023; Tanous et al. 2023; Mahomed 2023; Agha et al. 2024). The critical moment came with the victory of Hamas in 2006, which provided both a focus for international criticism and a rationale for Israeli policies that have persisted. But to see Gaza's predicament exclusively as Hamas' doing would be to ignore the intricate historical and geopolitical circumstances that have shaped the area since 1948.

Netanyahu has framed the Palestinian right to self-defense under Article 51 of the UN Charter and their right to resistance as acts of terror. Although Israel withdrew from Gaza and relocated 9000 settlers, the Strip remained under full control of Israeli occupation authorities. According to Ariel Sharon's aide, Dov Weisglass, the 2005 disengagement was not intended to grant Gaza freedom but to ensure that "there will not be a political process with the Palestinians." Today, Israel is using Hamas as a fig leaf for its annexation efforts, and the rockets of the resistance are labeled terror. Hamas relies on rocket fire as a negotiating tactic to pressure Israel to ease access of goods and people and loosen the blockade. On the other hand, Israel employs military might to deter Hamas. These tactics of Hamas and Israel enabled short-term victories for both at the expense of a long-term resolution. Uncontested, Palestine lives under a prolonged Israeli military occupation that must end (Iriqat 2023a, 2024a; Boisen 2024).

Israel and its allies portrayed the 2005 unilateral withdrawal from Gaza as a significant step toward peace, framing it as a move to liberate Gaza. However, this action was executed without coordination with the Palestinian Authority (PA) or the Palestinian Liberation Organization (PLO), which undermined Palestinian unity and governance. Instead of fostering stability, the withdrawal deepened internal divisions by empowering Hamas, a group initially established as a resistance movement against Israeli occupation, while marginalizing the PA. This uncoordinated disengagement solidified the geographic and political split between Gaza and the West Bank, entrenching the division between Fatah and Hamas. Over 17 years of discord have been described as a "Nakba" for Palestinians, further weakening their leadership's ability to unite or effectively govern. This fragmentation has also been used as a justification for denying Palestinians the political cohesion necessary for self-determination (Heer 2023).

Occupied Palestine territories in general and Gaza Strip in specific has been grappling with repeated conflicts and systemic violence is plagued by deep-rooted social and economic inequalities. High unemployment, limited job prospects, and a heavy reliance on foreign aid have exacerbated economic disparities, leading to significant differences in the well-being of its citizens. Moreover, the rapidly growing population has highly increased competition for limited resources and jobs. On top of that, the political landscape has also transformed Gaza's economy, leading to a heavy reliance on aid and the emergence of a tunnel economy, which has disrupted traditional economic structures (Tannira 2021; Alagha and Hussein 2024).

16.4 Implications of War on Gaza for Society and Development

Past and recent wars on Gaza have caused extensive destruction to critical infrastructure, severely impacting essential civilian services, including healthcare, education, food production, and water systems. The massive destruction of civilian infrastructure during prolonged war and occupation has far-reaching consequences. It disrupts essential services like water and energy supply, leading to lasting harm to both people and the environment (Weinthal and Sowers 2019; Hassoun et al. 2024).

16.4.1 Impact on Critical Infrastructure

The collapse of food production facilities and contamination of water resources have significantly exacerbated food and water insecurity in Gaza. These crises, combined with a severe lack of fuel and an energy crisis that cripples hospitals, water pumps, and essential services, have pushed the region to the brink of famine. Large portions of the population face severe malnutrition, hunger, and deteriorating living conditions, further deepening the humanitarian catastrophe (Alnabih et al. 2024; de Waal 2024). Even at the early stages of the war, significant damage has been inflicted on critical infrastructure, including healthcare and educational facilities, alongside severe food, water, and energy crises (Asi et al. 2024; Iriqat et al. 2025). For example, hospitals such as Al-Shifa have been severely damaged, with critical units like the intensive care unit (ICU) rendered inoperable due to structural damage, lack of fuel, and medical supplies. This has led to devastating consequences, including the deaths of patients reliant on life-saving technologies such as ventilators and incubators, with over 70% of ICU patients succumbing to these conditions (Shellah 2024a). The housing sector and population centers have borne the brunt of the destruction, accounting for approximately 69% of the total damaged structures in the Gaza Strip, with an estimated 245,123 housing units affected (The World Bank 2024; UNITAR-UNOSAT 2024).

Access to education is essential for reducing inequalities, equipping individuals with the tools for upward mobility and societal progress. In Gaza, the tight siege and unprecedented violence not only restrict educational opportunities but also deepen socioeconomic disparities, entrenching cycles of poverty and inequality. International collaboration is needed to provide digital resources, professional training, and mental health support, aligning with SDG 10 to reduce inequalities. Empowering Gaza's youth through education fosters resilience, counters poverty, and contributes to a more equitable future. Even before the recent war, higher education in Gaza faced significant challenges due to repeated conflicts and a prolonged siege, resulting in loss of life among staff and students, as well as damage to university infrastructure (Milton et al. 2023).

Since 2007, the siege and military actions have severely impacted schools and educational facilities, rendering them unsafe and affecting students and teachers economically, socially, and psychologically. The Israeli occupation and siege have crippled Gaza's education system, leading to critical shortages of supplies and technology, such as computers and science labs, while isolating educators from global peers due to permit restrictions for professional exchanges (Hussein et al. 2024). The destruction of Gaza's educational system denies individuals their right to education and undermines society's intellectual and human capital. This systematic targeting of teachers and students perpetuates oppression, transitioning from spatial cleansing and de-development to "Educide" a deliberate strategy to dismantle environments fostering critical thinking and knowledge exchange. By eroding cultural heritage and intellectual potential, these actions suppress resistance and limit Palestinians' access to knowledge and progress, erasing the future in Gaza (Rabaia and Habash 2024; Shellah 2024b).

The widespread destruction of Gaza's infrastructure, including healthcare, education, and water systems, calls for sustainable development initiatives to rebuild resilient structures. Priorities include sustainable water and energy systems to reduce dependence on foreign aid and enhance self-reliance. Resilient infrastructure supports SDG 10 by creating equal access to essential resources, promoting economic and social mobility, and enabling communities to recover more effectively from crises. By prioritizing infrastructure development, Gaza can address inequalities that hinder growth, aligning its recovery with global goals for equitable development.

16.4.2 Impact on Women, Children, and Persons with Disabilities

The war on Gaza has disproportionately affected women, children, and persons with disabilities, highlighting severe humanitarian crises. Reports of early 2024 indicate that over 25,000 people were killed in less than 4 months, with nearly half being children, and many more injured (Wells et al. 2024). As per Elnakib et al. (2024), 70% of the fatalities were women and children, exacerbating the plight of these vulnerable populations. After a year of war, over 42,000 lives have been killed in Gaza (Stamatopoulou-robbins 2024; Khalifa Aleghfeli 2024). A higher percentage is expected to be observed due to failure to manage obstetrical complications. The intersection of gender, age, and disability creates compounded vulnerabilities, as children with disabilities are often unable to escape danger and face neglect (Ćerimović 2023). This can alter Gaza population's demographics and leave communities there with fewer women, further diminishing their social power and economic opportunities.

16.4.2.1 Impact on Women

The war in Gaza has had devastating consequences for women, exacerbating gender inequalities and severely impacting their physical and mental health. Pregnant women face heightened risks due to limited access to prenatal care, essential resources like food and clean water, and functional medical facilities. The destruction of healthcare infrastructure has forced many women to give birth in unsanitary and overcrowded conditions, with some undergoing cesarean sections without anesthesia. The lack of resources has also contributed to a 300% increase in miscarriage rates, with malnutrition, dehydration, and stress being key factors in these adverse outcomes (Ben Saad 2024; Jangda et al. 2024; Sabet et al. 2024; Irfan et al. 2024).

Displacement and overcrowded shelters have worsened conditions for women, depriving them of privacy and increasing their vulnerability to sexual and gender-based violence. Many women are unable to access menstrual hygiene supplies or even basic necessities like clean water, further affecting their dignity and health. Surveyed displaced women report showering as infrequently as once every 3 weeks, while nearly 96% lack access to basic hygiene products, amplifying the risks of infections and other health complications (UNDP 2024).

The impact of the war extends beyond health, stripping women of social and economic agency. A majority of displaced women have lost access to livelihoods, and many are now the sole providers within their families. However, they lack sufficient financial resources to meet basic needs such as food, shelter, and healthcare. Addressing these challenges requires urgent international intervention to deliver essential services, safeguard the rights of women, and ensure their safety. Support for programs tailored to gender-specific needs is crucial to restoring dignity and resilience among women in Gaza (UNDP 2024; Zaiter 2024).

16.4.2.2 Impact on Children

Children are disproportionately affected by wars and conflicts, with recent data showing that one in six children globally are impacted by armed violence, a sharp increase from one in ten. These children face severe challenges, including displacement, loss of access to education and healthcare, and exposure to violence. The destruction of infrastructure leaves many vulnerable to malnutrition, disease, exploitation, and long-term psychological trauma such as anxiety and post-traumatic stress disorder (PTSD) (Bürgin et al. 2022; Hazer and Gredebäck 2023; Molloy and Bearer 2024; Jain et al. 2025).

The physical and mental health consequences of war are particularly acute among children. Educational disruptions caused by school closures and constant violence impede cognitive development and learning, making it difficult for children to keep up academically and develop critical thinking skills. Prolonged exposure to trauma leads to emotional and psychological disorders, including PTSD, anxiety, and depression, affecting emotional regulation and causing long-term mental health challenges (Buheji and Buheji 2024; D'Andrea et al. 2024; Elnakib et al. 2024).

The destruction of family and community structures, essential for the social development for children, hinders their ability to form healthy relationships, develop empathy, and learn social norms. Speech and language development, crucial for communication and learning, is also impacted by the stress of war and disrupted family interactions. Research highlights that during war, children's education significantly suffers, leading to higher rates of delayed language skills acquisition, lower IQ scores, and poorer overall academic achievement. These cognitive impairments further limit the overall health and socioeconomic opportunities of affected children (Bürgin et al. 2022; Cozza 2024; Ferrara et al. 2024).

Child labor in Gaza and the West Bank remains a critical issue, with many children involved in hazardous activities such as agriculture and street vending in conflict zones. These activities expose children to dangerous chemicals, unsafe tools, and life-threatening risks, often resulting in injuries or death (Jabr and Berger 2022). It was reported that 79.1% of surveyed children experienced work-related injuries, with younger children, boys, and those attending UNRWA schools being most affected (Jildeh et al. 2014). Child labor exacerbates inequality by limiting access to education and hindering social mobility, trapping families in cycles of poverty. Policies and interventions should safeguard children's rights by eliminating exploitation through labor, addressing their interconnected health and social needs holistically, and promoting protective factors such as family income, peer relationships, and access to education. For children with severe psychiatric disorders, targeted therapeutic interventions within integrated care plans are essential, alongside further research to address their specific mental health needs in diverse socioeconomic contexts (Thabet et al. 2011).

The economic devastation in Gaza due to prolonged conflict has significantly worsened job scarcity and reduced income opportunities, particularly for youth and marginalized groups. To mitigate these challenges, economic interventions such as microloans, vocational training, and local employment programs are recommended to support job creation and financial stability. These initiatives align closely with SDG 10 to reduce economic inequalities, as they help provide vulnerable populations with income-generating opportunities. By investing in programs that stimulate economic activity, stakeholders can contribute to a more stable and self-sufficient economy in Gaza, facilitating a pathway out of poverty and ensuring that economic growth benefits all segments of society.

16.4.2.3 Impact on Persons with Disabilities

The lack of readily available medicine and tools that help people with disabilities, such as wheelchairs, glasses, and hearing aid, makes manageable health problems much harder to manage and leads to increased disability and higher mortality. Persons with disabilities and medically vulnerable Palestinians in a region experiencing genocide, occupation, and a long-standing siege are at a higher risk of dying due to both sudden acts of violence and the ongoing, slower harm caused by these conditions (Mosleh et al. 2018; ACAPS 2024).

Palestinian adults in Gaza with disabilities reported significant needs, including feelings of insecurity (69.6%), dissatisfaction with quality of life (39%), difficulty adapting to the demands of their circumstances (53.3%), a strong desire for love and social recognition (69.4%), and heightened awareness of their personal potential and abilities (69.4%) (Thabet et al. 2018). Men with disabilities reported higher self-esteem than women with disabilities, while those with no income experienced lower social adaptation, life status, social role, and self-concept. These findings highlight the crucial need to focus on improving social support, self-esteem, and well-being for disabled Palestinian adults and their families. A previous similar study found that 42.6% of adults with physical disabilities were diagnosed with psychiatric morbidity, compared to 37.3% of those with other disabilities. Additionally, 10.1% of participants with physical disabilities experienced moderate to severe depression, while 6.6% of those with other disabilities reported similar levels of depression (Thabet et al. 2015).

The compounded challenges faced by persons with disabilities in Gaza, due to the lack of medical supplies and assistive tools, demand immediate and sustained attention. Recommendations include partnerships with nongovernmental organizations to deliver essential aids and medical care, ensuring that the needs of persons with disabilities are met despite the constraints of conflict. This approach is deeply aligned with the commitment of SDG 10 to leave no one behind. By addressing health inequalities and enhancing support for individuals with disabilities, these actions foster social equity and ensure that all individuals have the resources necessary for a dignified life. This commitment is vital for achieving SDG 10 targets of inclusive development.

16.5 International Response to the Recent War on Gaza

The war is not solely a war against Hamas but raises broader concerns about the Palestinian right to self-determination and significant violations of international law. The displacement of over 1.9 million civilians in Gaza (Kerbaj et al. 2024) constitutes forced displacement, which qualifies as a war crime under Article 49 of the Fourth Geneva Convention. These actions resemble the events of the Nakba and highlight the urgent need for accountability, as stipulated under Article 146 of the convention, which mandates the prosecution of individuals responsible for war crimes. Despite numerous condemnations from international organizations, these efforts have failed to produce tangible outcomes. The influence of a well-organized and well-funded Israeli lobby has played a significant role in obstructing accountability and meaningful action, exacerbating systemic inequalities and injustices faced by Palestinians. This situation underscores the urgent need for the global community to enforce international law and ensure justice.

Israel targeted children, women, hospitals, churches, mosques, and clinics. The narrative that delegitimizes Palestinian resistance, and that attributes the unprecedented crisis in Gaza to Hamas is advantageous to a number of geopolitical players, most notably Israel, which uses it to defend its atrocities in Gaza (Iriqat 2023b). However, this perspective overlooks the historical and geopolitical realities that have defined the region since 1948. In particular, it disregards the confiscation of land by the Israeli regime, the expansion of illegal settlements, arbitrary detentions and extrajudicial killings, severe restrictions on mobility, deliberate suppression of economic activity, and frequent violations of human rights (Browne 2023; Nijim 2024).

Despite public outrage all around the capitals calling for ceasefire and to a halt of the genocidal war in Gaza, multilateralism has not served justice. The UN and peaceful international legal and diplomatic instruments could not serve humanity in 2024. However, while diplomatic and legal approaches demand patience, relying on peaceful and lawful means does not diminish the undeniable truth that Palestinians have the right to resist and the right to self-determination. It is essential to recall UN General Assembly Resolutions 3314 (1974) and 37/43 (1982), which affirm the right of Palestinians and all occupied nations to resist using all legitimate means, including armed resistance, in their struggle for liberation, dignity, and freedom from military occupation. Now Israel is being tried at the International Court of Justice (ICJ) that had ruled for plausibility of Genocide after South Africa filed against the state of Israel, many states joined South Africa in its case afterwards (Boisen 2024; Iriqat 2024a; Whyte 2024).

Investigating diplomacy and the role of the international community is pertinent to our discussion of Gaza. International interventions and diplomatic attempts have mostly failed to alleviate the basic injustices that Palestinians must endure or to improve the situation in Gaza. Many international condemnations of the conflict have proven ineffective, highlighting the shortcomings of existing diplomatic frameworks and the outsized influence of pro-Israel lobbying in shaping global policy on the issue. This lack of accountability perpetuates impunity, posing a significant obstacle to achieving justice and lasting resolution.

The Saudi-led international alliance, launched in September 2024 during the UN General Assembly to advance the implementation of the two-state solution, has recently gained significant momentum, with 94 states now supporting the initiative. This coalition, driven by Arab and European countries with full UN backing, represents a crucial international effort toward resolving the conflict. However, it is imperative to caution against reconstruction plans for Gaza that perpetuate the rhetoric of development under occupation. Even with millions of dollars continuing to fund Israeli military operations, the Marshall Plan for Gaza cannot address the deeper structural issues of occupation and inequality. A comprehensive political framework must take precedence over economic measures to ensure a just and sustainable resolution for the region.

16.6 Mitigation Measures and Recommendations

Addressing systemic inequalities and the humanitarian crisis in Gaza requires comprehensive, multi-level mitigation measures and actionable recommendations. These approaches must aim to alleviate immediate impacts of war while fostering long-term resilience and equity.

The immediate priority is to provide humanitarian aid that meets basic needs of the population in Gaza. Delivering essential medical supplies, including medications, surgical kits, and assistive devices is crucial. Collaborative efforts with international organizations like the World Health Organization (WHO) and Médecins Sans Frontières can enhance the effectiveness of these interventions (Ahmed et al. 2024; Elnakib et al. 2024). Addressing hunger through food distribution programs in partnership with agencies such as the World Food Programme (WFP) is vital. Long-term strategies must focus on enhancing local food production through sustainable agricultural practices, which also align with SDG Target 10.1 by supporting income growth for disadvantaged groups and reducing reliance on external aid. Restoring safe housing and repairing water sanitation systems are also urgent to ensure secure living conditions and prevent disease outbreaks (Asi et al. 2024; Asi 2020; Hassoun et al. 2024).

Extensive destruction of infrastructure in Gaza necessitates targeted efforts to rebuild healthcare facilities, schools, and essential utilities. Reconstructing hospitals and clinics must prioritize emergency preparedness, maternal and child health, and rehabilitation services for individuals with disabilities (Ben Saad 2024; Shellah 2024a). Rebuilding educational institutions and integrating digital learning tools can bridge learning gaps caused by prolonged closures, while efforts to make education more inclusive align with SDG 10.3 by eliminating discriminatory barriers (Milton et al. 2023; Hussein et al. 2024). Sustainable energy solutions, such as solar power systems, and water desalination technologies are required to address chronic energy and water shortages, ensuring access to essential services and supporting SDG 10.2 by fostering social inclusion through improved infrastructure (Weinthal and Sowers 2019; ESCWA/UNDP 2024).

Economic recovery in Gaza must focus on creating sustainable livelihoods. Vocational training programs tailored to market demands can enhance employability, particularly for youth, women, and individuals with disabilities, contributing to SDG 10.4 by addressing economic disparities through improved labor policies (UNCTAD 2024; UNEP 2024). Microfinance initiatives and support for small and medium-sized enterprises can stimulate local economic activity and create jobs, while investments in modern farming techniques can boost agricultural productivity and enhance food security (Tannira 2021; Alnabih et al. 2024).

Ensuring accountability for war crimes and upholding international law are vital for achieving justice and long-term peace. Strengthened efforts to prosecute war crimes through the International Criminal Court (ICC) and the International Court of Justice (ICJ) must be supported. Advocacy campaigns to pressure governments and institutions to lift the blockade on Gaza and recognize the right to self-determination for Palestinians are essential steps toward justice and equality, aligning with SDG 10.6, which promotes representation of all nations in decision-making processes (Heer 2023; Boisen 2024; Iriqat 2024b). Multilateral diplomacy should focus on advancing peace initiatives, such as the Saudi-led international alliance, emphasizing a two-state solution rooted in justice and equality.

Rebuilding community resilience is critical to fostering long-term recovery. Expanding mental health services to address trauma among children, women, and individuals with disabilities is essential. Community-based and school-based interventions can promote healing and emotional well-being, supporting SDG 10.2 by fostering social inclusion and reducing inequalities (Bürgin et al. 2022; D'Andrea et al. 2024). Gender-responsive programs should provide women with economic opportunities and ensure access to reproductive healthcare and hygiene supplies. Addressing child labor through education incentives and social protections is vital, aligning with SDG 10.1 by improving opportunities for the most vulnerable. For individuals with disabilities, rebuilding efforts must prioritize accessibility and inclusivity, providing assistive technologies and fostering social inclusion through community programs (Thabet et al. 2018; UNDP 2024; Elnakib et al. 2024).

Reconstruction and recovery efforts in Gaza must align with SDGs, particularly SDG 10 on reducing inequalities. Policies should ensure equitable access to resources such as healthcare, education, and economic opportunities, focusing on marginalized groups. Strengthening governance structures in Gaza to enhance transparency, accountability, and inclusivity is essential. Empowering local organizations and civil society groups to lead development initiatives can foster ownership and sustainability. Promoting international solidarity through partnerships with global stakeholders can provide the technical and financial support needed for sustainable recovery (Tartir et al. 2021; Tanous et al. 2023; Bhandari 2024).

Mitigating impacts of war in Gaza and addressing systemic inequalities requires an integrated approach that combines humanitarian relief, infrastructure rebuilding, economic empowerment, and legal accountability. By aligning efforts with the SDGs, particularly SDG 10, stakeholders can create pathways for resilience, equity, and long-term peace. International solidarity and sustained commitment are essential to transforming the future of the region and ensuring residents of Gaza can live with dignity and hope.

On 15 January 2025, an agreement was reached between Hamas and Israel to end the war in Gaza. However, Israel may continue using the presence of Hamas as a pretext for launching new wars if the root causes of the conflict remain unaddressed. The success of any agreement to end the conflict in Gaza largely depends on the details of the deal, the commitment of all parties, and the guarantees of enforcement and monitoring mechanisms. Without a comprehensive approach, any resolution risks being temporary.

Future outcomes hinge on the scope and objectives of the agreement. A deal that includes clear steps to rebuild Gaza and improve living conditions, such as easing the blockade and enhancing access to essential services like electricity, water, and healthcare, could lead to gradual relief for the population. However, agreements limited to prisoner exchanges or temporary ceasefires, without tackling the underlying issues, will likely result in a fragile peace that paves the way for renewed violence. Repeated wars and temporary truces have shown that without addressing core issues like the blockade, the occupation, and guarantees of political rights for the Palestinian people, any deal will be prone to collapse. Clear mechanisms for monitoring compliance and ensuring accountability are essential. Additionally, internal Palestinian unity and dialogue are critical, as divisions undermine stability and allow strategies aimed at eroding Palestinian rights to persist.

A sustainable solution requires a comprehensive political framework that prioritizes economic development, political rights, and the management of Gaza's governance, alongside a complete withdrawal of the occupation. Without these elements, any agreement risks perpetuating the cycle of conflict and suffering.

16.7 Conclusion

This chapter highlights the critical intersection of SDG 10 and the realities of the war on Gaza, stressing the need for integrated approaches that promote stability and equitable conflict resolution. Tackling the root causes of prolonged assaults, particularly the military occupation and its associated impacts on Palestinian society, is vital to advancing SDG 10 and reducing inequalities. True progress requires addressing structural injustices, fostering sustainable development, and supporting Palestinian self-determination to build a society founded on equality and resilience. Achieving SDG 10 in Gaza means implementing peace-oriented solutions that support infrastructure rebuilding, facilitate access to education and healthcare, and enable economic empowerment. These approaches offer pathways toward dismantling entrenched inequalities and fostering inclusive growth. The role of international community and solidarity is pivotal in this endeavor, providing targeted assistance and encouraging diplomatic engagement to end the occupation and support the right of Palestinian people to self-determination. Through such efforts, stakeholders can contribute meaningfully to SDG 10, not only by reducing inequalities but by establishing a framework for sustainable peace and resilience, enabling Gazans to strive toward a future of dignity and empowerment despite immense challenges.

References

Abdelnour S (2023) Making a killing: Israel's military-innovation ecosystem and the globalization of violence. Organ Stud 44:333–337. https://doi.org/10.1177/01708406221131938

ACAPS (2024) Impact of the conflict on people with disabilities in the Gaza strip

Agha Z, Esson J, Griffiths M, Joronen M (2024) Gaza: A decolonial geography. Trans Inst Br Geogr 49:e12675. https://doi.org/10.1111/tran.12675

Ahmed F, Alhajeeli N, Badheeb M et al (2024) A call for global intervention: the pressing requirement to address the health emergency in Gaza health crisis. Avicenna 2024:1–6. https://doi.org/10.5339/avi.2024.5

Alagha M, Hussein N (2024) The political economy of Israeli economic warfare on the Gaza strip. In: Elkahlout G (ed) Gaza's cycle of destruction and rebuilding. Palgrave Macmillan, Cham, pp 125–145

Alnabih A, Alnabeh N, Aljeesh Y, Aldabbour B (2024) Food insecurity and weight loss during wartime: a mixed-design study from the Gaza strip. J Health Popul Nutr 43:222

Ashiagbor D (2021) Race and colonialism in the construction of labour markets and precarity. Ind Law J 50:506–531. https://doi.org/10.1093/indlaw/dwab020

Asi YM (2020) Achieving food security through localisation, not aid: "De-development" and food sovereignty in the Palestinian territories. J Peacebuild Dev 15:205–218. https://doi.org/10.1177/1542316620918555

Asi Y, Mills D, Greenough PG et al (2024) 'Nowhere and no one is safe': spatial analysis of damage to critical civilian infrastructure in the Gaza strip during the first phase of the Israeli military campaign, 7 October to 22 November 2023. Confl Heal 18:1–13. https://doi.org/10.1186/s13031-024-00580-x

Belle D, Bullock HE (2023) The psychology of poverty, wealth, and economic inequality. Cambridge University Press

Ben Saad H (2024) Urgent humanitarian appeal: protecting the lives of women and children in the Gaza strip (Palestine). J Public Health Res 13:1–2. https://doi.org/10.1177/22799036241229312

Bhandari MP (2024) What is next for the sustainable development goals, what are the challenges concerning SDG 10—reduced inequalities? Sustain Earth Rev 7:2–5. https://doi.org/10.1186/s42055-024-00093-8

Bin-Nashwan SA, Hassan MK, Muneeza A (2024) Russia–Ukraine conflict: 2030 Agenda for SDGs hangs in the balance. Int J Ethics Syst 40:3–16. https://doi.org/10.1108/IJOES-06-2022-0136

Boisen C (2024) Israel's punitive war on Palestinians in Gaza. J Genocide Res 1–22. https://doi.org/10.1080/14623528.2024.2406098

Browne BC (2023) Transitional justice and enforcing the 'peace' on Palestine. In: Transitional (in)justice and enforcing the peace on Palestine. Rethinking peace and conflict studies. Palgrave Macmillan, Cham, pp 23–40

Buheji M, Buheji B (2024) Mitigating risks of slow children development due to war on Gaza 2023. Int J Psychol Behav Sci 14:11–21. https://doi.org/10.5923/j.ijpbs.20241401.02

Bürgin D, Anagnostopoulos D, Doyle M et al (2022) Impact of war and forced displacement on children's mental health—multilevel, needs-oriented, and trauma-informed approaches. Eur Child Adolesc Psychiatry 31:845–853. https://doi.org/10.1007/S00787-022-01974-Z/FIGURES/1

Ćerimović E (2023) At risk and overlooked: children with disabilities and armed conflict. Int Rev Red Cross 105:192–216. https://doi.org/10.1017/S181638312200087X

Cojocaru TM, Ionescu GH, Firoiu D et al (2022) Reducing inequalities within and among EU countries—assessing the achievement of the 2030 agenda for sustainable development targets (SDG 10). Sustain For 14:7706. https://doi.org/10.3390/su14137706

Cozza SJ (2024) Children and war: exposures, outcomes, and interventions. J Am Acad Child Adolesc Psychiatry 63:S48. https://doi.org/10.1016/j.jaac.2024.07.225

D'Andrea W, Aboagye A, Lee KA et al (2024) Growing up on the edge: a community-based mental health intervention for children in Gaza. Res Child Adolesc Psychopathol 52:833–848. https://doi.org/10.1007/s10802-023-01124-2

de Waal A (2024) Famine in Gaza: an example of the global humanitarian crisis. Am J Clin Nutr 119:1383–1385. https://doi.org/10.1016/j.ajcnut.2024.04.015

Elkahlout G (2024) Gaza's cycle of destruction and rebuilding. Understanding the actors, dynamics, and responses. Palgrave Macmillan

Elnakib S, Fair M, Mayrhofer E et al (2024) Pregnant women in Gaza require urgent protection. Lancet 403:244. https://doi.org/10.1016/S0140-6736(23)02835-0

ESCWA/UNDP (2024) Gaza war: Expected socioeconomic impacts on the State of Palestine—October 2024

Ferrara P, Cammisa I, Zona M et al (2024) The global issue of violence toward children in the context of war. J Pediatr 269:114007. https://doi.org/10.1016/j.jpeds.2024.114007

Gauthier GR, Smith JA, García C et al (2021) Exacerbating inequalities: social networks, racial/ethnic disparities, and the COVID-19 pandemic in the United States. J Gerontol Ser B Psychol Sci Soc Sci 76:e88–e92. https://doi.org/10.1093/geronb/gbaa117

Gradín C, Leibbrandt M, Tarp F (2021) Inequality in the developing world. Oxford University Press

Hassoun A, Al-Muhannadi K, Hassan HF et al (2024) From acute food insecurity to famine: how the 2023/2024 war on Gaza has dramatically set back sustainable development goal 2 to end hunger. Front Sustain Food Syst 8:1402150. https://doi.org/10.3389/fsufs.2024.1402150

Hazer L, Gredebäck G (2023) The effects of war, displacement, and trauma on child development. Humanit Soc Sci Commun 10:1–19. https://doi.org/10.1057/s41599-023-02438-8

Heer J (2023) Why Netanyahu Bolstered Hamas. https://www.thenation.com/article/world/why-netanyahu-bolstered-hamas/. Accessed 29 Dec 2024

Hussein A, Wong S, Bright A (2024) History and impact of Israeli siege and attacks on education in Gaza, Palestine. Oxford Res Encycl Educ. https://doi.org/10.1093/ACREFORE/9780190264093.013.1903

Irfan B, Shammala AA, Saleh K (2024) Will there be a future for newborns in Gaza? Lancet 404:1725–1726. https://doi.org/10.1016/S0140-6736(24)02249-9

Iriqat D (2023a) 75 years Palestine Nakba/Israel independence. Palest J 28:6–15

Iriqat D (2023b) Nakba continued. https://shuddhashar.com/nakba-continued/. Accessed 29 Dec 2024

Iriqat D (2024a) International law: Israel's fig leaf or a route to Palestinian statehood? Polit Geogr 115:103208. https://doi.org/10.1016/j.polgeo.2024.103208

Iriqat D (2024b) Towards an inclusive political system, Palestinian perspectives on the reconstruction of Gaza. In: Antar U, Asseburg M, Shaban O (eds) Palestinian perspectives on the reconstruction of Gaza. Friedrich-Ebert-Stiftung (FES), Stiftung Wissenschaft und Politik (SWP)

Iriqat D, Alousi R, Aldahdouh TZ et al (2025) Educide amid conflict: the struggle of the Palestinian education system. Qual Educ All 2:81–99. https://doi.org/10.1108/QEA-10-2024-0120

Jabr S, Berger E (2022) The children of Palestine: struggle and survival under occupation. In: Tiliouine H, Benatuil D, Lau MKW (eds) Handbook of Children's risk, vulnerability and quality of life. International handbooks of quality-of-life. Springer, Cham, pp 279–296

Jain T, Cooper L, Khauli N et al (2025) Social connections related to health and well-being needs identified by children and adolescents affected by displacement in Lebanon: a participatory research study. J Migr Heal 11:100284. https://doi.org/10.1016/j.jmh.2024.100284

Jangda HN, Farhan K, Akilimali A (2024) Unheard voices: addressing sexual and reproductive health in Gaza's women. Int J Gynecol Obstet 166:139–140. https://doi.org/10.1002/ijgo.15596

Jildeh C, Abdeen Z, Al Sabbah H et al (2014) Labor and related injuries among schoolchildren in palestine: findings from the national study of Palestinian schoolchildren (HBSC-WBG2006). ISRN Pediatr 2014:729573. https://doi.org/10.1155/2014/729573

Kerbaj M, Traboulsi J, Maasri Z, Toukan H (2024) Gaza in plain sight: witnessing in solidarity. J Vis Cult 22:1–19. https://doi.org/10.1177/14704129241227847

Khalifa Aleghfeli Y (2024) Inclusive teaching and learning practices that promote and protect Reading and science literacy for Palestinian children. Educ Sci 14:1145. https://doi.org/10.3390/EDUCSCI14111145

Kuhn H (2020) Reducing inequality within and among countries: realizing SDG 10—a developmental perspective. In: Kaltenborn M, Krajewski M, Kuhn H (eds) Sustainable development goals and human rights. Interdisciplinary studies in human rights. Springer, Cham, pp 137–153

Little W (2023) Chapter 9. Social inequality. In: Introduction to sociology—3rd Canadian edition. BCcampus

Mahomed S (2023) When sanctuaries of humanity turn into corridors of horror: the destruction of healthcare in Gaza. South African J Bioeth Law 16:1–3. https://doi.org/10.7196/SAJBL.2023.V16I3.1732

Milton S, Elkahlout G, Barakat S (2023) Protecting higher education from attack in the Gaza strip. Compare 53:1024–1042. https://doi.org/10.1080/03057925.2021.1987192

Molloy EJ, Bearer CF (2024) Advocacy for children in global conflicts: the value of children in our world: how can we advocate for children as clinicians and researchers? Pediatr Res 95:1386–1387. https://doi.org/10.1038/s41390-024-03044-3

Mosleh M, Dalal K, Aljeesh Y, Svanström L (2018) The burden of war-injury in the Palestinian health care sector in Gaza strip. BMC Int Health Hum Rights 18:1–9. https://doi.org/10.1186/s12914-018-0165-3

Nijim M (2024) Gazacide: Palestinians from Refugeehood to ontological obliteration. Crit Sociol 1944:1–23. https://doi.org/10.1177/08969205241286530

Omoeva C, Moussa W, Hatch R et al (2018) The effects of armed conflict on educational attainment and inequality. Educ Policy Data Cent 18:1–33

Panneer S, Kantamaneni K, Palaniswamy U et al (2022) Health, economic and social development challenges of the COVID-19 pandemic: strategies for multiple and interconnected issues. Healthc 10:770. https://doi.org/10.3390/healthcare10050770

Rabaia I, Habash L (2024) The hidden war on higher education: unmasking the 'Educide' in Gaza. In: The war on Gaza and Middle East political science. George Washington University—POMEPS, pp 8–13

Remington TF (2023) Socio-economic and geographic inequality of the effects of the pandemic. In: Aspalter C (ed) Covid-19 pandemic: problems arising in health and social policy. Springer Nature, Singapore, pp 31–56

Rosenthal FS (2021) A comparison of health indicators and social determinants of health between Israel and the occupied Palestinian territories. Glob Public Health 16:431–447. https://doi.org/10.1080/17441692.2020.1808037

Roy S (2016) The Gaza strip: the political economy of de-development, 3rd edn. Institute for Palestine Studies, Washington, DC

Sabet C, Nguyen D, Nada SA et al (2024) Women and girls in Gaza face increasingly dire physical and mental health challenges. BMJ 384. https://doi.org/10.1136/BMJ.Q625

Shellah D (2024a) War on Gaza: the impossible duty to care for the critically ill. Intensive Care Med 50:311–313. https://doi.org/10.1007/S00134-023-07309-Z/METRICS

Shellah D (2024b) War on Gaza: attacks on science are erasing the future. Lancet 404:27–28. https://doi.org/10.1016/S0140-6736(24)00767-0

Simangunsong H, Stewart BP, Debortoli D (2023) The impact of economic inequality on social disparities: a quantitative analysis. J Sos Sains Terap dan Ris 11:159–167. https://doi.org/10.35335/mdjzzw52

Sinha AK (2019) Sustainable development and the problem of economic inequality. In: Bandyopadhyay S, Dutta M (eds) Opportunities and challenges in development: essays for Sarmila Banerjee. Springer, Singapore, pp 1–564

Stamatopoulou-robbins S (2024) The human toll: indirect deaths from war in Gaza and the West Bank, October 7, 2023 Forward

Tannira A (2021) The political economy ofthe Gaza strip under Hamas. In: Tartir A, Dana T, Seidel T (eds) Political economy of Palestine. Palgrave Macmillan, pp 135–154

Tanous O, Asi Y, Hammoudeh W et al (2023) Structural racism and the health of Palestinian citizens of Israel. Glob Public Health 18. https://doi.org/10.1080/17441692.2023.2214608

Tartir A, Dana T, Seidel T (2021) (Middle East today) Alaa Tartir, Tariq Dana, Timothy Seidel—political economy of Palestine. In: Critical, interdisciplinary, and decolonial perspectives. Palgrave Macmillan

Thabet AA, Matar S, Carpintero A et al (2011) Mental health problems among labour children in the Gaza strip. Child Care Health Dev 37:89–95. https://doi.org/10.1111/j.1365-2214.2010.01122.x

Thabet A, Tawahina A, Henley D et al (2015) Mental health of disabled Palestinian adults in the Gaza strip. Br J Educ Soc Behav Sci 10:1–10. https://doi.org/10.9734/bjesbs/2015/18268

Thabet AM, Vostanis P, Qama KA (2018) Social support of Palestinian adults with disabilities in the Gaza strip. Arab. J Psychiatry 29:145–156. https://doi.org/10.12816/0051279

The World Bank (2024) Gaza Strip—interim damage assessment

Ujunwa A, Okoyeuzu C, Nkwor N, Ujunwa A (2021) Potential impact of climate change and armed conflict on inequality in sub-Saharan Africa. South African J Econ 89:480–498. https://doi.org/10.1111/saje.12271

UNCTAD (2024) Preliminary assessment of the economic impact of the destruction in Gaza and prospects for economic recovery

UNDP (2024) Impact of the 2023 Gaza war on displaced women in the Gaza strip

UNEP (2024) Environmental impact of the conflict in Gaza: preliminary assessment of environmental impacts. Nairobi

UNITAR-UNOSAT (2024) UNOSAT Gaza Strip comprehensive damage assessment—December 2024

United Nations (2023) The sustainable development goals report 2023: Special Edition

Veenstra G (2011) Race, gender, class, and sexual orientation: intersecting axes of inequality and self-rated health in Canada. Int J Equity Health 10:1–11. https://doi.org/10.1186/1475-9276-10-3

Weinthal E, Sowers J (2019) Targeting infrastructure and livelihoods in the West Bank and Gaza. Int Aff 95:319–340. https://doi.org/10.1093/IA/IIZ015

Wells K, Cortés-Morales S, Esson J et al (2024) Israel's war on Gaza and the violation of children's rights. Child Geogr 1–4. https://doi.org/10.1080/14733285.2024.2316752

Whyte J (2024) A "tragic humanitarian crisis": Israel's Weaponization of starvation and the question of intent. J Genocide Res:1–15. https://doi.org/10.1080/14623528.2024.2339637

Yılmaz E, Bağçe S (2022) Social identity and economic inequalities. Soc Ident 28:400–416. https://doi.org/10.1080/13504630.2022.2029740

Zaiter WA (2024) Women at the heart of Gaza's rebirth: Palestinian perspectives on the reconstruction of Gaza. In: Antar U, Asseburg M, Shaban O (eds) Palestinian perspectives on the reconstruction of Gaza. Friedrich-Ebert-Stiftung (FES), Stiftung Wissenschaft und Politik (SWP)

17 Disrupted Systems: The War's Impact on Sustainable Production and Consumption in Gaza

Abdo Hassoun, Iman Dankar,
and Hussein F. Hassan

Abstract

This chapter explores the profound impact of the recent war (October 2023–January 2025) on Gaza on Sustainable Development Goal 12 (SDG 12), which focuses on ensuring sustainable consumption and production patterns. This Israeli genocidal war has severely disrupted consumption and production systems in Gaza, leading to unsustainable practices that hinder long-term development and environmental stability. The chapter delves into the extensive damage inflicted by the war on key infrastructure, including manufacturing facilities, supply chains, and waste management systems. The destruction has led to reliance on inefficient and unsustainable practices, exacerbating resource scarcity and environmental degradation. Key issues addressed include the collapse of local industries, increased waste from damaged goods and buildings, and challenges in managing hazardous materials. Social and economic implications are also discussed, such as increased unemployment and poverty, which drive unsustainable consumption patterns. This chapter illustrates how the war has reversed progress toward achieving SDG 12 in Gaza and highlights the need for collaborative efforts by local and international organizations to mitigate these impacts. Recommendations are provided for rebuilding systems in Gaza to promote sustainable production and consumption. This comprehensive examination aims to provide a deeper understanding of the critical challenges faced by the population of Gaza in achieving sustainable consumption and production, offering insights into the urgent actions needed to address these issues and promote long-term environmental and economic stability.

Keywords

Economic instability · Environmental degradation · Poverty · Resilience · SDG 12 · Socioeconomic · Sustainable practices · Sustainable recovery · The Gaza Strip · Unemployment

A. Hassoun
Sustainable AgriFoodtech Innovation & Research (SAFIR), Arras, France

Faculty of Agricultural Engineering, University of Aleppo, Aleppo, Syria
e-mail: a.hassoun@saf-ir.com

I. Dankar
Department of Liberal Education, School of Arts and Sciences, Lebanese American University, Beirut, Lebanon
e-mail: Imane.Dankar@balamand.edu.lb

H. F. Hassan (✉)
Department of Nutrition and Food Science, School of Arts and Sciences, Lebanese American University, Beirut, Lebanon
e-mail: hussein.hassan@lau.edu.lb

A. Hassoun (ed.), *War on Gaza*, Sustainable Development Goals Series,
https://doi.org/10.1007/978-3-031-88500-6_17

17.1 Introduction

Wars and conflicts worldwide create significant challenges for achieving sustainable development, particularly with respect to Sustainable Development Goal 12 (SDG 12), which emphasizes responsible consumption and production. These challenges stem from the destruction of critical infrastructure, the collapse of governance systems, and the disruption of essential services such as waste management and supply chains. Prolonged conflicts often push affected regions into cycles of dependency on unsustainable practices, leading to environmental degradation, resource inefficiency, and long-term economic instability. The absence of effective recovery mechanisms exacerbates these issues, making the pursuit of sustainability nearly unattainable in war-torn areas (Mhlanga and Ndhlovu 2023; Bin-Nashwan et al. 2024; Gupta and Shukla 2024).

The Gaza Strip illustrates the broader global trends where wars and conflicts significantly hinder efforts to achieve sustainable development. The region faces severe political, economic, and geographical constraints, which have been intensified by prolonged conflict. This has led to widespread destruction of infrastructure and critical systems, including waste management, supply chains, and production facilities. The disruptions have necessitated the reliance on resource-intensive and informal alternatives, further exacerbating environmental degradation. The cyclical nature of destruction, compounded by the blockade and military operations, has left Gaza in a persistent state of rebuilding (United Nations 2017; European Union/World Bank Group/United Nations 2021; Elkahlout 2024a).

Over the past two decades, a series of hostilities, starting in 2008 and followed by escalations in 2012, 2014, and 2021, have caused extensive damage to infrastructure and ecosystems in Gaza. These events severely impacted sustainable production and consumption patterns by destroying industrial facilities, agricultural lands, and waste management systems. The damage forced the adoption of inefficient and environmentally harmful coping mechanisms, leaving little room for recovery or sustainable development. Repeated targeting of critical infrastructure has entrenched unsustainable practices into the daily life in this region, making them a necessity rather than a choice. The challenges faced by Gazans exemplify the complex intersection of conflict and sustainability, where immediate survival imperatives consistently override long-term development goals (UNDP 2021; World Bank Group 2023; AbuHamed et al. 2023).

The recent war on Gaza has introduced another level of destruction, severely damaging already fragile systems and further stretching limited resources. This situation not only delays recovery but also deepens reliance on unsustainable practices. The redirection of resources and attention away from long-term development goals, including sustainability, toward immediate reconstruction efforts intensifies enduring challenges. This shift exacerbates issues such as the depletion of natural resources, the deepening of economic instability, and the escalation of environmental degradation (Shaheen et al. 2024; UNCTAD 2024; UNEP 2024). Without effective recovery mechanisms and coordinated strategies, the systems of production and consumption remain incapable of functioning sustainably. This underscores the urgent need for solutions that address both immediate and long-term challenges, paving the way for sustainable development in conflict-affected regions.

This chapter explores how the recent war on Gaza has undermined progress toward achieving SDG 12. It focuses on the critical systems disrupted by this war, including waste management, production facilities, and supply chains, examining structural challenges that hinder sustainable development. Additionally, it addresses the broader implications of these disruptions for environmental and socioeconomic stability. The chapter concludes with actionable recommendations for fostering sustainable consumption and production in the face of these challenges, highlighting the importance of international collaboration, innovative approaches, and comprehensive recovery strategies.

17.2 Overview of Sustainable Development Goal 12 (SDG 12): Responsible Consumption and Production

Sustainable Development Goal 12 (SDG 12) aims to ensure sustainable consumption and production patterns by promoting resource efficiency, reducing environmental degradation, and improving the quality of life for present and future generations. Central to this goal are the sustainable management and efficient use of natural resources, the reduction of waste through prevention, recycling, and reuse, and the adoption of sustainable practices across public and private sectors. SDG 12 also seeks to raise public awareness about sustainable lifestyles and encourage organizations to integrate sustainability principles into their policies and practices (Gasper et al. 2019; Arora and Mishra 2023).

Globally, SDG 12 addresses critical issues such as climate change, water and sanitation, and food security, all of which are interconnected with consumption and production systems. Unsustainable consumption patterns contribute to resource depletion, biodiversity loss, and increased greenhouse gas emissions. Transitioning to sustainable practices mitigates environmental harm, enhances resource efficiency, and fosters economic opportunities through green innovations and job creation (Mensah et al. 2023; Raman et al. 2024).

However, in conflict-affected regions, the challenges to achieving SDG 12 are particularly acute. Economic instability caused by high unemployment and poverty compels reliance on unsustainable consumption patterns. Governance systems, often weakened or collapsed in conflict zones, are unable to enforce regulatory frameworks that support sustainability (Meaza et al. 2024). Moreover, wars directly contribute to environmental harm, including pollution, habitat destruction, and biodiversity loss, which further impede sustainable resource management. The destruction of critical infrastructure, such as waste management systems, production facilities, and supply chains in conflict zones disrupts sustainable practices, further exacerbating environmental degradation (Rawtani et al. 2022; Abu-Rmeileh and Iriqat 2024; Dardona et al. 2024; Rahiman et al. 2024).

Several SDG 12 targets are particularly relevant in conflict contexts. Target 12.2, which emphasizes the sustainable management and efficient use of natural resources (Trummer et al. 2022), is directly impacted by the infrastructure destruction and resource inefficiency in war zones. Target 12.3, which focuses on reducing food waste and losses along supply chains, becomes nearly unattainable as conflicts disrupt agricultural systems, storage facilities, and transportation networks (Ardra and Barua 2022; Jacob-John et al. 2023). Target 12.4, which advocates for the environmentally sound management of chemicals and waste, is undermined by the collapse of waste management systems in conflict areas. This leads to unregulated dumping of hazardous materials, chemical spills, and pollution of natural ecosystems. Finally, Target 12.c, aimed at rationalizing fossil fuel subsidies to encourage sustainable consumption, is challenged as conflicts force reliance on inefficient, outdated energy systems and fossil fuels (Jacob-John et al. 2021).

Addressing these barriers requires a comprehensive approach that integrates sustainable practices into the recovery and reconstruction efforts in conflict-affected regions. By embedding sustainable consumption and production patterns into the rebuilding process, these regions can establish a foundation for long-term stability, environmental recovery, and resilience against future disruptions.

17.3 Impacts of the Recent War on Gaza on Sustainable Production and Consumption

17.3.1 Destruction of Production and Supply Systems

Israel has imposed a strict blockade on Gaza, severely restricting the movement of goods such as food, agricultural inputs, medical supplies, construction materials, and fuel, resulting in a severe humanitarian crisis. This blockade has deeply affected Gaza's economy, infrastructure, and social fabric, leading to widespread shortages and skyrocketing prices of essential goods. Food security has been particularly impacted, with limited access to agricultural supplies and fuel disrupting local food production and distribution systems. Furthermore, the war has exacerbated these challenges by halting the import of critical goods and services necessary for survival. The interconnectedness of water, energy, and food systems has been weaponized, leaving Gaza's population struggling to access even the most basic necessities (ESCWA 2023; Alagha and Hussein 2024; Anera 2024; Vos 2024).

17.3.2 Disruption of Sustainable Food Production Systems

Gaza, under siege and amidst the recent war, has been denied access to food supplies through its crossings to the outside world. Families forced to abandon their homes and flee northern Gaza were left without food, transportation, water, or shelter, compounding their already dire humanitarian situation. Those who remained in the north saw their bakeries, flour stores, and food markets bombed and destroyed. The war has disrupted access to eco-friendly farming inputs and technologies, such as organic fertilizers, pest control methods, and efficient irrigation systems. This disruption has forced farmers to adopt less sustainable practices, increasing the environmental impact of agriculture and reducing the overall sustainability of food production systems (Alnabih et al. 2024; Elkahlout 2024b; Hassoun et al. 2024).

Since the onset of this war, Gaza has experienced unprecedented destruction of all food production facilities, including agriculture, fishing, and related food processing and distribution ones. This mass destruction has direct consequences on the food security of Gaza's population and will severely compromise the Palestinian food system's ability to feed future generations. The destruction of agricultural assets, food infrastructure, and the contamination and degradation of soil and land have drastically limited the functionality of Gaza's food system, necessitating large-scale investment and considerable time for recovery. Economically, the widespread destruction of farms, livestock, and other food production facilities in Gaza has had disastrous consequences on both food systems and the economy, with recovery likely requiring decades to return to pre-October 2023 welfare levels (Nassar et al. 2023; Appiah-Kusi and Al-Madbouh 2024). As highlighted by previous conflicts, such as the May 2021 escalation, which resulted in significant rubble and hazardous waste impacting agricultural lands, similar environmental and economic repercussions are being observed now. Addressing these challenges necessitates coordinated international efforts to rebuild and stabilize Gaza's food systems, ensuring sustainable recovery and resilience (AbuHamed et al. 2023).

Before October seventh, Gaza maintained a flourishing agricultural and fisheries sector, achieving self-sufficiency in fruits and vegetables. However, the war has decimated homes, agricultural lands, and vital resources. As of May 20, 2024, Israeli-imposed restrictions on farmland and the sea, combined with targeted destruction, have resulted in the loss of 28% of Gaza's cropland, the destruction of over 70% of its fishing fleet, damage to 488 agricultural wells, and the obliteration of 21% of the Strip's greenhouses. Even before this devastating war, 80% of Gaza's population depended on humanitarian aid, highlighting the preexisting vulnerabilities now exacerbated by this crisis (Irfan et al. 2024; Wakilzada and Husseini 2024).

This widespread destruction has severely undermined progress toward achieving SDG 12, particularly Target 12.3, which aims to halve food waste and reduce food losses along production and supply chains. The inability to maintain sustainable food production systems in Gaza highlights the critical need for international support and long-term strategies to rebuild agricultural and food systems in line with sustainability goals.

17.3.3 Strain on Sustainable Energy Use in Production

The damage to energy infrastructure has led to a decrease in energy efficiency. The reliance on outdated and inefficient energy systems has increased due to the destruction of all energy infrastructure, including solar power plants and rooftop solar photovoltaic systems (Elnaggar 2023; Amnesty International 2024; The World Bank 2024a). This shift is contrary to both SDG 7, which focuses on ensuring access to affordable, reliable, sustainable, and modern energy for all, as well as SDG 12, which aims to ensure sustainable consumption and production patterns, including reducing inefficiencies in energy use and promoting sustainability. The long-term impact includes higher energy costs, increased greenhouse gas emissions, and greater environmental degradation (Rettig 2024).

Roads, power plants, water treatment facilities, and public buildings are routinely damaged or destroyed during intense military escalations, forcing ongoing repairs that deplete scarce resources and impede development. For example, frequent damage to Gaza's electrical infrastructure has led to severe power outages, disrupting schools, hospitals, businesses, and other essential services. This instability in electrical supply further restricts the possibilities for renewable energy alternatives to address the energy issue, leaving the region reliant on outside sources. Similarly, damage to water and sanitation infrastructure exacerbates Gaza's environmental challenges, hindering attempts to implement sustainable resource management methods, which are critical for long-term rehabilitation. Following a year of Israeli operations, Gaza lies in ruins, and estimates suggest that complete reconstruction could take several decades (Elnaggar 2023; ESCWA/UNDP 2024).

17.3.4 Collapse of Waste Management and Recycling Systems

The recent war has severely crippled local capacities and infrastructure for solid waste management, making the two main sanitary landfills (i.e., Al-Fukhary/Sofa and Johr Al-Deek) inaccessible. This has led to the proliferation of temporary waste dumping sites and the accumulation of solid waste, including hazardous medical waste, in the streets and around internally displaced people shelters and makeshift camps. These unregulated dumping sites, which contain nearly half a million tons of solid waste often contaminated with hazardous materials, have become breeding grounds for disease vectors such as mosquitoes and rats. Additionally, the destruction of medical waste treatment facilities during the war has resulted in hazardous medical waste contaminating the solid waste, posing further serious risks (PAX 2024; UNDP 2024; UNDP/PAPP 2024).

By the end of May 2024, UNDP/PAPP's team of experts and consultants estimated that the solid waste management sector had incurred approximately $60 million in damages. These costs are expected to rise as the war continues, causing further damage and destruction to infrastructure. The affected assets include waste collection and transfer vehicles, waste containers, landfill equipment, landfill components and units, waste transfer stations, solid waste management facilities (including furniture and spare parts), operational vehicles used in waste management, as well as the medical waste collection and treatment system, and waste sorting facilities. Waste collection vehicles represent the highest share of damage costs at 35%, followed by waste management facilities at 17%, and waste containers at 16% (UNDP/PAPP 2024; UNEP 2024).

This collapse of waste management and recycling systems severely undermines SDG 12, particularly Target 12.4, which aims to achieve environmentally sound management of chemicals and waste, and Target 12.5, which seeks to substantially reduce waste generation through prevention, reduction, and recycling. The lack of proper waste disposal and recycling mechanisms in Gaza not only poses immediate health and environmental risks but also creates long-term challenges for sustainable consumption and production patterns in the region.

17.4 Socioeconomic Implications of the War

The relentless fighting has catastrophically devastated Gaza's essential infrastructure, crippling the foundation for socioeconomic stability and long-term development, with far-reaching consequences for public services, economic recovery, and the overall well-being of the population. The cost of damage to critical infrastructure in Gaza is estimated at around $18.5 billion according to a report by the World Bank and the United Nations released in March 2024. That is equivalent to 97% of the combined GDP of the West Bank and Gaza in 2022. Power networks, solar-generated systems, and nearly all electricity supply have been disrupted since the first week of the war. With 92% of primary roads destroyed or damaged and the communications infrastructure seriously impaired the delivery of basic humanitarian aid to people has become exceedingly difficult (The World Bank 2024a).

17.4.1 Impact on Local Industries, Employment, and Poverty

The war on Gaza has had devastating effects on local industries, employment, and poverty levels. The war has led to widespread destruction of infrastructure, severely disrupting local businesses and economic activities. Many factories, shops, and offices have been destroyed or rendered inoperable, leading to significant job losses and a sharp decline in economic activity. The destruction of physical assets and the disruption of supply chains have crippled local industries. Businesses that managed to survive the initial onslaught face severe challenges in maintaining operations due to damaged infrastructure and unsafe conditions. This has led to a significant reduction in production capacity and a loss of local and international market share (Butter 2024; ILO 2024; PFI/FPCCIA 2024).

Moreover, the war has caused a dramatic increase in unemployment rates. With many businesses shutting down or scaling back operations, countless individuals have lost their jobs. This surge in unemployment has exacerbated poverty levels, as families struggle to make ends meet without a steady income (UNCTAD 2024; UNEP 2024). The economic instability caused by the war has plunged many families into poverty. The loss of income, coupled with rising prices for basic goods due to supply chain disruptions, has made it difficult for many to afford necessities. This has led to increased reliance on humanitarian aid and has strained local resources (ILO 2024). Unemployment rates in Gaza are now among the highest in the world, with nearly 80% of the population without stable employment following a year of intensified war. This economic devastation has also led to a dramatic contraction in GDP, which has shrunk by almost 85%, leaving Gaza's economy in a state of severe crisis (Elkahlout 2024a; The World Bank 2024b).

17.4.2 Increased Waste Generation and Challenges in Hazardous Material Management

The war has caused a substantial surge in waste generation, particularly hazardous materials, posing severe environmental and health risks. The widespread destruction of buildings and infrastructure has produced massive quantities of construction and demolition waste, including hazardous substances such as asbestos, lead, and other toxic materials. Additionally, the war has generated vast amounts of military debris, destroyed vehicles, and remnants of damaged infrastructure, compounding the waste management crisis in the region. This waste poses significant environmental and health risks, as it often contains hazardous materials that require specialized handling and disposal. The sheer volume of waste has overwhelmed local waste management systems, leading to improper disposal and increased environmental contamination (AbuHamed et al. 2023; UNDP/PAPP 2024; UNEP 2024).

Managing hazardous waste in a war zone presents unique challenges. The presence of unexploded ordnance, chemical contaminants, and other dangerous materials complicates cleanup efforts and poses risks to workers and residents alike. Additionally, the destruction of waste management infrastructure has made it difficult to safely collect, transport, and dispose of hazardous materials. This has increased environmental pollution and health hazards for the local population (UNEP 2024).

17.4.3 Link Between Socioeconomic Instability and Unsustainable Consumption Patterns

The socioeconomic instability induced by the war has significantly contributed to the proliferation of unsustainable consumption patterns. The disruption of local economies, coupled with widespread poverty, has compelled many individuals to adopt short-term survival strategies that often result in environmental degradation. This cycle of economic hardship has hampered community recovery efforts, creating an enduring challenge for sustainable development. The loss of employment opportunities and income sources has forced a growing dependence on unsustainable practices, such as the overexploitation of natural resources and illegal activities like logging. While these practices may offer immediate financial relief, they inflict long-term damage on the environment and undermine local economic resilience (UNDP/PAPP 2024; UNEP 2024).

The war and its devastating impacts have significantly altered consumption behaviors, driving a shift toward unsustainable practices. For example, prior to the recent war, Gazans heavily relied on solar energy to address the unreliable electrical supply caused by Israel's blockade. Solar power served as a cost-effective and sustainable alternative for households and businesses. However, this devastating war has destroyed much of this infrastructure, with solar equipment lying in ruins amidst the rubble of thousands of demolished buildings. The debris includes not only shattered solar panels but also pulverized building materials containing hazardous substances, such as asbestos, which pose significant environmental and health risks (El-Khozenadar et al. 2022; London et al. 2024).

A lack of access to affordable energy sources has also forced many households to resort to alternative, unsustainable options for basic needs such as cooking. With limited availability of cooking gas and electricity, people have turned to burning wood, plastic, and other waste materials. While these methods offer immediate relief, they contribute to severe environmental pollution and expose families to harmful toxins (UNEP 2024). This shift underscores the intersection of poverty, resource scarcity, and environmental harm, further complicating efforts to rebuild sustainable livelihoods in the region.

The crisis has further entrenched unsustainable consumption patterns, such as the increased use of disposable goods, reliance on inexpensive yet environmentally harmful products, and neglect of sustainable practices. These behaviors are exacerbated by the lack of access to affordable and sustainable alternatives, making it increasingly difficult for affected communities to transition toward more sustainable consumption habits (ESCWA/UNDP, 2024; UNEP, 2024) (ESCWA/UNDP 2024; UNEP 2024). This entrenchment underscores the critical need for targeted interventions to rebuild infrastructure and promote sustainable consumption practices in Gaza.

17.5 The Environmental Consequences of the Recent War

Amidst the humanitarian disaster and extensive bombings that have claimed thousands of lives, the war on Gaza has triggered a catastrophic environmental crisis, casting a long shadow over both the region's ecological and human health. Beyond the immediate toll on human life, the war has devastated Gaza's natural resources, raising concerns about the long-term sustainability of the environment. The war has led to the unprecedented destruction of agricultural land, which is vital to Gaza's economy and food security. Satellite imagery and ground reports indicate that between 38% and 48% of Gaza's tree cover and farmland have been destroyed. The decimation of olive groves, farms, and greenhouses not only halts food production but also poses a significant threat to the region's biodiversity and ecological balance (Fontaine 2024). The loss of vegetation cover exacerbates climate change effects, reducing the area's capacity to sequester carbon and regulate the local climate. Additionally, the heavy use of munitions and the resulting debris have contaminated soil and groundwater with hazardous materials, including heavy metals and unexploded ordnance, affecting long-term soil health. This presents immediate dangers to human health, and threatens future agricultural productivity, perpetuating a cycle of environmental degradation and food insecurity (Shaheen et al. 2024; UNEP 2024).

Water pollution has become a critical issue in Gaza, directly affecting human health and leading to a complex humanitarian crisis. The destruction of waste management and sewage treatment infrastructure, exacerbated by siege conditions, has resulted in over 100,000 m^3 of untreated sewage being discharged daily into the Mediterranean Sea. This not only endangers marine life but also human health, as polluted waters contaminate local food chains and drinking water sources. The public health crisis worsens as contaminated water becomes a breeding ground for pathogens, increasing the occurrence of diseases such as cholera, dysentery, and typhoid. Moreover, the widespread burning of waste, use of alternative fuels, and emissions from military actions have significantly deteriorated air quality. This toxic air has led to an increase in respiratory illnesses among Gaza's residents, exacerbating the public health crisis. Long-term exposure to air pollution poses serious health risks, underscoring the need for immediate actions to improve air quality and protect public health (Abd El Hay 2024; Human Rights Watch 2024a; UNEP 2024).

The war has generated millions of tons of debris, much of which is contaminated with hazardous substances like asbestos and heavy metals, posing a massive waste management challenge. The cleanup and safe disposal of this debris are crucial to preventing further environmental contamination and ensuring the safety of Gazans. The war in Gaza underscores the urgent need for international cooperation and strict adherence to environmental and humanitarian laws, highlighting how environmental and climate protections are deeply intertwined with the laws of armed conflict. The classification of environmental destruction as "ecocide" and its investigation as a potential war crime highlight the gravity of the situation (Gordon and Haddad 2024; Human Rights Watch 2024b; Qumsiyeh 2024; UNEP 2024). Comprehensive assessments and robust remediation efforts are vital for mitigating the environmental impacts of this war and laying the groundwork for the sustainable recovery and resilience of Gaza's environment and its communities.

17.6 Rebuilding Gaza: Collaborative Efforts and Pathways for Sustainable Recovery

17.6.1 Role of International and Local Collaboration

Rebuilding Gaza requires coordinated efforts among international organizations, local authorities, NGOs, and community stakeholders. Effective collaboration enables the pooling of resources, knowledge, and expertise, providing a comprehensive approach to tackling the multifaceted challenges of post-conflict recovery. International organizations can provide funding and technical support, while local stakeholders ensure culturally and regionally appropriate solutions. The decision by the Israeli government to ban UNRWA, the United Nations agency providing aid to Palestinian refugees, creates significant obstacles to these efforts. UNRWA plays an indispensable role in providing essential services such as education, healthcare, and emergency relief, which are crucial for immediate recovery and long-term sustainability (Hamad et al. 2024;

Nijim 2024; Studies et al. 2024; Iriqat et al. 2025). The ban disrupts access to these critical resources, exacerbating the strain on already overburdened local institutions and NGOs. It is imperative that the ban be lifted immediately to ensure the continuity of essential services and to enable meaningful progress toward rebuilding and resilience.

17.6.2 Sustainable Waste Management and Recycling

The recent war in Gaza has caused widespread destruction, generating an estimated 50 million tons of debris and waste (OCHA 2024). This immense volume poses a critical environmental challenge, demanding urgent attention and sustainable solutions. Rebuilding efforts must include modernized waste management systems that prioritize recycling and environmentally sound disposal practices, aligned with the targets of SDG 12, particularly Target 12.4. Utilizing war-damaged materials in reconstruction can reduce waste while promoting sustainability. Case studies from conflict-affected regions, such as Russia-Ukraine war, show that debris recycling can simultaneously support reconstruction and minimize environmental degradation (Hossain et al. 2024).

17.6.3 Sustainable Energy Use and Efficiency

Energy infrastructure in Gaza has been severely impacted, forcing reliance on fossil fuels and outdated systems. However, reconstruction provides an opportunity to transition to renewable energy sources, such as solar or wind power, reducing greenhouse gas emissions and enhancing energy efficiency, in line with SDG 12.c. Renewable energy projects can also help stabilize energy supply for essential services, industries, and households, fostering long-term resilience and sustainability (Juaidi et al. 2022; Rantissi et al. 2024).

17.6.4 Sustainable Resource Management and Production

The war has significantly disrupted production systems in Gaza, leading to inefficient and unsustainable resource use. Recovery efforts must prioritize rebuilding industries using green technologies and resource-efficient methods (PFI/FPCCIA 2024; UNEP 2024). Adopting circular economy principles, such as reusing materials, recycling, and reducing waste, can promote sustainable production systems, aligned with SDG 12.2. For instance, integrating water-saving technologies and energy-efficient manufacturing practices can rebuild industries while conserving resources. Such measures not only address immediate recovery needs but also lay the foundation for a more sustainable and resilient industrial sector in Gaza (Rantissi et al. 2024).

17.6.5 Sustainable Food Systems

The destruction of agricultural lands and food supply chains in Gaza has worsened food insecurity and reliance on imports (FAO/UN 2024). Reconstruction must focus on rebuilding agricultural infrastructure and adopting eco-friendly farming practices to reduce food losses, in alignment with SDG 12.3. Initiatives to strengthen local food systems, such as supporting smallholder farmers and introducing sustainable farming inputs, can improve food security while fostering sustainability (FAO 2018).

17.6.6 Addressing Economic and Social Stability

Economic instability and high unemployment caused by this war have exacerbated unsustainable consumption patterns in Gaza. Reconstruction efforts should focus on creating green jobs and fostering local industries that align with sustainable practices. Training programs in renewable energy, waste management, and sustainable agriculture can empower local populations to participate in the recovery process. Such efforts can reduce poverty and build a more stable, resilient, and sustainable economy (Ameer 2024; ESCWA/UNDP 2024).

17.6.7 Governance and Policy Development

Reconstruction in Gaza requires robust governance frameworks to implement and oversee sustainable recovery efforts. Policies that prioritize sustainability, such as incentives for renewable energy projects and stricter waste management regulations, are essential. Establishing monitoring and evaluation mechanisms can ensure that progress toward SDG 12 targets is tracked and adjusted as needed. Effective governance not only ensures accountability and transparency but also attracts international funding and partnerships, further supporting long-term recovery goals (Iriqat 2024).

Rebuilding Gaza presents an opportunity to integrate sustainable consumption and production practices into recovery efforts, fostering resilience and long-term development. Aligning recovery with SDG 12 targets ensures that reconstruction not only addresses immediate needs but also promotes environmental and economic stability. By prioritizing sustainability in reconstruction, Gaza can emerge from conflict with a stronger foundation for a sustainable future.

17.7 Conclusion

The recent war on Gaza has created profound disruptions in sustainable consumption and production systems, severely impeding progress toward achieving SDG 12. The destruction of critical infrastructure, including waste management systems, production facilities, and supply chains, has forced the region into dependence on inefficient and unsustainable practices. These challenges have been further exacerbated by widespread poverty and unemployment, which perpetuate unsustainable consumption patterns and deepen the vulnerability of the Gaza Strip.

Addressing these challenges requires urgent and coordinated strategies that prioritize the health and safety of the population in Gaza while integrating sustainability into reconstruction efforts. Immediate action should focus on mitigating the potential toxicity of debris from materials of the war, which poses severe health and environmental risks. Policies should emphasize the adoption of advanced waste management systems to handle and neutralize hazardous materials safely. Establishing specialized facilities for sorting, recycling, and disposing of debris will reduce the risk of toxic exposure. Transitioning to renewable energy sources, such as solar and wind power, reduces reliance on fossil fuels and supports the environmentally sound rebuilding of Gaza's energy systems. These efforts, combined with sustainable agricultural practices to restore food security and promote ecological balance, will protect the population and create a foundation for safer and more resilient recovery.

Effective governance frameworks are central to the success of these initiatives. Robust policies must be developed to oversee reconstruction efforts, enforce environmental regulations, and promote resource efficiency. Establishing monitoring and evaluation mechanisms will ensure that recovery programs remain aligned with SDG 12 targets and can adapt to evolving challenges. Such governance can also build confidence among international donors, attracting the funding and partnerships necessary to support large-scale recovery projects.

Social and economic stability must also be a core focus of the recovery process. Investments in green job creation, such as training programs in renewable energy installation, sustainable agriculture, and waste management, can empower local communities to actively participate in rebuilding efforts. These initiatives not only provide employment opportunities but also help break the cycle of poverty and foster a culture of sustainability.

Finally, fostering collaboration between local authorities, international organizations, and community stakeholders is essential. International expertise and funding can provide critical support, while local knowledge ensures that recovery solutions are culturally appropriate and regionally effective. Together, these partnerships can address immediate needs while laying the foundation for long-term resilience.

The reconstruction of Gaza presents a pivotal opportunity to reimagine and rebuild systems in alignment with sustainable development principles, fostering resilience and long-term sustainability. By embedding sustainability into every aspect of recovery, the Gaza Strip can move beyond survival to create a future that is more resilient, equitable, and environmentally stable. These efforts will not only benefit the people of Gaza but also serve as a model for integrating sustainability into recovery processes in conflict-affected regions worldwide.

References

Abd El Hay M (2024) The environmental-humanitarian impacts of the Israel Hamas war in Gaza. Arav Inst Environ Stud:1–28

AbuHamed H, Al Bursh W, Abu Mfarreh S, Yoshida M (2023) Managing post-conflict demolition wastes in Gaza strip: a case study on May 2021 conflict. J Mater Cycles Waste Manag 25:684–693. https://doi.org/10.1007/S10163-022-01570-Y/FIGURES/4

Abu-Rmeileh Q, Iriqat I (2024) Navigating economic development in Palestine: analysis of key challenges. Int J Sustain Dev Plan 19:591–597. https://doi.org/10.18280/ijsdp.190216

Alagha M, Hussein N (2024) The political economy of Israeli economic warfare on the Gaza strip. In: Elkahlout G (ed) Gaza's cycle of destruction and rebuilding. Palgrave Macmillan, Cham, pp 125–145

Alnabih A, Alnabeh N, Aljeesh Y, Aldabbour B (2024) Food insecurity and weight loss during wartime: a mixed-design study from the Gaza strip. J Health Popul Nutr 43:222

Ameer S (2024) Rebuilding Gaza: Navigating the challenges of post-war reconstruction. https://www.orfonline.org/expert-speak/rebuilding-gaza-navigating-the-challenges-of-post-war-reconstruction?utm_source=chatgpt.com. Accessed 25 Dec 2024

Amnesty International (2024) 'You feel like you are subhuman'—Israel's genocide against Palestinians in Gaza

Anera (2024) Food production systems under attack in Gaza. https://www.anera.org/blog/food-production-systems-under-attack-in-gaza/. Accessed 30 Mar 2024

Appiah-Kusi E, Al-Madbouh S (2024) Adapting to climate change in a politically complex environment: agricultural and food security resilience strategies in Hebron governorate, Palestine. In: Singh P, Yadav N (eds) The climate-health-sustainability nexus. Springer Nature, Cham, pp 127–160

Ardra S, Barua MK (2022) Halving food waste generation by 2030: the challenges and strategies of monitoring UN sustainable development goal target 12.3. J Clean Prod 380:135042. https://doi.org/10.1016/j.jclepro.2022.135042

Arora NK, Mishra I (2023) Responsible consumption and production: a roadmap to sustainable development. Environ Sustain 6:1–6. https://doi.org/10.1007/s42398-023-00266-9

Bin-Nashwan SA, Hassan MK, Muneeza A (2024) Russia–Ukraine conflict: 2030 Agenda for SDGs hangs in the balance. Int J Ethics Syst 40:3–16. https://doi.org/10.1108/IJOES-06-2022-0136

Butter D (2024) Economic impact of the Gaza war. IEMed Mediterr Yearb:273–275

Dardona Z, Amane M, Dardona A, Boussaa S (2024) Health and environmental impacts of Gaza conflict (2023–2024): a review. One Heal Bull. https://doi.org/10.4103/ohbl.ohbl

Elkahlout G (2024a) Gaza's cycle of destruction and rebuilding. Understanding the actors, dynamics, and responses. Palgrave Macmillan

Elkahlout G (2024b) War on Gaza: an unceasing humanitarian tragedy. In: Elkahlout G (ed) Gaza's cycle of destruction and rebuilding. Palgrave Macmillan, Cham, pp 209–223

El-Khozenadar HJ, Khatib T, Attaee B, El-Khozondar RJ (2022) Assessment of solar e-cookers social acceptance in Gaza strip. Sci Rep 12:17226. https://doi.org/10.1038/s41598-022-22326-6

Elnaggar M (2023) Useful energy, economic and reduction of greenhouse gas emissions assessment of solar water heater and solar air heater for heating purposes in Gaza, Palestine. Heliyon 9:e16803. https://doi.org/10.1016/J.HELIYON.2023.E16803

ESCWA (2023) War on Gaza: weaponizing access to water, energy and food

ESCWA/UNDP (2024) Gaza war: Expected socioeconomic impacts on the State of Palestine—October 2024

European Union/World Bank Group/United Nations (2021) Gaza rapid damage and needs assessment, June 2021

FAO (2018) Transforming food and agriculture to achieve the SDGs

FAO/UN (2024) Agricultural Damage Assessment in the Gaza Strip from October 7th 2023 to September 1st 2024

Fontaine R (2024) Strengthening environmental resilience in conflict zones: analysis of UNEA-6 resolution and the PERAC principles

Gasper D, Shah A, Tankha S (2019) The framing of sustainable consumption and production in SDG 12. Glob Policy 10:83–95. https://doi.org/10.1111/1758-5899.12592

Gordon N, Haddad M (2024) The road to famine in Gaza. New York Rev Books 1:1–12

Gupta P, Shukla DP (2024) Implications of Russia–Ukraine war on land surface temperature and air quality: long-term and short-term analysis. Environ Sci Pollut Res 31:46357–46375. https://doi.org/10.1007/s11356-024-32800-5

Hamad BA, Jamaluddine Z, Aly S et al (2024) UNRWA's work is at risk again. Lancet 404:2157. https://doi.org/10.1016/S0140-6736(24)02476-0

Hassoun A, Al-Muhannadi K, Hassan HF et al (2024) From acute food insecurity to famine: how the 2023/2024 war on Gaza has dramatically set back sustainable development goal 2 to end hunger. Front Sustain Food Syst 8:1402150. https://doi.org/10.3389/fsufs.2024.1402150

Hossain MA, Ferdous N, Ferdous E (2024) Crisis-driven disruptions in global waste management: impacts, challenges and policy responses amid COVID-19, Russia-Ukraine war, climate change, and colossal food waste. Environ Challeng 14:100807. https://doi.org/10.1016/J.ENVC.2023.100807

Human Rights Watch (2024a) Extermination and acts of Genocide Israel deliberately depriving palestinians in Gaza of water

Human Rights Watch (2024b) "Hopeless, starving, and besieged" Israel's forced displacement of Palestinians in Gaza

ILO (2024) A year of war in Gaza: impacts on employment and livelihoods in the West Bank and Gaza strip. Bulletin No 5

Irfan B, Lulu I, Hamawy A et al (2024) Combating infections under siege: healthcare challenges amidst the military assault in Gaza. World Med Heal Policy:1–26. https://doi.org/10.1002/wmh3.642

Iriqat D (2024) Towards an inclusive political system, Palestinian perspectives on the reconstruction of Gaza. In: Antar U, Asseburg M, Shaban O (eds) Palestinian Perspectives on the reconstruction of Gaza. Friedrich-Ebert-Stiftung (FES), Stiftung Wissenschaft und Politik (SWP)

Iriqat D, Alousi R, Aldahdouh TZ et al (2025) Educide amid conflict: the struggle of the Palestinian education system. Qual Educ All 2:81–99. https://doi.org/10.1108/QEA-10-2024-0120

Jacob-John J, D'souza C, Marjoribanks T, Singaraju S (2021) Synergistic interactions of SDGs in food supply chains: a review of responsible consumption and production. Sustain For 13:8809. https://doi.org/10.3390/SU13168809

Jacob-John J, D'Souza C, Marjoribanks T, Singaraju S (2023) Sustainable development goals: a review of SDG 12.3 in food supply chain literature. Benchmarking 30:3465–3481. https://doi.org/10.1108/BIJ-12-2021-0736/FULL/XML

Juaidi A, Anayah F, Assaf R et al (2022) An overview of renewable energy strategies and policies in Palestine: strengths and challenges. Energy Sustain Dev 68:258–272. https://doi.org/10.1016/j.esd.2022.04.002

London L, Watterson A, Mergler D et al (2024) A call from 40 public health scientists for an end to the continuing humanitarian and environmental catastrophe in Gaza. Environ Heal A Glob Access Sci Source 23:1–6. https://doi.org/10.1186/S12940-024-01097-9/PEER-REVIEW

Meaza H, Ghebreyohannes T, Nyssen J et al (2024) Managing the environmental impacts of war: what can be learned from conflict-vulnerable communities? Sci Total Environ 927:171974. https://doi.org/10.1016/j.scitotenv.2024.171974

Mensah K, Wieck C, Rudloff B (2023) Sustainable food consumption and sustainable development goal 12: conceptual challenges for monitoring and implementation. Sustain Dev. https://doi.org/10.1002/SD.2718

Mhlanga D, Ndhlovu E (2023) The Implications of the Russia–Ukraine War on Sustainable Development Goals in Africa. Fudan J Humanit Soc Sci 16:435–454. https://doi.org/10.1007/S40647-023-00383-Z/FIGURES/2

Nassar S, Tóth ZN, Vasa L (2023) Economic empowerment as a result of achieving SDGs with resource access: a comparative research between Gaza strip and Hungary. J Int Stud 16:9–33. https://doi.org/10.14254/2071-8330.2023/16-2/1

Nijim M (2024) Gazacide: Palestinians from Refugeehood to ontological obliteration. Crit Sociol 1944:1–23. https://doi.org/10.1177/08969205241286530

OCHA (2024) Gaza Humanitarian Response Update | 24 November—7 December 2024. https://www.ochaopt.org/content/gaza-humanitarian-response-update-24-november-7-december-2024?_gl=1*1egyngz*_ga*NDE3Nzg4ODEuMTczMzQ2NTkzMQ..*_ga_E60ZNX2F68*MTczNTE0MjExMi4xMC4xLjE3MzUxNDIxMTguNTQuMC4w. Accessed 25 Dec 2024

PAX (2024) War and garbage in Gaza, The public health and environmental crisis from widespread solid waste pollution

PFI/FPCCIA (2024) Impact of the Gaza war on private sector and pathways for recovery

Qumsiyeh MB (2024) Impact of the Israeli military activities on the environment. Int J Environ Stud 81:977–992. https://doi.org/10.1080/00207233.2024.2323365

Rahiman HU, Sarea A, Kodikal R (2024) Russia-Ukraine conflict: will attainment of sustainable development goals be a dream?—owing to increasing risk in global supply chain. Int J Bus Emerg Mark 16:389–410. https://doi.org/10.1504/IJBEM.2024.139430

Raman R, Lathabai HH, Nedungadi P (2024) Sustainable development goal 12 and its synergies with other SDGs: identification of key research contributions and policy insights. Discov Sustain 5:150. https://doi.org/10.1007/s43621-024-00289-0

Rantissi T, Gitis V, Zong Z, Hankins N (2024) Transforming the water-energy nexus in Gaza: a systems approach. Glob Chall 8:2300304. https://doi.org/10.1002/GCH2.202300304

Rawtani D, Gupta G, Khatri N et al (2022) Environmental damages due to war in Ukraine: a perspective. Sci Total Environ 850:157932. https://doi.org/10.1016/j.scitotenv.2022.157932

Rettig E (2024) Relieving Gaza's electricity burden after the war. BESA Geo Energ Insigh 27:1–5

Shaheen A, Dajani R, Zinszer K et al (2024) The war on the Gaza strip and its consequences on global warming. Front Hum Dyn 6:1463902

Studies P, Perspectives N, Studies A (2024) The Israeli war against UNRWA and its impact on Palestinian refugees. Al-Muntaqa New Perspect Arab Stud 7:91–98

The World Bank (2024a) Gaza Strip—interim damage assessment

The World Bank (2024b) Impacts of the conflict in the middle east on the Palestinian economy-December 2024

Trummer P, Ammerer G, Scherz M (2022) Sustainable consumption and production in the extraction and processing of raw materials—measures sets for achieving SDG target 12.2. Sustain For 14:10971. https://doi.org/10.3390/su141710971

UNCTAD (2024) Preliminary assessment of the economic impact of the destruction in Gaza and prospects for economic recovery

UNDP (2021) Gaza 2021 infrastructure damage assessment report

UNDP (2024) A silent threat: Gaza's struggle with solid waste management . https://stories.undp.org/a-silent-threat-gazas-struggle-with-solid-waste-management. Accessed 9 Jan 2025

UNDP/PAPP (2024) Impact of the war on solid waste management in the Gaza Strip

UNEP (2024) Environmental impact of the conflict in Gaza: preliminary assessment of environmental impacts. Nairobi

United Nations (2017) Gaza ten years later—UN Country team in the occupied Palestinian territory report

Vos R (2024) One year of war in Gaza: food emergency continues with no end in sight. IFPRI. https://www.ifpri.org/blog/one-year-of-war-in-gaza-food-emergency-continues-with-no-end-in-sight/. Accessed 9 Jan 2025

Wakilzada S, Husseini S (2024) How can Gaza achieve food security and self-sufficiency during post-conflict reconstruction? Agrilinks. https://agrilinks.org/post/how-can-gaza-achieve-food-security-and-self-sufficiency-during-post-conflict-reconstruction. Accessed 9 Jan 2025

World Bank Group (2023) West Bank and Gaza: country climate and development report

Global Unfulfilled Promises: The Recent War on Gaza and the Erosion of Global Partnerships

18

Rasha Abousalem

Abstract

This chapter examines the adverse effects of the recent war on Gaza on Sustainable Development Goal 17 (SDG 17), which seeks to strengthen global partnerships for sustainable development. This war has not only caused extensive human and material damage but has also severely undermined international cooperation and the achievement of shared global goals. The war on Gaza has caused a crisis of credibility and erosion of trust, which has complicated diplomatic efforts, peace negotiations, and projects aimed at fostering sustainable development. The destruction of infrastructure and disruption of economic activities in Gaza also have long-term repercussions on regional and global economic stability, hindering efforts to build resilient economies and inclusive societies. This chapter provides a comprehensive examination of how the war on Gaza undermines global partnerships and cooperation and suggests recommendations for mitigating these negative impacts and revitalizing international efforts to achieve SDG 17. The chapter emphasizes the need for sustained international engagement and peacebuilding initiatives, highlighting the critical importance of solidarity and collaboration in overcoming the challenges posed by this war.

Keywords

Gaza · Global partnerships · Humanitarian crises · Human rights · International cooperation · Peacebuilding · SDG 17 · Sustainable development · War

18.1 Introduction

The concept of sustainable development is defined as "the development that meets the needs of the present without compromising the ability of future generation to meet their own needs" (Gusmão Caiado et al. 2018). Encompassing 17 key goals and supported by 169 specific targets, the SDG Agenda was designed to guide international efforts focused on sustainable development to the goal year of 2030 (UN 2024a, b, c, d). According to Ryan and Hauser (2020), partnerships, which are central to realizing the SDG 2030 Agenda, are "enshrined as an integral mechanism for the achievement of the other 16 goals." One of the primary challenges in achieving the SDGs is the persistence of inequality within and between societies in the form of economic disparities, conflict, unequal access to education and healthcare, and discrimination,

R. Abousalem (✉)
Humanitarian Aid Worker and Independent Researcher
Columbia, MO, USA

A. Hassoun (ed.), *War on Gaza*, Sustainable Development Goals Series,
https://doi.org/10.1007/978-3-031-88500-6_18

which hinder progress towards individual goals. Additionally, geopolitical strain and nationalism are likely to thwart attempts at international cooperation and coordination, as economic challenges such as high inflation have exacerbated the risk of debt distress and aggressive competition (rather than cooperation) between states (United Nations 2023a). According to the UN, these issues highlight the importance of a "collective surge in action to provide developing countries with the necessary financing and technologies to accelerate the implementation of the SDGs" (United Nations 2023b).

Sitting atop of the SDG pyramid is Goal #17 (Global Partnerships), which is essential for the success of the overall SDGs for a more inclusive, connected international community. Partnerships and networking between UN entities, governments, and public and private sectors are increasingly crucial to help establish collaborations on SDGs that can aid in pragmatic transitions. In her report on "Enhanced cooperation between the United Nations and all relevant partners, in particular the private sector," Sanda Ojiambo, Assistant Secretary-General of the United Nations Global Compact, remarked that cooperation frameworks are essential starting points for formulating robust global partnership strategies. The report also affirmed that stronger relationships with a diversified portfolio of enterprises, whether they are micro or macro, along with the role of diplomatic partnerships can support the acceleration of meeting the SDGs. The importance lies in the promotion of partnerships that aid in achieving international peace and security, as well as provide humanitarian, development, education, and health needs assistance in countries that are impacted by instability.

For Gaza, however, it begs the question of whether authentic, effective global partnerships are possible when there is an occupying power controlling every aspect of life—socially, financially, politically, and physically. Due to the numerous obstacles that the Gazan economy, and the overall Palestinian economic sphere, has endured, it is not surprising that international ties remain stagnant, as the ability to build global partnerships requires a minimum level of agency and independence which have been largely denied by the Israeli government. Moreover, with the scope of the most recent, unrelenting destruction of Gaza, this has set in motion a path of economic annihilation that will take Palestinians generations to recuperate from, assuming they are given the proper opportunity to do so, as Israel has a history of physical and economic aggressions against Gaza through the use of blockades and numerous military operations that have intentionally aspired to cripple its economy and livelihood. Therefore, if we are to analyze the role SDG 17 has in the Palestinian sphere, in particular to Gaza, then the ultimate question is not how long it will take Gaza to recover, but whether or not it is meant to (in terms of its Occupier).

18.2 Sustainable Development Goal (SDG) 17: A Brief Breakdown

According to the UN, SDG 17 aims to "strengthen the means of implementation and revitalize the Global Partnership for sustainable development and recognizes the relevancy multi-stakeholder partnerships as important vehicles for mobilizing and sharing knowledge, expertise, technologies and financial resources to support the achievement of the sustainable development goals in all countries, particularly developing countries" (UNDESA 2023). The 2030 Agenda calls for action by all countries—developed and developing—requiring partnerships between governments, the private sector, and civil society (UN 2023a, b). This ambitious goal can only be realized with a resilient commitment to global partnership and cooperation.

There are seven primary components to SDG 17 that form a comprehensive approach to global sustainable development: capacity building, multi-stakeholder partnerships, trade, finance, science, financial inclusion, and technology (Bulmer and Yáñez-Araque 2023). According to the UN, the objective of the SDGs is to end poverty, safeguard the environment, and ensure that by the year 2030 there is international peace and prosperity to allow societies to flourish. While the purpose of SDG 17 is a righteous one, in that global partnerships will make sure no society is left behind in its goal of a better future, the reality is sadly far from it, as Gaza faces immeasurable obstacles that will be impossible to overcome by the 2030 deadline.

18.3 Gaza: A History of De-Development

The ethical peace dimension of the SDG Agenda—concerning values such as freedom, justice, equality, and human rights—is essential to the overall achievement of the goals, as a lack of any one of these values will substantially impact a society's ability to prosper (Ibrahim 2020).

Gaza's uniqueness unfortunately lies in its decades of occupation, apartheid, and blockades, all of which have paralyzed genuine economic development. The persistent deprivation has resulted in stagnating local and international economic growth, de-development, and dependence on international aid. While Israel claims that blockades and other restrictive measures are necessary for its security objectives, over the years its strategies have eroded the productive capacity of Gaza, "depriving it from its growth potential and turning into a captive market" fully dependent on its occupier for survival (UNESCWA 2022). As a result, the infrastructure of Gaza remained undeveloped, and, in the absence of an integrated market, the economy stagnated (Roy 2016; Iqtait 2024). This put the Gazan existence in a perpetual state of Israeli dependency, rendering it vulnerable to the political, economic, and social constraints of its occupier.

SDG 17.10 and 17.11 discuss trade and reference the World Trade Organization (WTO), the Doha Agreements, and the increase of exports for developing states, all of which require international coordination and partnerships that are absent from Gaza's structure (UN 2023a, b). Since its military occupation of Palestinian territories began in June 1967, Israel has deployed a matrix of policies and practices that seek to control and monopolize every aspect of Palestinian life, in what is referred to as the "matrix of control and domination." The "matrix of control," coined by Israeli political activist and writer Jeff Halper, is a term used to describe the complex set of Israeli policies and practices designed and implemented to establish control over all aspects of Palestinian life (for more on this see Halper 2000, 2009). Since then, the term has been referenced by the United Nations Economic and Social Commission for Western Asia (ESCWA 2022) when describing Israeli policies and practices to emphasize the suppressive and domination dimension of the occupation. Sustainable development requires a certain level of capacity, agency, and autonomy, but in situations where the state is financially insecure and administratively fractured, its ability to fulfill its social obligations is compromised. It was Sara Roy who coined the term "de-development" to describe the continuous degradation of Gaza's structures and capacities over time, which ensures a total dependency on Israel in every aspect of existence. De-development is defined as a process which undermines or weakens the ability of an economy to grow and expand by preventing it from accessing and utilizing critical inputs needed to promote internal growth beyond a specific structural level (Roy 2016). Gaza's intentional de-development is the result of numerous interconnected factors resulting from decades of Israeli territorial occupation, suffocating economic restraints, and societal fragmentation. Decades of physical distance between the West Bank and Gaza Strip led to strained social relationships which ultimately resulted in the erosion of social cohesion, while institutional decline created the conditions for "state collapse" to a level that has become nearly irreversible under such conditions (Jebril and Deakin 2022).

While Targets 17.1 (strengthening domestic resource mobilization), 17.3 (additional financial resources), 17.4 (long-term debt sustainability), and 17.5 (investment promotion practices) highlight the financial portion of the goal (von Schnurbein 2020; UN 2023a, b), the stagnated economy inside Gaza is a result of specific and intentionally damaging Israeli policies which have aimed to impede and sabotage the ability of the Gazan economy to create the necessary infrastructure and global partnerships needed for sustainable development. According to Roy (1987), Gaza's economy has undergone the following processes that have contributed to its de-development over the years:

(1) Low levels of governmental investment in social and economic infrastructure and development inside Gaza;
(2) Discriminatory tax laws against Palestinian producers;
(3) A lack of financial incentives and guarantees for Palestinian producers (which are available for Israeli producers);
(4) Strict control over Gaza's terms of trade, resulting in highly restricted access to foreign markets, with minimal protection from the importation of Israeli goods into Gaza;
(5) Restrictions on certain forms of economic activity;
(6) The growing dispossession of land; and
(7) The lack of political, economic, and social linkages between foreign groups.

For economies, and thereby societies, to thrive, basic levels of industrialization and investments are needed to accumulate capital. For a society to achieve this, it must have the ability to freely access the financial, political, and technological resources required to achieve growth, meaning some level of agency and sovereignty are also required. For this step to evolve, political and economic alliances are necessary. In the case of Gaza, it has de-evolved vis-à-vis the trifecta effect: (1) The internal Gazan economy is suppressed, but with just enough activity to allow for the most basic living necessities to be acquired; (2) All borders are under Israeli control, meaning nothing or no one can go in or out without a permit from Israeli authorities; and (3) In general, Palestinian society has been alienated from international investments, as investors are discouraged or banned from creating such professional relationships, especially in the case with Hamas' political takeover of Gaza (Fischer et al. 2001).

While there are numerous restrictions on trade, imports, and exports applied against Palestinians, Israeli producers, on the other hand, have unlimited access to Gazan, Israeli, and international markets (Roy 1987). This one-sided trade structure has effectively left Palestinian producers unable to compete under such constraining and disadvantageous conditions. In turn, this has left Gaza completely dependent on Israel for even the most minimal-structured global partnerships (Sayigh 1986).

Under such conditions, a viable economy and sustainable societal structure are not feasible. Statistics from 2021 highlighted lackluster economic growth rates, stagnant per capita income levels, eroded production capacity, persistent budget and trade deficits, sharp fall in foreign aid, rising public debt, and high rates of poverty and unemployment, all of which have had devastating impacts on the quality of life for Palestinians (ESCWA 2022). As a result of hindering Israeli practices, the long-sustained physical and economic fragmentation of Palestinian society has culminated in the inability of Palestinians to build beneficial global partnerships that would allow Gaza to demonstrate its ability to have a sustainable, sovereign society, even if under a minimally invasive occupation.

18.4 Impact of Blockade and Israeli Repeated Assaults

18.4.1 Challenges Facing Gaza/Occupied Palestinian Territory (Pre-October 2023)

Since the West Bank (WB) and Gaza are under occupation, they remain highly dependent on Israeli supplies and trade. Almost all Palestinian imports and exports must go through Israel or Israeli-controlled checkpoints, while any international trade must complete stringent Israeli paperwork. Just as damaging are the frequent short-notice adjustments implemented by the Israeli government in response to security incidents. These lengthy delays and at times complete isolation from external markets have resulted in damaging effects on Gaza's already limited private sector, economic growth, and international partnerships. Insufficient facilitation of public goods such as road infrastructures, utility networks, waste collection, electricity, water facilities, and health amenities affect the overall well-being of the Palestinian economy (Lovatt 2018), while it also inhibits its ability to form critical global partnerships that would allow for a prosperous society in Gaza to foster (Sayigh 1986; Garson 2020; Alkhaldi et al. 2024).

What is also important to note is the decline in foreign aid to Palestinians over the years due to continuous political instability and strained diplomatic relations. Foreign aid to the Palestinian economy substantially declined from $2 billion in 2008 to $550 million in 2022, which has "really serious economic ramifications," according to Mutasim Elagraa, economist at the United Nations Conference on Trade and Development (UNCTAD). The UN agency also stated that the Palestinian economy is in a state of "forced economic dependence" on Israel, in which "excessive production and transaction costs and barriers to trade with the rest of the world have resulted in a chronic trade deficit and a pervasive, lopsided dependence on Israel" (Millard 2023).

18.4.2 The Aftermath: Global Impacts of Genocide in Gaza

In its annual report on the Palestinian economy, UNCTAD stated that "Gaza has experienced 16 years of de-development and suppressed human potential and the right to development" (Millard 2023). Richard Kozul-Wright, the director of UNCTAD's Globalization and Development Strategies Division, claimed that the "economic consequences of the current and ongoing humanitarian crisis in Gaza are impossible to determine," but expects it would take "tens of billions of dollars" to repair the Gazan economy to what it was before October 2023 (Millard 2023). The recent Israeli military campaign has debilitated all of Gaza's sectors—financial, industrial, agricultural, commercial, technological, and communication, leaving no conventional economy to exist in Gaza (even at a primal level), as the black-market has completely taken over all aspects of life in the besieged strip.

Since the October 7th, 2023 attack by Hamas and the subsequent Israeli military campaign in Gaza that followed, political polarization has deepened, as the ramifications of the recent onslaught of violence have drastically shifted once reliable diplomatic ties, as well as impacted global trading partnerships. In their report on the "economic impact of the Israeli military operation in Gaza from October 2023 to May 2024," UNCTAD (2024a, b, c) detailed the profound economic destruction that has squeezed Gaza to the brink of total collapse, highlighting the staggering scale of economic devastation and unprecedented decline, far exceeding the impact of all previous military confrontations in 2008, 2012, 2014, and 2021. By early 2024, UNCTAD said up to 96% of Gaza's agricultural assets—including farms, orchards, irrigation systems, and machinery—had been "decimated," with 82% of businesses in Gaza had also been damaged or destroyed (Al Jazeera 2024; Soussi and Tahhan 2024).

As a result of the post-October 2023 Israeli military campaign, Gaza's economy and infrastructure have been left in "utter ruin" as the UN warns that it would take an estimated 350 years to return to its pre-conflict levels (Elliott 2024). The report, presented to UN general assembly, said economic activity across Gaza—which had been subpar before the war—had come to an almost complete standstill, apart from the modicum of humanitarian aid that Israel has permitted to enter under extreme conditions of severe water, fuel, and electricity shortages. The months of unrelentless attacks on Gaza have led to a significant decline in many of its industries, as construction output has been cut by 96%, agriculture output by 93%, manufacturing by 92%, and services sector output by 76% (Elliott 2024).

Israel, on the other hand, has been highly criticized over their offensive in Gaza, as they too are feeling the repercussions of their actions in Gaza. Accusations of ethnic cleansing, genocide, war crimes, and disregard of International Humanitarian Law (IHL) have resulted in not only strained global partnerships, but boycotts, calls for sanctions, and pending cases with the International Court of Justice (ICJ) and International Criminal Court (ICC). Although Israel has carried out repressive and oppressive acts against Palestinians for decades, it has enjoyed a fairly popular reputation as the "only democracy in the Middle East" while greatly benefiting from its most loyal global partner—the United States. The ongoing situation in Gaza, though, has not only significantly impacted Israel's own global partnerships and reputation, but unsurprisingly has led to a ripple effect for other states and affiliations as well. Once reliable international partnerships have now fractured, separating on the lines of those who either support Israel or support Palestinians. This strained escalation is particularly noticeable in the Global South, where states have led efforts to hold Israel accountable under IHL. Most noticeably is the case in the ICJ accusing Israel of genocide, filed by South Africa in December 2023. By November 2024, the ICC issued arrest warrants for Israeli Prime Minister Benjamin Netanyahu and Knesset member Yoav Gallant for crimes against humanity and war crimes in Gaza (Kretzmer 2005; Baruch 2019; HRW 2024; ICRC 2024).

These current accusations and actions against Israel by international courts indicate a significant shift in diplomatic attitudes, with some governments re-evaluating their once-reliable alliances with Israel. Though some traditional allies, such as the USA and several European states, have largely continued to maintain not only their unwavering support to Israel, but have refused to comply with the ICC arrest warrants. Yet rapidly growing domestic opposition within the Israeli loyalist nations coupled with international protests against Israeli actions in Gaza indicate a growing polarization of public opinion that could influence future state-to-state interactions and global partnerships with Israel, especially impacting trade and investments. While some will see this is as a positive, in that international pressure will result in the dismantling of Israel's oppressive regime just as it did in the case of the apartheid in South Africa, one must consider the opposite outcome—if Israel continues to occupy the WB and Gaza, severed global partnerships with Israel will have the biggest impact on the Palestinians, as they not only depend on Israel for survival, but will most likely be prohibited from forming partnerships with those who have boycotted Israel.

It is imperative to be mindful of the occupier-occupied relationship that exists between Israel-Palestine, as it plays a significant role when discussing Gaza and its abilities to fulfill the SDGs, especially #17, which serves as the central element to the overall Agenda. Gaza, along with the WB, have been intentionally devoid of self-determination for decades. This absence of agency for the Palestinians means that they have been deprived of the ability to form global partnerships that would lead to trading and investment opportunities, technological advances, and general societal and political growth, all of which are necessary to formulate sustainable development for their society. For many states, such as the USA, the Palestinian territories are not recognized as destinations separate and distinct from Israel, meaning any importing/exporting to Gaza will have to first include proper paperwork under the Israeli government, followed by a rigorous and overly restrictive slew of paperwork and permits required to enter said products. Decades of politicized and strategically driven strategies taken against the Palestinians have entirely inhibited their ability to formulate actions and envision plans for the benefit of their own society. Under such circumstances, it is practically impossible for Gaza to formulate the global partnerships that Goal 17 calls for that are required to meet the overall SDG Agenda (Al-Mughrabi 2023; Bureau of Industry and Security 2023; ITA 2023, 2024).

18.5 Impacts on Global Partnerships

Israel's campaign in Gaza could have serious, worldwide implications that transcend beyond direct partnerships with Israel or Palestine. As previously mentioned, the ripple effect of the situation in Gaza is being felt far and wide, with possible unforeseen consequences to the entire SDG 2030 Agenda. Criticisms at the micro-level come from everyday citizens and grassroots organizations in regard to the international community's failure to mediate a ceasefire or humanitarian pause, while the United Nations' reputation as a leading international multilateral organization has come under substantial scrutiny, mainly due to its inability to adhere to its own resolutions, conventions, and international laws (Darcy and Reynolds 2010; Lynk 2023; Akram 2024).

Israel's rapidly declining international standing is palpable, as worldwide condemnation continues to swiftly spread. What is a safe assumption to make is that Israel's military campaign in a post-October 2023 Gaza will undoubtedly leave behind decades-worth of damage to its global reputation, or as President Biden recently stated—Israel is "facing major, possibly generational damage to their reputation" (Serhan 2024). Since the offensive campaign in Gaza started over 1 year ago, the divide between North-South, East-West, Developed-Developing has steadily widened, straining not only diplomatic ties and global partnerships, but has invited widespread and fierce criticism from the public. In an interview with TIME magazine several months after Israel began its campaign in Gaza, Dylan Williams, Vice President for Government Affairs at the US-based Center for International Policy, stated: "In five and a half months of war, Israel has done more damage to its global standing than the preceding five and a half decades of occupation" (Serhan 2024). Failure from the USA and Western nations to denounce Israeli violations of IHL after months of violence has not only undermined their own moral compass, but has now transformed from complacency to complicitness, especially in the case for the USA who has continued sending billions of dollars and weapons to Israel. As a result, SDG 17, along with the entire SDG Agenda, hang by a thread.

18.6 Challenges to Effective Partnerships in Gaza

The 2030 SDG pillars encourage collaboration between states and actors that are integral to ending worldwide injustices and propelling nations into a more connected and sustainable world. According to researchers Al Koni and Mahamda (2023), the achievement of the SDGs has garnered significant attention from business experts, researchers, and policymakers worldwide, as "sustainable development is essential for promoting economic growth while preserving the environment, promoting social equality, and improving the quality of life for all people." Yet an endurable future cannot, and more importantly *should not*, be built on the pillars of occupation, apartheid, and injustices. Fundamental human rights, such as the right to self-determination, that are enshrined in the UN Charter, are universal and not to be practiced on an "a-la-carte" basis. There are no social or security situations that should have a bearing on the application of basic human rights, as they are to be pertinent under all circumstances. Gaza, with its inadequate resources, lack of political support, and occupation that robs it from all forms of freedom, undermines the SDG Agenda as it chips away at its core values of equity and sustainability.

If one were to examine the separate components of SDG 17, it would become immediately apparent that Gaza—whether it be pre- or post-October 2023—had no legitimate prospect of achieving the goal, let alone any portion of the SDG Agenda. Decades of Israeli occupation, apartheid, and ethnic cleansing have alienated, and more recently annihilated, Gaza's ability to secure global partnerships that would help form a society with economic independence and social prosperity. Additionally, due to political polarization and stigmatization, many foreign investors have no desire to implement projects in the OPTs, especially in a Gaza under Hamas leadership. Gaza seems to be the exception to the rule of international customary norms and human rights laws, which are, at their core, intertwined. With the overwhelming and blatant disregard for international order, Israel's recent actions in Gaza have possibly permanently altered future diplomatic relations that will impact the success (or failure) of the SDG 2030 Agenda (Ibrahim 2020; Leal Filho et al. 2024).

SDGs 17.13–17.19 address systemic issues within policy, civil society, and global partnerships and discuss the steps needed to achieve the overall SDGs (UN 2023a, b). But the protracted and magnified violence in Gaza presents substantial challenges to attaining these goals due to issues like instability, economic hardship, blockades, and limited access to the external stage that, in an otherwise healthy environment, global partnerships can formulate from. However, this is crucial in addressing Gaza's future humanitarian and developmental needs if it is to be given any fair chance at not only survival, but sustainability. One of the most essential core components of any global partnership is referenced in SDG 17.15: "Respect each country's policy space and leadership to establish and implement policies for poverty eradication and sustainable development" (UN 2023a, b), which is all but impossible for Gaza considering not only Israel's stance against Hamas, but the predominate anti-Hamas sentiment shared by many international states and their unwillingness to recognize the militant group as a legitimate governing body. Additional challenges that are hindering Gaza's ability to transform into a prosperous society include aid restrictions, donor fatigue, and political polarization, all of which discourage or prohibit international and diplomatic partnerships from investing their time and funds into such a volatile region (Garson 2020; Jebril and Deakin 2022; Buheji and Marouf 2024).

What should also not be overlooked when discussing meeting the SDG Agenda year of 2030 is the amount of time it will take for Gaza to rebuild in order to fairly take part in such an endeavor. UNCTAD (2024a, b, c) reports that if the current military operation in Gaza were to end immediately and reconstruction were to start right away, it would take Gaza until 2092 to restore its economy to 2022 GDP levels. With Gaza reduced to 50 million tons of rubble, it would cost an estimated $80 billion to rebuild (OCHA 2024; Hodali et al. 2024).

Although SDGs 17.2, 17.3, and 17.4 stress the need for developed countries to provide financial assistance and debt relief support towards developing countries (UN 2023a, b), this will require tremendous effort on the international community's part and will demand a serious financial commitment from global actors including states, NGOs, the UN, and other stakeholders. Yet again, it is important to stress the previously mentioned point that we cannot and should not be discussing pre-October 2023 Gaza as a successful goal. Rather, it is imperative that we break the cycle of oppressive destruction that has rendered 80% of the population dependent on international aid even prior to the latest military assaults of the last year (UNICEF 2023). UNCTAD's assessment warns that a post-hostilities phase of economic rehabilitation should not return to the previous status quo, as the vicious cycle of destruction and oppression needs to be broken, or else risk periodic outbreaks of violence and destruction (Bulmer and Yáñez-Araque 2023; UNCTAD, January 2024; UNCTAD, July 2024).

18.7 Key Strategies for Global Partnerships to Support Gaza

The current SDGs, which target stakeholders, build on their predecessor, the UN Millennium Development Goals (MDGs). But unlike the MDGs, which concentrate primarily on the governments, the 2030 SDG Agenda embraces the crucial role of private actors in steering sustainable development (Marx and Depoorter 2020). Current SDGs are more aligned with people-centered values along with the principals of universality, transparency, participation, equality, nondiscrimination, and liability (Frey and MacNaughton 2016; Gusmão Caiado et al. 2018), all of which expand well beyond the confines of the MDGs. Decades of research on business relationships, public-private partnerships, cross-sector partnerships, and multi-stakeholder partnerships, have proven that the quality of interaction between actors is fundamental in achieving progressive outcomes (Ryan and Hauser 2020). While SDGs 17.9, 17.16, and 17.17 discuss enhancing and encouraging capacity-building, multi-stakeholder partnerships, and private-civil society partnerships, respectively (UN 2023a, b), to share expertise and resources, under the Gazan-context these are restricted due to the refusal or inability of external actors to interact directly with Hamas leadership or the professional sector or the civilian population of Gaza.

A lack of motivation to coordinate and unwillingness to engage among global partners is exacerbated by political agendas, conflicting ideologies, and differing priorities. In the research article by Sondermann and Ulbert (2021), it was argued that "policy documents are replete with statements on the necessity of 'meaningful' engagement…without clarifying what 'meaningful' stands for," and that SDG 17 simply states the "right" way to collaborate between different actors. The authors made in interesting suggestion: operationalizing the concept of partnership according to different levels of accountability and participation, allowing for a gradual enhancement of the quality of partnership in terms of "meaningfulness."

To generate the types of partnerships needed for Gaza to transform into a sustainable and prosperous society, two types of partnerships are essential: (1) Global partnerships that embrace civil society, industry, and the UN (related to SDGs 17.9, 17.16, and 17.17); and (2) Multi-stakeholder partnerships that are inclusive of local, regional, and global affiliations (Castillo-Villar 2020). Sincere efforts to create meaningful and lasting change in Gaza will require the commitment of diverse stakeholders, including governments, the private sector, grassroots/civil society organizations (CSO), and multilateral organizations. These efforts via collaborations and global partnerships must be a kaleidoscope of approaches that concentrate on:

- **Development:** To help rebuild and support a sustainable economy and environment within Gaza. Global investments to fund the building of proper infrastructures that help establish a stable trade partner, leading to a prosperous and self-sustaining society that will not depend on indefinite humanitarian assistance. Developmental needs will include rebuilding healthcare and educational facilities roads, and buildings, as well as a seaport and functioning airport to support trade within Gaza and connect it to external markets.
- **Capacity building:** To help support governance and institutional infrastructures in Gaza, ensuring that local authorities and organizations are equipped with the skills, knowledge, and resources necessary to implement sustainable development initiatives effectively.
- **Humanitarian assistance:** To help advocate for and mobilize increased essential humanitarian and medical aid into Gaza to assist in its recovery and reconstruction. This assistance should run concurrently with assistance in development.
- **Access to resources:** Global partnerships can help ensure Gaza has equitable access to resources that will help boost its trading, economy, and involvement at a global level in an unrestricted manner.

The UN (2023a, b) stresses the vital role multi-stakeholder partnerships play to help "leverage the inter-linkages between the Sustainable Development Goals to enhance their effectiveness and impact and accelerate progress in achieving the Goals." In the case for Gaza, the SDGs can only transpire with a dynamic commitment to global partnerships and cooperation to ensure that the Palestinian society has a fair opportunity to succeed and prosper. It will require diverse sectors and actors working together in an integrated manner to merge financial resources, knowledge, and expertise (UNDESA 2020) that will be accessible to Gaza and customized to its unique needs in the post-occupation restructuring phase. The World Bank implored the international community to rapidly increase funding to help maintain fundamental public services and to initiate planning for long-term recovery and reconstruction. Additionally, "measures to facilitate trade and strengthen private sector activities in both the WB and Gaza are crucial to stimulate income generation" were stressed (World Bank 2024).

A report released in 2022 by ESCWA discussed a policy framework for Palestine and identified three crucial pillars that are essential in changing the dynamic of the existence in Gaza: The first pillar centers on reducing the chronic dependence of the Palestinian economy on Israel through the implementation of policies that redefine the exceedingly discriminatory and irregular economic ties with Israel. The second pillar focuses on alleviating the continuous de-development of the territories through policy measures that empower the private sector and facilitate active participation in sectors that can assist in re-establishing the productive capabilities of the Palestinian economy. The third pillar, which should be evident but in Palestine's case is all too often overlooked, promotes a rights-based approach to Palestinian development, through which the right to self-determination is focused on as the priority and the central goal which all development policies and programs seek to achieve (ESCWA 2022).

To do this, there needs to be a reinforcement of global support that is in conjunction with a reaffirmation to international human rights in order to secure dignity and opportunity for the Palestinian people. There is a crucial and mandatory correlation between basic human rights and the SDGs, as human rights are an essential prerequisite in providing states and societies with the proper leverage to take part in this Agenda (Winkler and Williams 2017; Alkhaldi et al. 2024). Again, the situation with Gaza is incredibly unique in that it is a modern-day apartheid state in addition to a prolonged violent military occupation that strips away every basic human right of the Palestinians—freedom of movement, freedom of self-determination, and discriminatory laws based on ethnicity, among many others. When we discuss a sustainable global future, we must envision a future in which Gaza (and the WB) are freed from the shackles of oppression and can seek out opportunities for growth and prosperity. This calls for unprecedented collaborations among Members States in adopting ambitious global and national commitments for SDG-supportive partnerships.

Again, we return to the concept of self-determination, which, according to the United Nations, is the "focal precondition for development and must be the driving force behind the key policy goals addressing Palestinian dependence on the Israeli economy and empowering the Palestinian private sector, ensuring that policies are contextualized and aligned with the Charter of the United Nations and the 2030 Agenda for Sustainable Development" (ESCWA 2022). Securing international support for the Palestinian right to development in the context of universal human rights, including the right to self-determination, is imperative in the application of the SDG framework and its anticipated success. Endowing the Palestinian people with their due agency and ownership of their institutions can contribute significantly to gaining the international backing necessary for developing Gaza (ESCWA 2022).

For Gaza's economy to proliferate, a foundation for autonomy and territorial integrity that allows unrestricted access to the open global market is imperative to achieve long-term economic viability and sustainability. Unfortunately, the matrix of oppressive Israeli policies and practices in Gaza remains the single largest obstacle to Palestinian development. The recent military operation has significantly impacted Israel's current global partnerships, creating a polarized international environment that will undoubtedly have a bearing on trade agreements and global partnerships, leaving many to question if Israel will be able to mitigate the damage and salvage its reputation.

A 2024 UNCTAD report stressed the urgency of providing immediate and dynamic budget support to the Palestinian government by the international community, while urging donors and partners to "recognize that the constraints on the Palestinian economy, particularly in Gaza, extend beyond the recent confrontation." As ESCWA stated in its 2022 report on Palestine: "An eviscerated economy operates far below its full potential and becomes increasingly dependent on other countries and actors…and the vast asymmetry in power between Israel and the occupied Palestinian territory, finding sustainable solutions to the chronic economic challenges will require a shift in the approach adopted by relevant development actors" (ESCWA 2022). Rami Alazzeh, an economist who works on the Occupied Palestinian Territories at UNCTAD reminds the international community that, "Gaza needs to be part of the development agenda rather than being treated as a humanitarian case" (Tétrault-Farber and Mantovani 2024). Reforming current global alliances by involving influential actors from within Gaza is a priority if Palestinian society is expected to take part in achieving the SDGs. These partnerships should prioritize accountability, advocacy, and diplomacy to support human rights and promote sustainable development in Palestine (Alkhaldi et al. 2024).

18.8 Conclusion

The scope of death and destruction in Gaza has been catastrophic and unprecedented, resulting in the killings of tens-of-thousands of people, infrastructural annihilation, dependency on international aid for the entire population, and a total collapse of the economic system. International institutions and markets, along with global partnerships, are feeling the ripple effect of the unrelenting violence in Gaza. Rising voices from the Global South are challenging traditional diplomatic alliances, economic partnerships, and multilateral institutions that have connections with Israel. How this will impact future global partnerships is yet to be known, but if current trends are a prelude, Israel will be facing the dissolution of global partnerships, boycotts, and sanctions, while Gaza will witness global solidarity in the form of foreign investors and trading partners. The United Nations, although undergoing an existential crisis due to polarizing loyalties between Member States over stances on Israel and Gaza, continues to play a crucial role in supporting countries in delivering their SDGs. Palestinians, on the other hand, will never be able to achieve sustainable development or discover their role in the global market without the freedom and sovereignty to do so, as practically every aspect of the SDG components requires some rudimentary characteristic of sovereignty (even if under an occupying power).

As the 2030 deadline to fulfill the SDGs draws near, political leaders, public institutions, and private sectors will need to rally all stakeholders to wholeheartedly commit to not only rebuilding Gaza but irrevocably concluding its decades-long dependency on international aid. For Gaza, the role of the global community is to act assertively to uphold the fundamental rights embedded in the SDGs which seek to ensure an accessible, equitable, and sustainable future for all. According to the UN, the "defining principle of the 2030 Agenda for Sustainable Development is a shared promise by every country to work together to secure the rights and well-being of everyone...." But that promise is in peril, as the genocide in Gaza has put a spotlight on global double-standards—the international community, primarily the Global North and Western states, advocate for universal human rights, while all-too-often ignoring Gaza's suffering.

But with these recent acts of aggression, alliances are rearranging, adversely impacting global markets and partnerships. A fundamental shift is needed—in commitment, solidarity, finance, and action—to put Gaza on a better path, one that it has been systematically denied for generations. The international community is now compelled to confront decades of neglecting Israeli injustices and violations against Gaza. The status quo of deprivation, which should have never been allowed to exist in the first place, has now reached an impasse—either the oppressive regime persists, or the international community collectively breaks the chains that bind Gaza. Although the future success of the SDG Agenda seems bleak, there is still hope. The prospect lies within SDG 17 and in the collective action of global partnerships that can help pave the way for Gaza towards a just future. With coordinated international support that confronts the occupation and blockade, Gaza could overcome its current challenges and make strides toward achieving all SDGs.

References

Akram SM (2024, January 22) The failures of the UN in the Israel-Palestine conflict. OpenGlobalRights. https://www.openglobalrights.org/failures-un-israel-palestine-conflict/

Al Jazeera (2024, September 12) UN warns of "staggering" economic devastation in Gaza, occupied West Bank. https://www.aljazeera.com/news/2024/9/12/un-warns-of-staggering-economic-devastation-in-gaza-occupied-west-bank

Al Koni SR, Mahamda MK (2023) The role of business schools in achieving sustainable Development goals: Palestine as a case study. Preprints. https://doi.org/10.20944/preprints202305.0823.v2
Alkhaldi M, Akilan B, Hoteit M, Elhadidy M, Morrison S (2024) Suboptimal attainment of global goals of human rights, universal health coverage and sustainable development goals in Gaza during 2023–2024. Risk Manage Healthcare Policy 17:3207–3216. https://doi.org/10.2147/RMHP.S488944
Al-Mughrabi N (2023, September 5) Israeli ban on Gaza exports deals blow to long-suffering economy. Reuters. https://www.reuters.com/world/middle-east/israeli-ban-gaza-exports-deals-blow-long-suffering-economy-2023-09-05/
Baruch PS (2019) The international criminal court on Israel: another step toward investigation. Institute for National Security Studies http://www.jstor.org/stable/resrep23504
Buheji M, Marouf M (2024) Mapping reconstruction of Gaza to overcome donors' fatigue. Gradiva 63(07):116–130. https://www.researchgate.net/publication/382889087_Mapping_Reconstruction_of_GAZA_to_Overcome_Donors'_Fatigue
Bulmer E, Yáñez-Araque B (2023) Tackling climate change through multi-stakeholder partnerships: promoting SDG 17 to combat climate change. Energies 16(9):3777. https://doi.org/10.3390/en16093777
Bureau of Industry and Security (2023, October) FAQs related to exports to Israel. Bureau of Industry and Security. https://www.bis.gov/exports-to-israel#:~:text=For%20export%20control%20purposes%2C%20the,Gaza%20and%20the%20West%20Bank
Castillo-Villar RG (2020) Identifying determinants of CSR implementation on SDG 17 partnerships for the goals. Cogent Busi Manage 7(1). https://doi.org/10.1080/23311975.2020.1847989
Darcy S, Reynolds J (2010) An enduring occupation: the status of the Gaza strip from the perspective of international humanitarian law. J Confl Secur Law 15(2):211–243. http://www.jstor.org/stable/26294629
Economic and Social Commission for Western Asia (ESCWA) (2022, October) Countering economic dependence and de-development in the Occupied Palestinian Territory. United Nations. https://www.un.org/unispal/wp-content/uploads/2023/03/ESCWAREPORT_280223.pdf
Elliott L (2024, October 22) "Utter ruin": Gaza economy would take 350 years to return to pre-conflict level, UN says. The Guardian https://www.theguardian.com/world/2024/oct/22/ruined-gaza-economy-un-report-israel
Fischer S, Alonso-Gamo P, Von Allmen UE (2001) Economic developments in the West Bank and Gaza since Oslo. Econ J 111:254–275. https://doi.org/10.1111/1468-0297.00629
Frey DF, MacNaughton G (2016) A Human Rights Lens on Full Employment and Decent Work in the 2030 Sustainable Development Agenda. SAGE Open, 6(2). https://doi.org/10.1177/2158244016649580 (Original work published 2016)
Garson M (2020) Are we really aiding Gaza? New Zealand Int Rev 45(1):18–21. https://www.jstor.org/stable/48552248
Gusmão Caiado RG, Leal Filho W, Quelhas OL, de Mattos L, Nascimento D, Ávila LV (2018) A literature-based review on potentials and constraints in the implementation of the sustainable development goals. J Clean Prod 198:1276–1288. https://doi.org/10.1016/j.jclepro.2018.07.102
Halper J (2000) The 94 percent solution: a matrix of control. Middle East Rep 216. https://doi.org/10.2307/1520209
Halper J (2009) "Dismantling the matrix of control," Middle East Report Online, September 11, 2009
Hodali F, Kao J, Haque J, Lin JCF (2024, August 16) Gaza reduced to 42 million tonnes of rubble. What will it take to rebuild? Bloomberg.com. https://www.bloomberg.com/graphics/2024-gaza-who-will-pay-to-rebuild/
Human Rights Watch (HRW) (2024, January 26) Gaza: World Court orders Israel to prevent genocide. Human Rights Watch. https://www.hrw.org/news/2024/01/26/gaza-world-court-orders-israel-prevent-genocide
Ibrahim M (2020, August) Implementing the 2030 agenda for sustainable development in Palestine: an innovation-centric economic growth perspective. United Nations. https://www.un.org/development/desa/dspd/wp-content/uploads/sites/22/2020/09/Maysoun-Ibrahim_UN-EGM-2020-Innovation-and-SDGs.pdf
International Committee of the Red Cross (ICRC) (2024, July 26) What does the law say about the responsibilities of the occupying power in the Occupied Palestinian Territory?. International Committee of the Red Cross. https://www.icrc.org/en/document/ihl-occupying-power-responsibilities-occupied-palestinian-territories
International Trade Administration (ITA) (2023, October 6) West Bank and Gaza - Market Challenges. West Bank and Gaza Country Commercial Guide. https://www.trade.gov/country-commercial-guides/west-bank-and-gaza-market-challenges
International Trade Administration (ITA) (2024, September 26) Israel raw materials supply chain affected by Israel-Hamas conflict. International Trade Administration. https://www.trade.gov/market-intelligence/israel-raw-materials-supply-chain-affected-israel-hamas-conflict
Iqtait A (2024) The state of the financial system in the West Bank and Gaza. Strateg Comments 30(7):x–xiii. https://doi.org/10.1080/13567888.2024.2425247
Jebril M, Deakin S (2022, April 2) The political economy of health in the Gaza strip: reversing de-development. J Glob Heal. https://jogh.org/2022/jogh-12-03014
Kretzmer D (2005) The advisory opinion: the light treatment of international humanitarian law. Am J Int Law 99(1):88–102. https://doi.org/10.2307/3246092

Leal Filho W, Dibbern T, Pimenta Dinis MA, Coggo Cristofoletti E, Mbah MF, Mishra A, Clarke A, Samuel N, Castillo Apraiz J, Abubakar IR, Aina YA (2024) The added value of partnerships in implementing the UN sustainable development goals. J Clean Prod 438:140794. https://doi.org/10.1016/j.jclepro.2024.140794

Lovatt H (2018) Gaza's fragile calm: the search for lasting stability. European Council on Foreign Relations http://www.jstor.org/stable/resrep21611

Lynk M (2023, November 13) Can international law prevail in Gaza and Israel? DAWN. https://dawnmena.org/can-international-law-prevail-in-gaza-and-israel/

Marx A, Depoorter C (2020) Achieving the global 2030 agenda: what role for voluntary sustainability standards? In: von Schnurbein G (ed) Transitioning to strong partnerships for the sustainable development goals, Transitioning to sustainability series 17. MDPI, Basel

Millard R (2023, October 25) Gaza has gone through 16 years of "de-development": UN. Barron's. https://www.barrons.com/news/gaza-has-experienced-16-years-of-de-development-un-5e93cca8

Paris, C. D., & Murillo, C. (2024, August 5). *Palestine: Mid-year metrics 2024*. ACLED. https://acleddata.com/2024/08/05/palestine-mid-year-metrics-2024/

Roy S (1987) The Gaza strip: a case of economic de-development. J Palest Stud 17(1):56–88. https://doi.org/10.2307/2536651

Roy SM (2016) The Gaza Strip: The Political Economy of de-development (Third). Institute for Palestine Studies USA, Inc

Ryan A, Hauser C (2020) Reflecting on the role of academia–private sector partnerships in moving forward with the SDGs. In: von Schnurbein G (ed) Transitioning to strong partnerships for the sustainable development goals, Transitioning to sustainability series 17. MDPI, Basel

Sayigh YA (1986) The Palestinian economy under occupation: dependency and pauperization. J Palest Stud 15(4):46–67. https://doi.org/10.2307/2536611

Serhan Y (2024, April 4) How Israel and its allies lost global credibility in Gaza. Time. https://time.com/6963032/israel-netanyahu-allies-global-standing/

Sondermann E, Ulbert C (2021) Transformation through 'meaningful' partnership? SDG 17 as Metagovernance norm and its Global Health implementation. Polit Govern 9(1):152–163. https://doi.org/10.17645/pag.v9i1.3656

Soussi A, Tahhan ZA (2024, January 23) How Israel controls $188 million of Palestinian money every month. Al Jazeera. https://www.aljazeera.com/news/2024/1/23/why-is-israel-sending-palestinian-taxes-to-norway

Tétrault-Farber G, Mantovani C (2024, January 31) Gaza's economy could take decades to recover, UN trade body says. Reuters. https://www.reuters.com/world/middle-east/gazas-economy-could-take-decades-recover-un-trade-body-says-2024-01-31/

UNESCWA (2022, October) Countering economic dependence and de-development in the Occupied Palestinian Territory. United Nations Economic and Social Commission for Western Asia (UNESCWA). https://www.unescwa.org/sites/default/files/pubs/pdf/palestine-economic-dependency-development-arabic.pdf

UN Trade and Development (UNCTAD) (2024a, January 31) Gaza: unprecedented destruction will take tens of billions of dollars and decades to reverse. UNCTAD. https://unctad.org/news/gaza-unprecedented-destruction-will-take-tens-billions-dollars-and-decades-reverse

UN Trade and Development (UNCTAD) (2024b, January) Preliminary assessment of the economic impact of the destruction in Gaza and prospects for economic recovery. UNCTAD rapid assessment. https://unctad.org/publication/preliminary-assessment-economic-impact-destruction-gaza-and-prospects-economic-recovery

UN Trade and Development (UNCTAD) (2024c, September 12) Economic crisis worsens in occupied Palestinian territory amid ongoing Gaza conflict. UN Trade and Development (UNCTAD). https://unctad.org/news/economic-crisis-worsens-occupied-palestinian-territory-amid-ongoing-gaza-conflict

United Nations (2022, October) Countering economic dependence and de-development in the Occupied Palestinian Territory. United Nations Economic and Social Commission for Western Asia (UNESCWA). https://www.unescwa.org/sites/default/files/pubs/pdf/palestine-economic-dependency-development-arabic.pdf

United Nations (2023a) Special edition 2023—UN statistics division—the United Nations. The Sustainable Development Goals Report Special Edition (2023). https://unstats.un.org/sdgs/report/2023/The-Sustainable-Development-Goals-Report-2023.pdf

United Nations (2023b) Global partnerships—United Nations sustainable development. United Nations Sustainable Development Goals https://www.un.org/sustainabledevelopment/globalpartnerships/

United Nations. (2024a, September 10). Economic costs of the Israeli occupation for the Palestinian people: The economic impact of the Israeli military operation in Gaza from October 2023 to May 2024—UNCTAD Report (A/79/343)—question of Palestine. The Question of Palestine. https://www.un.org/unispal/document/unctad-report-10sep24/#:~:text=Once%20a%20ceasefire%20is%20reached,to%20its%20level%20in%202022

United Nations (2024b, July 17) Humanitarian situation in Gaza "a moral stain on us all", secretary-general tells security council, stressing international law must be respected by all | meetings coverage and press releases. United Nations https://press.un.org/en/2024/sgsm22310.doc.htm

United Nations (2024c, September 26) World Bank report: impacts of the conflict in the Middle East on the Palestinian economy—September 2024 update—question of Palestine. United Nations. https://www.un.org/unispal/document/world-bank-report-26sep24/#:~:text=Official%20data%20reveals%20a%2035,percent%20contraction%20in%20Q1%2D2024

United Nations (2024d, October 16) Amid global volatility, partnerships between organization, private sector increasingly critical, Speakers Tell Second Committee. United Nations. https://press.un.org/en/2024/gaef3606.doc.htm

United Nations Children Fund (UNICEF) (2023, July 3) Fifteen years of the blockade of the Gaza Strip. UNICEF. https://www.unicef.org/mena/press-releases/fifteen-years-blockade-gaza-strip#:~:text=%22Largely%20due%20to%20the%20blockade,of%20the%20youth%20are%20unemployed

United Nations Department of Economic and Social Affairs (UNDESA) (2020) The SDG partnership guidebook. THE SDG partnership guidebook: a practical guide to building high impact multi-stakeholder partnerships for the sustainable development goals. https://sdgs.un.org/sites/default/files/2022-02/SDG Partnership Guidebook 1.11.pdf

United Nations Department of Economic and Social Affairs (UNDESA). (2023). Goal 17 | Department of economic and social affairs. United Nations. https://sdgs.un.org/goals/goal17

United Nations Office for the Coordination of Humanitarian Affairs (OCHA). (2024, December 10). Gaza humanitarian response update: 24 November–7 December 2024. OCHA. https://www.unocha.org/publications/report/occupied-palestinian-territory/gaza-humanitarian-response-update-24-november-7-december-2024

Von Schnurbein G (2020) Achieving the global 2030 agenda: what role for voluntary sustainability standards? In: von Schnurbein G (ed) Transitioning to strong partnerships for the sustainable development goals, Transitioning to sustainability series 17. MDPI, Basel

Winkler IT, Williams C (2017) The sustainable development goals and human rights: a critical early review. Int J Human Right 21(8):1023–1028. https://doi.org/10.1080/13642987.2017.1348695

World Bank (2024, September) Impacts of the conflict in the Middle East on the Palestinian economy (September 2024 update). World Bank Economic Monitoring Report. https://thedocs.worldbank.org/en/doc/c25061ab26d14d7acc0330d5a7b4d496-0280012024/world-bank-economic-monitoring-report-impacts-of-the-conflict-in-the-middle-east-on-the-palestinianeconomy-september-2024-update

Wright, R. (2023, October 31). *The Five Global Dangers from the Gaza War*. Wilson Center. https://www.wilsoncenter.org/article/five-global-dangers-gaza-war

Correction to: 'Waiting for a Better Time Is a Waste of Time': Motivations of Learning Amid the Genocide in Gaza in Light of Sustainable Development Goal 4

Tahani Aldahdouh, Khalid Dader, Alaa Aldahdouh, Rawia Awadallah, and Imed Romdhani

Correction to:
Chapter 5 in: A. Hassoun (ed.), *War on Gaza*, Sustainable Development Goals Series, https://doi.org/10.1007/978-3-031-88500-6_5

The chapter "Waiting for a Better Time Is a Waste of Time: Motivations of Learning Amid the Genocide in Gaza in Light of Sustainable Development Goal 4," written by Tahani Aldahdouh, Khalid Dader, Alaa Aldahdouh, Rawia Awadallah, and Imed Romdhani, was originally published without open access. Following the authors' decision to opt for open access, the copyright of the chapter changed to © The Author(s) 2025 and the chapter is now distributed under the terms of the Creative Commons Attribution 4.0 International License (http://creativecommons.org/licenses/by/4.0/), which permits use, sharing, adap-tation, distribution, and reproduction in any medium or format, as long as you give appropriate credit to the original author(s) and the source, provide a link to the Creative Commons license, and indicate if changes were made.

The updated version of this chapter can be found at
https://doi.org/10.1007/978-3-031-88500-6_5

A. Hassoun (ed.), *War on Gaza*, Sustainable Development Goals Series,
https://doi.org/10.1007/978-3-031-88500-6_19

The manufacturer's authorised representative in the EU is Springer Nature Customer Service Centre GmbH, Europaplatz 3, 69115 Heidelberg, Germany. If you have any concerns regarding our products, please contact ProductSafety@springernature.com

Printed and bound by CPI Group (UK) Ltd, Croydon, CR0 4YY
01/10/2026
02225960-0020